Handbook of Beta Distribution and Its Applications

Additional Volumes in Preparation

Handbook of Beta Distribution and Its Applications

edited by
Arjun K. Gupta
Bowling Green State University
Bowling Green, Ohio, U.S.A.

Saralees Nadarajah
University of South Florida
Tampa, Florida, U.S.A.

CRC Press
Taylor & Francis Group
Boca Raton London New York

CRC Press is an imprint of the
Taylor & Francis Group, an **informa** business

First published 2004 by Marcel Dekker, Inc.

Published 2018 by CRC Press
Taylor & Francis Group
6000 Broken Sound Parkway NW, Suite 300
Boca Raton, FL 33487-2742

First issued in paperback 2020

© 2004 by Taylor & Francis Group, LLC
CRC Press is an imprint of Taylor & Francis Group, an Informa business

No claim to original U.S. Government works

ISBN 13: 978-0-367-57832-9 (pbk)
ISBN 13: 978-0-8247-5396-2 (hbk)

**Visit the Taylor & Francis Web site at
http://www.taylorandfrancis.com**

**and the CRC Press Web site at
http://www.crcpress.com**

Library of Congress Cataloging-in-Publication Data
A catalog record for this book is available from the Library of Congress.

Preface

This handbook is devoted to a most flexible family of distributions. The Beta family, whose origin can be traced to 1676 in a letter from Sir Isaac Newton to Henry Oldenberg, has been utilized extensively in statistical theory and practice for over one hundred years. Originally defined on the unit interval [0, 1] but easily extended to any finite interval, the Beta distribution can take an amazingly great variety of forms. Thus it can be fitted to practically any data representing a phenomenon in almost any field of application. Its crucial importance is in Bayesian statistical analyses. This diversity and the apparent ease of application require careful analysis of the properties of the distribution to avoid pitfalls and misrepresentation.

It is therefore somewhat surprising that a handbook devoted to the Beta distributions has not been available while a number of handbooks dealing with other, less prominent distributions have been compiled and published in the last twenty years. Perhaps one of the reasons is that compilation of such a handbook is a daunting task, as our table of contents indicates. Indeed, the scope of the Beta distributions is unsurpassed, and although there are books dealing with continuous distributions and review articles in the *Encyclopedia of Statistical Sciences* and the *Encyclopedia of Biostatistics* the material presented in these sources is at best the tip of an iceberg.

Our goal in this handbook is to present in an organized and user-friendly manner all the information available in the literature worthy of publication. For this purpose we have enlisted several experts in the field and left no stone unturned to coordinate their efforts so that the handbook would be accessible to novices and be valuable to seasoned researchers. A proper balance between theory and applications has been maintained.

We are very thankful to the contributors for their cooperation in timely completion of the project. Finally, we wish to extend our thanks to Maria Allegra, Acquisitions Editor at Marcel Dekker, Inc., for her help in bringing this work to publication.

Arun K. Gupta
Saralees Nadarajah

Contents

v

Contributors

Kyoung-Gee Ahn University of Michigan, Ann Arbor, Michigan, U.S.A.

Barry C. Arnold University of California, Riverside, California, U.S.A.

Yontha Ath California State University, Carson, California, U.S.A.

Manish C. Bhattacharjee Center for Applied Mathematics and Statistics, New Jersey Institute of Technology, Newark, New Jersey, U.S.A.

Enrique Castillo University of Cantabria, Santander, Spain

Can Cogun Gazi University, Maltepe, Ankara, Turkey

Gerrit H. de Rooij Wageningen University, Wageningen, The Netherlands

Alan H. Feiveson National Aeronautics and Space Administration, Lyndon B. Johnson Space Center, Houston, Texas, U.S.A.

Krzysztof S. Frankowski University of Minnesota, Minneapolis, Minnesota, U.S.A.

Steven T. Garren James Madison University, Harrisonburg, Virginia, U.S.A.

Narayan Giri University of Montreal, Montreal, Quebec, Canada

Arjun Gupta Bowling Green State University, Bowling Green, Ohio, U.S.A.

Thomas A. Mazzuchi The George Washington University, Washington, D.C., U.S.A.

Kosto Mitov Airforce Academy "G. Benkovski," Pleven, Bulgaria

Dirk F. Moore Temple University, Philadelphia, Pennsylvania, U.S.A.

Saralees Nadarajah University of South Florida, Tampa, Florida, U.S.A.

Truc T. Nguyen Bowling Green State University, Bowling Green, Ohio, U.S.A.

Sudhir R. Paul University of Windsor, Windsor, Ontario, Canada

T. Pham-Gia University of Moncton, Moncton, Canada

José M. Sarabia University of Cantabria, Santander, Spain

Woollcott Smith Temple University, Philadelphia, Pennsylvania, U.S.A.

R. T. Smythe Oregon State University, Corvallis, Oregon, U.S.A.

Milton Sobel University of California, Santa Barbara, California, U.S.A.

Frank Stagnitti School of Ecology and Environment, Deakin University, Warrnambool, Victoria, Australia

J. René van Dorp The George Washington University, Washington, D.C., U.S.A.

Handbook of Beta Distribution and Its Applications

Beta Function and the Incomplete Beta Function

Saralees Nadarajah[1] and Arjun Gupta[2]

[1]Department of Mathematics
University of South Florida
Tampa, Florida 33620
[2]Department of Mathematics and Statistics
Bowling Green State University
Bowling Green, Ohio 43403

I.   Beta Function

The beta function is defined by

$$B(a, b) = \int_0^1 z^{a-1}(1 - z)^{b-1}dz, \qquad a > 0, \quad b > 0.$$

It can be represented in several different forms. In this section we provide a comprehensive list of integral representations, series representations, functional relations and closed form expressions of the beta function.

A.   Integral Representations

The integral representations of B are:

$$B(a, b) = 2 \int_0^1 w^{2a-1} \left(1 - w^2\right)^{b-1} dw,$$
$$a > 0, \quad b > 0,$$

$$B(a, b) = 2 \int_0^{\pi/2} \sin^{2a-1} w \cos^{2b-1} w dw,$$
$$a > 0, \quad b > 0,$$

$$B(a, b) = \int_0^\infty \frac{z^{a-1}}{(1 + z)^{a+b}} dz$$
$$= 2 \int_0^\infty \frac{z^{2a-1}}{(1 + z^2)^{a+b}} dz,$$
$$a > 0, \quad b > 0,$$

$$B(a, b) = 2^{2-a-b} \int_{-1}^1 \frac{(1 + z)^{2a-1}(1 - z)^{2b-1}}{(1 + z^2)^{a+b}} dz,$$
$$a > 0, \quad b > 0,$$

$$B(a, b) = \int_0^1 \frac{z^{a-1} + z^{b-1}}{(1 + z)^{a+b}} dz$$
$$= \int_1^\infty \frac{z^{a-1} + z^{b-1}}{(1 + z)^{a+b}} dz,$$
$$a > 0, \quad b > 0,$$

$$B(a, b) = z^{1-a-b} \int_0^1 \left\{ (1 + w)^{a-1}(1 - w)^{b-1} \right.$$
$$\left. + (1 + w)^{b-1}(1 - w)^{a-1} \right\} dw,$$
$$a > 0, \quad b > 0,$$

$$B(a, b) = z^b(1 + z)^a \int_0^1 \frac{w^{a-1}(1 - w)^{b-1}}{(w + z)^{a+b}} dw,$$
$$a > 0, \quad b > 0, \quad 1 < z < 0, \quad a + b < 1,$$

$$B(a,b) = z^b(1+z)^a \int_0^1 \frac{\cos^{2a-1} w \sin^{2b-1} w}{(z + \cos^2 w)^{a+b}} dw,$$

$$a > 0, \quad b > 0, \quad -1 < z < 0, \quad a + b < 1,$$

$$B(a,b) = \frac{1}{(d-c)^{a+b-1}} \int_c^d (z-c)^{a-1}(d-z)^{b-1} dz,$$

$$d > c, \quad a > 0, \quad b > 0,$$

$$B(a,b) = (c+e)^a(d+e)^b \int_0^1 \frac{w^{a-1}(1-w)^{b-1} dw}{\{cw + d(1-w) + e\}^{a+b}},$$

$$c \geq 0, \quad d \geq 0, \quad e > 0, \quad a > 0, \quad b > 0,$$

$$B(a,b) = \frac{(d-e)^a(c-e)^b}{(d-c)^{a+b-1}} \int_c^d \frac{(w-c)^{a-1}(d-w)^{b-1} dw}{(w-e)^{a+b}},$$

$$a > 0, \quad b > 0, \quad e < d < c,$$

$$B(a,b) = \int_0^1 \left[c^a w^{a-1}(1-cw)^{b-1} \right.$$

$$\left. + (1-c)^b w^{b-1} \{1-(1-c)w\}^{a-1} \right] dw,$$

$$a > 0, \quad b > 0, \quad |c| < 1,$$

$$B(a,b) = \frac{|c-e|^a |d-e|^b}{(c-d)^{a+b-1}} \int_c^d \frac{(d-z)^{a-1}(z-c)^{b-1} dz}{|z-e|^{a+b}},$$

$$a > 0, \quad b > 0, \quad 0 < e < c < d,$$

$$0 < c < d < e,$$

$$B\left(\frac{a}{b}, c\right) = b \int_0^1 w^{a-1} \left(1 - w^b\right)^{c-1} dw,$$

$$a > 0, \quad b > 0, \quad c > 0,$$

$$B\left(\frac{a}{2}, 1-b-\frac{a}{2}\right) = 2 \int_0^\infty z^{a-1} \left(1 + z^2\right)^{b-1} dz,$$

$$a > 0, \quad b + \frac{a}{2} < 1,$$

$$B\left(1 - b - \frac{a}{c}, b\right) = c \int_1^\infty z^{a-1} (z^c - 1)^{b-1} \, dz,$$

$$c > 0, \quad b > 0, \quad a < c(1 - b),$$

$$B\left(\frac{a}{c}, b - \frac{a}{c}\right) = c d^{a/c} \int_0^\infty z^{a-1} (1 + dz^c)^{-b} \, dz,$$

$$c > 0, \quad 0 < a < bc,$$

$$B(a, b) = 2^{2-a-b} \int_{-1}^1 \frac{(1 + z)^{2a-1}(1 - z)^{2b-1} dz}{(1 + z^2)^{a+b}},$$

$$a > 0, \quad b > 0,$$

$$B(ca, b) = \frac{1}{c} \int_0^\infty \{1 - \exp(-z/c)\}^{b-1} \exp(-az) dz,$$

$$a > 0, \quad b > 0, \quad c > 0,$$

$$B\left(a + \frac{b}{c}, a - \frac{b}{c}\right) = c 4^{1-a} \int_0^\infty \frac{\cosh 2bz}{\cosh^{2a} cz} dz,$$

$$a \pm b > 0, \quad c > 0, \quad b > 0,$$

$$B\left(\frac{a+1}{2}, \frac{b-a}{2}\right) = 2 \int_0^\infty \frac{\sinh^a z}{\cosh^b z} dz,$$

$$a > -1, \quad a - b < 0,$$

$$B\left(\frac{a}{2c} - \frac{b}{2}, b + 1\right) = c 2^{b+1} \int_0^\infty \sinh^b cz \exp(-az) dz,$$

$$c > 0, \quad b > -1, \quad a > bc,$$

$$B\left(\frac{a}{c} - b, 2b + 1\right) = c 2^b \int_0^\infty (\cosh cz - 1)^b \exp(-az) dz,$$

$$c > 0, \quad b > -\frac{1}{2}, \quad a > bc,$$

$$B\left(\frac{a}{2}, b - \frac{a}{2}\right) = 2 \int_0^{\pi/2} \tan^{a-1} w \cos^{2b-2} w dw$$

$$= 2 \int_0^{\pi/2} \cot^{a-1} w \sin^{2b-2} w dw,$$

$$0 < a < 2b,$$

$$B\left(\frac{a+b+1}{2}, \frac{a-b+1}{2}\right)$$

$$= \frac{a2^{a-1}}{\pi \sin(b\pi/2)} \bigg/ \int_0^\pi \sin^{a-1} w \sin bw \, dw$$

$$= \frac{a2^{a-1}}{\pi \cos(b\pi/2)} \bigg/ \int_0^\pi \sin^{a-1} w \cos bw \, dw$$

$$= \frac{a2^a}{\pi} \bigg/ \int_0^{\pi/2} \cos^{a-1} w \cos bw \, dw, \qquad a > 0,$$

$$B\left(\frac{a+b+1}{2}, \frac{a-b+1}{2}\right)$$

$$= \frac{\pi \sin(b\pi/2)}{a2^{a-1}} \bigg/ \int_{-\pi/2}^{\pi/2} \cos^{a-1} w \sin\left\{b\left(w + \frac{\pi}{2}\right)\right\} dw,$$

$$a > 0,$$

$$B\left(\frac{a+b}{2} + 1, \frac{a-b}{2} + 1\right)$$

$$= \frac{b\pi}{a(a+1)2^{a+1}} \bigg/ \int_0^{\pi/2} \cos^{a-1} w \sin bw \sin w \, dw,$$

$$\frac{1}{B(a+1, b+1)}$$

$$= \frac{(a+b+1)2^{a+b+2}}{\pi} \int_0^{\pi/2} \cos^{a+b} w \cos aw \cos bw \, dw - 1,$$

$$a + b > -1,$$

$$B(a, b) = \frac{1}{\sin(a\pi/2)} \int_0^{\pi/2} \sin^{a-1} w \cos^{b-1} w \sin(a+b)w \, dw$$

$$= \frac{1}{\cos(a\pi/2)} \int_0^{\pi/2} \sin^{a-1} w \cos^{b-1} w \cos(a+b)w \, dw$$

$$a > 0, \quad b > 0,$$

$$B(a + b - 1, 1 - a)$$

$$= -\frac{1}{\cos\left((a + b)\pi/2\right)} \int_0^{\pi/2} \tan^a w \sin^{b-2} w \sin bw dw$$

$$= \frac{1}{\sin\left((a + b)\pi/2\right)} \int_0^{\pi/2} \tan^a w \sin^{b-2} w \cos bw dw$$

$$= \frac{1}{\cos\left(a\pi/2\right)} \int_0^{\pi/2} \cot^a w \cos^{b-2} w \sin bw dw$$

$$= \frac{1}{\sin\left(a\pi/2\right)} \int_0^{\pi/2} \cot^a w \cos^{b-2} w \cos bw dw,$$

$$a + b > 1 > a,$$

$$B(a, b) = 2c^{2a} d^{2b} \int_0^{\pi/2} \frac{\sin^{2a-1} w \cos^{2b-1} w dw}{\left(c^2 \sin^2 w + d^2 \cos^2 w\right)^{a+b}},$$

$$a > 0, \quad b > 0,$$

$$B(a, b) = \int_0^{\pi/2} \frac{\sin^{a-1} w \cos^{b-1} w dw}{(\sin w + \cos w)^{a+b}},$$

$$a > 0, \quad b > 0,$$

$$B(a, b) = 2\left(1 - k^2\right)^a \int_0^{\pi/2} \frac{\sin^{2a-1} w \cos^{2b-1} w dw}{\left(1 - k^2 \sin^2 w\right)^{a+b}},$$

$$a > 0, \quad b > 0,$$

$$B(a, a) = 2^{2-2a} \int_0^1 \left(1 - w^2\right)^{a-1} dw$$

$$= 2^{1-2a} \int_0^1 \frac{(1 - w)^{a-1}}{\sqrt{w}} dw,$$

$$B\left(\frac{a}{2}, \frac{a}{2}\right) = 2^{2-a} \int_0^{\pi/2} \sin^{a-1} w dw$$

$$= 2^{2-a} \int_0^{\pi/2} \cos^{a-1} w dw,$$

$$B\left(\frac{n}{2},\frac{n}{2}\right) = 2(ab)^n \int_0^{\pi/2} \frac{\sin^{n-1} w \cos^{n-1} w\, dw}{\left(a^2 \cos^2 w + b^2 \sin^2 w\right)^n},$$
$$ab > 0,$$

$$B\left(\frac{a}{2},\frac{a}{2}\right) = 2^{1-a}\sqrt{(b^2-c^2)^a} \int_0^{\pi} \frac{\sin^{a-1} w\, dw}{(b+c\cos w)^a},$$
$$a > 0, \quad 0 < c < b,$$

$$B\left(\frac{a+1}{2},\frac{a+1}{2}\right) = \frac{\pi}{2^a} - \frac{a}{2^{a-1}} \int_0^{\pi/2} w \sin^a w \cot w\, dw,$$
$$a > -1,$$

$$B(a+b, a-b) = 4^{1-a} \int_0^{\infty} \frac{\cosh 2bz}{\cosh^{2a} z}\, dz,$$
$$a > |b|, \quad a > 0,$$

$$B\left(a,\frac{b}{c}\right) = z \int_0^1 (1-w^a)^{a-1}\, w^{b-1}\, dw,$$
$$z > 0, \quad \frac{b}{c} > 0, \quad a > 0,$$

$$\frac{1}{B(a,b)} = \frac{a+b-1}{2\pi(c+d)^{1-a-b}} \int_{-\infty}^{\infty} \frac{dz}{(c+iz)^a (d-iz)^b},$$
$$c > 0, \quad d > 0, \quad a,b \in \Re, \quad a+b > 1,$$

$$B(a+ib, a-ib) = 2^{1-2a}\alpha \exp\left(-2i\gamma b\right) \int_{-\infty}^{\infty} \frac{\exp\left(2i\gamma bz\right) dz}{\cosh^{2a}(\alpha z - \gamma)},$$
$$b,\gamma \in \Re, \quad \alpha > 0, \quad a > 0$$

and

$$\frac{1}{B(a,b)} = \frac{2^{a+b-1}(a+b-1)}{\pi} \int_0^{\pi/2} \cos\left((a-b)w\right) \cos^{a+b-2} w\, dw$$
$$= \frac{2^{a+b-2}(a+b-1)}{\pi \cos\left((a-b)\pi/2\right)} \int_0^{\pi} \cos\left((a-b)w\right) \sin^{a+b-2} w\, dw$$
$$= \frac{2^{a+b-2}(a+b-1)}{\pi \sin\left((a-b)\pi/2\right)} \int_0^{\pi} \sin\left((a-b)w\right) \sin^{a+b-2} w\, dw.$$

B.   Series Representations

Two infinite series representations of B are:

$$B(a,b) = \frac{1}{b}\sum_{k=0}^{\infty}(-1)^k b \frac{(b-1)\cdots(b-k)}{k!(a+k)},$$
$$b > 0,$$

$$B\left(a,\frac{1}{2}\right) = \sum_{k=1}^{\infty}\frac{(2k-1)!!}{2^k k!}\frac{1}{a+k} + \frac{1}{a}$$

and an infinite product representation is:

$$B(a+1,b+1) = \frac{1}{a+b+1}\prod_{i=1}^{\infty}\frac{i(a+b+i)}{(a+i)(b+i)},$$
$$a,b \neq -1,-2,\ldots.$$

C.   Functional Relations

The functional relations involving B are:

$$B(a,b) = \frac{\Gamma(a)\Gamma(b)}{\Gamma(a+b)} = B(b,a),$$

$$B(a+1,b) = \frac{a}{a+b}B(a,b),$$

$$B(a,b)B(a+b,c) = B(b,c)B(b+c,a),$$

$$\sum_{i=0}^{\infty}B(a,b+i) = B(a-1,b),$$

$$B(a,a) = 2^{1-2a}B\left(\frac{1}{2},a\right)$$

and

$$B(a,a)B\left(a+\frac{1}{2},a+\frac{1}{2}\right) = \frac{\pi}{2^{4a-1}a}.$$

D.   Special Cases

Finally, the closed form expressions of B which arise as special cases are:

$$B(a, 1-a) = \frac{\pi}{\sin a\pi},$$

$$\frac{1}{B(m,n)} = m\binom{m+n-1}{n-1}$$
$$= n\binom{m+n-1}{m-1},$$

$$B\left(\frac{1}{2}, \frac{n}{2}\right) = A\frac{(n-2)!!}{(n-1)!!},$$
$$A = 2 \text{ if } n \text{ is even and } A = \pi \text{ if } n \text{ is odd,}$$

$$B\left(\frac{3}{2}, \frac{n}{2}\right) = A\frac{(n-2)!!}{(n+1)!!},$$
$$A = 2 \text{ if } n \text{ is even and } A = \pi \text{ if } n \text{ is odd,}$$

and

$$B\left(\frac{a}{2}, \frac{a}{2}\right) = \sqrt{\pi}2^{1-a}\frac{\Gamma(a/2)}{\Gamma((a+1)/2)}$$

if a is a fraction with odd numerator and denominator.

II.   Incomplete Beta Function and Its Ratio

The incomplete beta function and the incomplete beta function ratio are defined by

$$B_x(a, b) = \int_0^x w^{a-1}(1-w)^{b-1}dw$$

and

$$I_x(a, b) = \frac{B_x(a, b)}{B(a, b)},$$

respectively. The incomplete beta function has the simple relationship

$$B_x(a, b) = \frac{x^a}{a} \, {}_2F_1(a, 1 - b; a + 1; x),$$ (II.1)

where ${}_2F_1$ is the well-known Gauss hypergeometric function defined by

$$ {}_2F_1(\alpha, \beta; \gamma; x) = \sum_{i=0}^{\infty} \frac{(\alpha)_i (\beta)_i}{(\gamma)_i} \frac{x^i}{i!},$$

where $(z)_i = z(z + 1) \cdots (z + i - 1)$ denotes the ascending factorial. The properties of the Gauss hypergeometric function are well established in the literature; see, for example, Section 9.1 of Gradshteyn and Ryzhik (2000). The properties of the incomplete beta function (or equivalently, the incomplete beta function ratio) can thus be deduced easily using (II.1). In this section we provide a comprehensive list of integral representations, series expansions, recurrence formulas, continued fractions, approximations, inequalities and closed form expressions.

A.   Integral Representations

The integral representations of I_x are:

$$I_x(a, b) = \frac{x^a}{aB(a, b)B(1 - b, a + b)}$$
$$\times \int_0^1 z^{-b}(1 - z)^{a+b-1}(1 - xz)^{-a}dz,$$

$$I_x(a, b) = \frac{2^{b-1}}{B(a, b)} \int_{-\log x}^{\infty} \sinh^{b-1}\left(\frac{z}{2}\right) \exp(-\gamma z)dz,$$
$$\gamma = a + (b - 1)/2,$$

$$I_x(a, a) = \frac{2^{1-2a}\Gamma(2a)}{\Gamma^2(a)} \int_{-\pi/2}^{\arcsin(2x-1)} \cos^{2a-1} w \, dw,$$

$$I_x(a, n-a+1) = 1 - \frac{1}{2\pi i} \int_O (1-x+zx)^n \frac{z^{-a}}{1-z} dz$$

if a is an integer,

$$I_x(a, b) = 1 - \frac{(1-x)^b}{2\pi i} \int_O (1-zx)^{-b} \frac{z^{-a}}{1-z} dz$$

$$= 1 - \frac{1}{2\pi i} \int_O \left\{ \frac{(1-x)z}{1-zx} \right\}^{-a} \left(\frac{1-x}{1-zx} \right)^{a+b} \frac{dz}{1-z}$$

if a is an integer,

$$I_x(a, b) = 1 - \frac{1}{2\pi i} \int_{-\infty}^{0+} \left(\frac{1-zx}{1-x} \right)^{-b} \frac{z^{-a}}{1-z} dz$$

and

$$I_x(a, b) = 1 - \frac{1}{2\pi i} \int_C \left\{ \frac{\sinh(z/2)}{\sinh\left(-\frac{1}{2}\log x\right)} \right\}^{-b}$$

$$\times \frac{x^{a+(b-1)/2} \exp\left\{(2a+b-1)z/2\right\}}{2\sinh\left\{(z+\log x)/2\right\}} dz,$$

where O denotes the contour of integration passing counter-clockwise round the origin while C denotes the contour going from $-\infty - \pi i$, across the real axis to the right of $-\log x$, and back to $-\infty + \pi i$.

B. Series Expansions

First consider the expansions of $I_x(a, b)$ in terms of elementary functions. Two expansions, due to Wishart (1927), for half-integer and integer values of a and b are:

$$I_x\left(\frac{n}{2}+1, \frac{n}{2}+1\right) = 1 - \frac{1}{w}\left(1 - \frac{w^2}{n}\right)^{n/2+1}\left[A_0 - \frac{A_1}{w^2+2}\right.$$

$$+ \frac{A_2}{(w^2+2)(w^2+4)}$$

$$\left. - \frac{A_3}{(w^2+2)(w^2+4)(w^2+6)} + \cdots \right],$$

where $w = 2\sqrt{n}(x - 1/2)$,

$$A_0 = \frac{1}{\sqrt{2\pi}} \left(1 - \frac{1}{4n} + \frac{1}{32n^2} + \frac{5}{128n^3} \right),$$

$$A_1 = \frac{1}{\sqrt{2\pi}} \left(1 + \frac{3}{4n} + \frac{25}{32n^2} + \frac{105}{128n^3} \right),$$

$$A_2 = \frac{1}{\sqrt{2\pi}} \left(1 + \frac{39}{4n} + \frac{1105}{32n^2} + \frac{13965}{128n^3} \right),$$

$$A_3 = \frac{1}{\sqrt{2\pi}} \left(5 + \frac{279}{4n} + \frac{20045}{32n^2} + \frac{522165}{128n^3} \right),$$

$$A_4 = \frac{1}{\sqrt{2\pi}} \left(9 + \frac{2151}{4n} + \frac{286185}{32n^2} + \frac{13207005}{128n^3} \right),$$

$$A_5 = \frac{1}{\sqrt{2\pi}} \left(129 + \frac{17907}{4n} + \frac{3895545}{32n^2} + \frac{283992345}{128n^3} \right)$$

and

$$I_x(l+1, m+1) = 1 - \left(1 + \sqrt{\frac{m}{nl}} u \right)^{l+1} \left(1 - \sqrt{\frac{l}{nm}} u \right)^{m+1}$$

$$\times \left[\frac{1}{u} \left\{ B_0 - \frac{B_1}{u^2 + r} \right. \right.$$

$$+ \frac{B_2}{(u^2 + r)(u^2 + 2r)}$$

$$\left. - \frac{B_3}{(u^2 + r)(u^2 + 2r)(u^2 + 3r)} + \cdots \right\}$$

$$- \frac{C_1}{(u^2 + r)(u^2 + 2r)}$$

$$\left. + \frac{C_2}{(u^2 + r)(u^2 + 2r)(u^2 + 3r)} - \cdots \right],$$

where $n = l + m$,

$$r = \frac{l}{m} + \frac{m}{l},$$

$$u = \sqrt{n(r+2)} \left(x - \frac{l}{n} \right),$$

$$B_0 = \frac{1}{\sqrt{2\pi}} \left(1 - \frac{1+r}{12n} + \frac{(1+r)^2}{288n^2} + \cdots \right),$$

$$B_1 = \frac{1}{\sqrt{2\pi}} \left(1 - \frac{11-r}{12n} + \frac{265 - 22r + r^2}{288n^2} + \cdots \right),$$

$$B_2 = \frac{1}{\sqrt{2\pi}} \left(3 - r - \frac{285 - 86r + r^2}{12n} \right.$$
$$\left. + \frac{31395 - 11059r + 169r^2 - r^3}{288n^2} + \cdots \right),$$

$$B_3 = \frac{1}{\sqrt{2\pi}} \left(15 - 9r + 2r^2 + \frac{4725 - 2430r + 247r^2 - 2r^3}{12n} \right.$$
$$+ \frac{1295175 - 696675r + 70607r^2 - 485r^3 + 2r^4}{288n^2}$$
$$\left. + \cdots \right),$$

$$B_4 = \frac{1}{\sqrt{2\pi}} \left(105 - 90r + 33r^2 - 6r^3 \right.$$
$$+ \frac{77175 - 57015r + 12585r^2 - 891r^3 + 6r^4}{12n}$$
$$+ \frac{41435625 - 30302328r + 6122214r^2 - 346494r^3 + 1749r^4 - 6r^5}{288n^2}$$
$$\left. + \cdots \right),$$

$$B_5 = \frac{1}{\sqrt{2\pi}} \left(945 - 1050r + 525r^2 - 105r^3 + 24r^4 \right.$$
$$+ \frac{1340955 - 130189r + 452025r^2 - 69375r^3 + 4014r^4 - 24r^5}{12n}$$
$$\left. + \cdots \right),$$

$$C_1 = \sqrt{\frac{2(r-2)}{n\pi}} \left(1 + \frac{35 - r}{12n} + \frac{1945 - 70r + r^2}{288n^2} + \cdots \right),$$

$$C_2 = \sqrt{\frac{2(r-2)}{n\pi}} \left(10 - 3r + \frac{1190 - 259r + 3r^2}{12n} \right.$$
$$\left. + \frac{184810 - 42487r + 508r^2 - 3r^3}{288n^2} + \cdots \right),$$

$$C_3 = \sqrt{\frac{2(r-2)}{n\pi}} \left(105 - 60r + 11r^2 \right.$$

$$+ \frac{28875 - 12789r + 1309r^2 - 11r^3}{12n}$$

$$+ \frac{9152385 - 3949698r + 356180r^2 - 2558r^3 + 11r^4}{288n^2}$$

By expanding a factor of the integrand in a binomial series and then integrating each term, Osborn and Madey (1968) obtained the following expansions for I_x for general a and b:

$$I_x(a,b) = \frac{x^a}{B(a,b)} \left\{ \frac{1}{a} + \frac{1-b}{a+1}x + \frac{(1-b)(2-b)}{2!(a+2)}x^2 \right.$$

$$\left. + \frac{(1-b)(2-b)(3-b)}{3!(a+3)}x^3 + \cdots \cdots \right\},$$

$$a \neq 0, 1, 2, \ldots \qquad (II.2)$$

and

$$I_x(a,b) = I_{1/2}(a,b) + \frac{1}{B(a,b)} \left\{ \frac{1-w^b}{b2^b} + \frac{(1-a)\left(1-w^{b+1}\right)}{1!(b+1)2^{b+1}} \right.$$

$$\left. + \frac{(1-a)(2-a)\left(1-w^{b+2}\right)}{2!(b+2)2^{b+2}} + \cdots \right\},$$

$$b \neq 0, 1, 2, \ldots, \qquad (II.3)$$

where $w = 2(1-x)$. Because of convergence considerations, (II.2) is useful when $0 < x \leq 1/2$ and (II.3) is useful when $1/2 < x \leq 1$.

Next consider expansions of $I_x(a,b)$ in terms of:

$$I(u,p) = \frac{1}{\Gamma(p+1)} \int_0^{u\sqrt{p+1}} z^p \exp(-z)dz, \qquad (II.4)$$

the incomplete gamma function, and

$$Q(u,p) = \frac{1}{\Gamma(p)} \int_u^{\infty} z^{p-1} \exp(-z)dz, \qquad (II.5)$$

the incomplete gamma function ratio. For integer values of a and b, an expansion of I_x in terms of (II.4) is:

$$I_x(l+1,m+1) = \frac{\Gamma(l+m+2)}{m^{l+1}\Gamma(m+1)}\left[I(w_0,l)\right.$$

$$-M_1\frac{\Gamma(l+3)}{\Gamma(l+1)}I(w_2,l+2)$$

$$-M_2\frac{\Gamma(l+4)}{\Gamma(l+1)}I(w_3,l+3)$$

$$+M_3\frac{\Gamma(l+5)}{\Gamma(l+1)}I(w_4,l+4)$$

$$\left.+M_4\frac{\Gamma(l+6)}{\Gamma(l+1)}I(w_5,l+5) - \cdots\right]$$

(assuming $m > l$) due to Wishart (1927), where $w_s = mx/(l+s+1)$,

$$M_1 = \frac{1}{2m},$$

$$M_2 = \frac{1}{3m^2},$$

$$M_3 = \frac{1}{8m^2}\left(1 - \frac{2}{m}\right),$$

$$M_4 = \frac{1}{6m^3}\left(1 - \frac{6}{5m}\right),$$

$$M_5 = \frac{1}{48m^3}\left(1 - \frac{26}{3m} + \frac{8}{m^2}\right),$$

$$M_6 = \frac{1}{24m^4}\left(1 - \frac{22}{5m} + \frac{24}{7m^2}\right),$$

$$M_7 = \frac{1}{384m^4}\left(1 - \frac{68}{3m} + \frac{348}{5m^2} - \frac{48}{m^3}\right),$$

$$M_8 = \frac{1}{144m^5}\left(1 - \frac{202}{45m} + \frac{892}{35m^2} - \frac{16}{m^3}\right),$$

$$M_9 = \frac{1}{3840m^5}\left(1 - \frac{140}{3m} + \frac{580}{3m^2} - \frac{4080}{7m^3} + \frac{384}{m^4}\right)$$

and an expansion of I_x in terms of (II.5) is:

$$I_x(a,b) = 1 - Q(w,a)$$
$$+ \frac{w^a \exp(-w)}{\Gamma(a)} \left\{ \frac{b_1}{2b} + \frac{b_2}{4b^2} + \frac{b_3}{8b^3} + \frac{b_4}{16b^4} \right\}$$

due to Hartley (1944), where $x = w/(w+b)$ and

$$b_1 = a - 1 - w,$$

$$b_2 = \frac{1}{6} \Big\{ (a-1)(a-2)(3a-1) - (a-1)(9a-2)w$$
$$+ (9a-1)w^2 - 3w^3 \Big\},$$

$$b_3 = \frac{1}{6} \Big\{ a(a-1)^2(a-2)(a-3) - a(a-1)(a-2)(5a-3)w$$
$$+ 2a(a-1)(5a-1)w^2 - 2a(5a+1)w^3$$
$$+ (5a+3)w^4 - w^5 \Big\},$$

$$b_4 = \frac{1}{360} \Big\{ (a-1)(a-2)(a-3)(a-4) \left(15a^3 - 30a^2 + 5a + 2 \right)$$
$$- (a-1)(a-2)(a-3) \left(105a^3 - 135a^2 + 10a + 8 \right) w$$
$$+ (a-1)(a-2) \left(315a^3 - 180a^2 + 5a + 12 \right) w^2$$
$$- (a-1) \left(525a^3 + 75a^2 + 50a + 8 \right) w^3$$
$$+ \left(525a^3 + 450a^2 + 175a + 2 \right) w^4$$
$$- 5 \left(63a^2 + 99a + 46 \right) w^5 + 15(7a+9)w^6 - 15w^7 \Big\}.$$

The second expansion is not suitable when a is large, because of slow convergence of $b_k/(2b)^k$. For small values of a and moderate or large values of b, however, the formula is quite useful. An expansion of I_x in terms of (II.4) for general a and b is:

$$I_x(a,b) = a_0 I \left(\frac{x(b-1)}{\sqrt{a}}, a-1 \right) + a_2 I \left(\frac{x(b-1)}{\sqrt{a+2}}, a+1 \right)$$
$$+ a_3 I \left(\frac{x(b-1)}{\sqrt{a+3}}, a+2 \right) + \cdots$$
$$+ a_s I \left(\frac{x(b-1)}{\sqrt{a+s}}, a+s-1 \right) + \cdots$$

due to Pearson and Pearson (1935), where

$$a_0 = \frac{\Gamma(a+b)}{\Gamma(b)} \frac{1}{(b-1)^a},$$

$$a_2 = -\frac{\Gamma(a+b)}{\Gamma(b)} \frac{(a+1)a}{2(b-1)^{a+1}},$$

$$a_3 = -\frac{\Gamma(a+b)}{\Gamma(b)} \frac{(a+2)(a+1)a}{3(b-1)^{a+2}},$$

$$a_4 = \frac{\Gamma(a+b)}{\Gamma(b)} \frac{(a+3)(a+2)(a+1)a}{8(b-1)^{a+2}} \left(1 - \frac{2}{b-1}\right),$$

$$a_5 = \frac{\Gamma(a+b)}{\Gamma(b)} \frac{(a+4)(a+3)(a+2)(a+1)a}{6(b-1)^{a+3}} \left(1 - \frac{6}{5(b-1)}\right),$$

$$a_6 = -\frac{\Gamma(a+b)}{\Gamma(b)} \frac{(a+5)(a+4)\cdots(a+1)a}{48(b-1)^{a+3}}$$
$$\times \left(1 - \frac{26}{3(b-1)} + \frac{8}{(b-1)^2}\right),$$

$$a_7 = -\frac{\Gamma(a+b)}{\Gamma(b)} \frac{(a+6)(a+5)\cdots(a+1)a}{24(b-1)^{a+4}}$$
$$\times \left(1 - \frac{22}{5(b-1)} + \frac{24}{7(b-1)^2}\right),$$

$$a_8 = \frac{\Gamma(a+b)}{\Gamma(b)} \frac{(a+7)(a+6)\cdots(a+1)a}{384(b-1)^{a+4}}$$
$$\times \left(1 - \frac{68}{3(b-1)} + \frac{696}{10(b-1)^2} - \frac{48}{(b-1)^3}\right),$$

$$a_9 = \frac{\Gamma(a+b)}{\Gamma(b)} \frac{(a+8)(a+7)\cdots(a+1)a}{144(b-1)^{a+5}}$$
$$\times \left(1 - \frac{472}{45(b-1)} + \frac{892}{35(b-1)^2} - \frac{16}{(b-1)^3}\right),$$

$$a_{10} = -\frac{\Gamma(a+b)}{\Gamma(b)} \frac{(a+9)(a+8)\cdots(a+1)a}{3840(b-1)^{a+5}}$$
$$\times \left(1 - \frac{140}{3(b-1)} + \frac{964}{3(b-1)^2} - \frac{23088}{35(b-1)^3}\right.$$
$$\left. + \frac{384}{(b-1)^4}\right),$$

$$a_{11} = -\frac{\Gamma(a+b)}{\Gamma(b)} \frac{(a+10)(a+9)\cdots(a+1)a}{1152(b-1)^{a+6}}$$

$$\times \left(1 - \frac{916}{45(b-1)} + \frac{3716}{35(b-1)^2} - \frac{6704}{35(b-1)^3}\right.$$

$$\left. + \frac{1152}{11(b-1)^4}\right),$$

$$a_{12} = \frac{\Gamma(a+b)}{\Gamma(b)} \frac{(a+11)(a+10)\cdots(a+1)a}{46080(b-1)^{a+6}}$$

$$\times \left(1 - \frac{250}{3(b-1)} + \frac{28828}{27(b-1)^2} - \frac{478024}{105(b-1)^3}\right.$$

$$\left. + \frac{155552}{21(b-1)^4} - \frac{3840}{(q-1)^5}\right).$$

More recently, there have been a number of expansions of $I_x(a,b)$ in terms of (II.5) for general a and b. Molina (1932) and Gnanadesikan et al. (1966) gave the following form useful when $2a + b > 1$:

$$I_x(a,b) = \sum_{k=0}^{6} \frac{A_k}{k!} \frac{\Gamma(b+k)}{N^{b+k}} \{1 - Q(y, b+k)\} + \cdots, \qquad (II.6)$$

where

$$N = a + \frac{b-1}{2},$$
$$y = -N \log x,$$
$$A_0 = 1,$$
$$A_1 = 0,$$
$$A_2 = \frac{b-1}{12},$$
$$A_3 = 0,$$
$$A_4 = \frac{(b-1)(5b-7)}{240},$$
$$A_5 = 0,$$
$$A_6 = \frac{(b-1)\left(35b^2 - 112b + 93\right)}{4032}.$$

An equivalent form of (II.6) was derived by Wise (1950):

$$I_x(a,b) = 1 - Q(b,y) - \frac{P_2(y)}{24N^2} + \frac{P_4(y)}{5760N^4} + \cdots,$$

where N and y are as defined above, and

$$P_2(y) = \frac{y^b \exp(-y)}{(b-2)!}(b+1+y),$$

$$P_4(y) = \frac{y^b \exp(-y)}{(b-2)!}\left\{(b-3)(b-2)(5b+7)(b+1+y)\right.$$
$$\left. -(5b-7)y^2(b+3+y)\right\}.$$

DiDonato and Morris (1992) derived another equivalent form of (II.6) suitable for large a, small b and $x > 1/2$, taking the form:

$$I_x(a,b) = \frac{1}{B(a,b)N^b} \sum_{k=0}^{\infty} \frac{c_k\Gamma(b+2k)}{N^{2k}}Q(b+2k,y), \qquad \text{(II.7)}$$

where N and y are as defined above, and c_k are the expansion coefficients of

$$\left(\frac{\sinh(t/2)}{t/2}\right)^{b-1} = \sum_{k=0}^{\infty} c_k t^{2k}.$$

These coefficients can be expressed in terms of the generalized Bernoul[l] polynomials as:

$$c_k = \frac{1}{(2k)!}B_{2k}^{1-b}\left(\frac{1-b}{2}\right).$$

Writing $d_k = 12^k c_k/z$, where $z = b - 1$, the first few d_k are:

$$d_1 = 1/2,$$
$$d_2 = \frac{z}{8} - \frac{1}{20},$$
$$d_3 = \frac{z^2}{48} - \frac{z}{40} + \frac{1}{105},$$
$$d_4 = \frac{z^3}{384} - \frac{z^2}{160} + \frac{101z}{16800} - \frac{3}{1400},$$

$$d_5 = \frac{z^4}{3840} - \frac{z^3}{960} + \frac{61z^2}{33600} - \frac{13z}{8400} + \frac{1}{1925},$$

$$d_6 = \frac{z^5}{46080} - \frac{z^4}{7680} + \frac{143z^3}{403200} - \frac{59z^2}{112000} + \frac{7999z}{19404000} - \frac{691}{5255250},$$

$$d_7 = \frac{z^6}{645120} - \frac{z^5}{76800} + \frac{41z^4}{806400} - \frac{11z^3}{96000} + \frac{5941z^2}{38808000} - \frac{2357z}{21021000} + \frac{6}{175175},$$

$$d_8 = \frac{z^7}{10321920} - \frac{z^6}{921600} + \frac{37z^5}{6451200} - \frac{73z^4}{4032000} + \frac{224137z^3}{6209280000} - \frac{449747z^2}{10090080000} + \frac{52037z}{1681680000} - \frac{10851}{1191190000}.$$

Doman (1996) provided the following cleaner version of (II.7):

$$I_x(a,b) = Q(b,y) + \frac{x^N}{B(a,b)} \sum_{k=0}^{\infty} \frac{T_k(b,x)}{N^{k+1}},$$

where

$$T_k(b,x) = \frac{d^k}{dt^k} \left(\sum_{l=[k/2]+1}^{\infty} c_l t^{2l+b-1} \right) \Bigg|_{t=-\log x}.$$

These quantities T_k satisfy the simple recurrence formulas:

$$T_{2k+1} = \frac{d}{dt} T_{2k},$$

$$T_{2k} = \frac{d}{dt} T_{2k-1} - c_k t^{b-1} \frac{\Gamma(2k+b)}{\Gamma(b)}.$$

Actually, T_k can be expressed directly in terms of b and x, for example,

$$T_0(b,x) = \left(\frac{1}{\sqrt{x}} - \sqrt{x} \right)^{b-1} - (-\log x)^{b-1}.$$

Finally, an expansion of I_x, due to Wishart (1927), in terms of

$$m_s(u) = \frac{1}{\sqrt{2\pi}} \frac{1}{(s-1)(s-3)\cdots 2 \text{ or } 1} \int_0^u z^s \exp\left(-z^2/2\right) dz,$$

the incomplete normal moment function, is:

$$I_x(l+1, m+1) = k_s\bigg\{ m_0(u) - k_3 m_3(u) - k_4 m_4(u)$$
$$-k_5 m_5(u) + k_6 m_6(u) + k_7 m_7(u)$$
$$+k_8 m_8(u) - k_9 m_9(u) - k_{10} m_{10}(u)$$
$$+k_{12} m_{12}(u) \bigg\},$$

where $n = l + m$,

$$r = \frac{l}{m} + \frac{m}{l},$$

$$u = \sqrt{(r+2)n}\left(x - \frac{l}{n}\right),$$

$$k_0 = \left(1 + \frac{1}{n}\right)\left(1 - \frac{1+r}{12n} + \frac{(1+r)^2}{288n^2}\right),$$

$$k_3 = \pm\frac{2}{3}\sqrt{\frac{r-2}{n}},$$

$$k_4 = \frac{3}{4}\frac{r-1}{n},$$

$$k_5 = \pm\frac{8r}{5n}\sqrt{\frac{r-2}{n}},$$

$$k_6 = \frac{5}{6}\left\{\frac{r-2}{n} - \frac{3\left(r^2 - r - 1\right)}{n^2}\right\},$$

$$k_7 = \pm\frac{4(r-1)}{n}\sqrt{\frac{r-2}{n}},$$

$$k_8 = \frac{7}{32}\frac{47r^2 - 94r + 15}{n^2},$$

$$k_9 = \pm\frac{64}{27}\frac{r-2}{n}\sqrt{\frac{r-2}{n}},$$

$$k_{10} = \frac{105}{8}\frac{(r-1)(r-2)}{n^2},$$

$$k_{12} = \frac{385}{72}\frac{(r-2)^2}{n^2}.$$

This formula is best suited for use when l and m are roughly of the same order and large.

C. Recursive Formulas

The recursive formulas involving I_x are:

$$I_x(a, b) = 1 - I_{1-x}(b, a),$$

$$I_x(a, b) = I_x(a + 1, b - 1) + \binom{a + b - 1}{a} x^a (1 - x)^{b-1},$$

$$I_x(a, b) = I_x(a + 1, b) + \binom{a + b - 1}{a} x^a (1 - x)^b,$$

$$I_x(a, b + 1) = I_x(a, b) + \binom{a + b - 1}{b} x^a (1 - x)^b,$$

$$I_x(a, b) = I_x(a + 1, b + 1) + \binom{a + b}{a} x^a (1 - x)^b \left(\frac{a}{a + b} - x\right),$$

$$x I_x(a, b) = I_x(a + 1, b) - (1 - x) I_x(a + 1, b + 1),$$

$$(a + b - ax) I_x(a, b) = b I_x(a, b + 1) + a(1 - x) I_x(a + 1, b - 1),$$

$$b I_x(a, b + 1) = (a + b) I_x(a, b) - a I_x(a + 1, b),$$

$$b I_x(a + 1, b + 1) = (a + b) x I_x(a, b) - \overline{(a + bx - b)} I_x(a + 1, b),$$

$$a I_x(a + 1, b + 1) = (a + b)(1 - x) I_x(a, b) - \overline{(b - a + bx)} I_x(a, b + 1),$$

$$(a + b - 1) x I_x(a - 1, b) = \overline{(a + b - 1x + a)} I_x(a, b) - a I_x(a + 1, b),$$

$$(a+b)(1-x)I_x(a+1,b-1) = \{(a+b)(1-x)+b\} I_x(a+1,b)$$
$$-aI_x(a+1,b+1),$$

$$(a)_n I_x(a+n,b) = \sum_{i=0}^{n}(-1)^i \binom{n}{i}(a+b+i)_{n-i}(b)_i I_x(a,b+i),$$

$$(b)_n I_x(a,b+n) = \sum_{i=0}^{n}(-1)^i \binom{n}{i}(a+b+i)_{n-i}(a)_i I_x(a+i,b)$$

and

$$(a+b)_n I_x(a,b) = \sum_{i=0}^{n}\binom{n}{i}(a)_{n-i}(b)_i I_x(a+n-i,b+i).$$

D. Continued Fractions

The continued fractions of I_x are:

$$I_x(a,b) = C \left[\frac{b_1}{1+}\frac{b_2}{1+}\frac{b_3}{1+}\frac{b_4}{1+}\cdots\right]$$

due to Müller (1930-1931), where

$$C = \frac{\Gamma(a+b)}{\Gamma(a+1)\Gamma(b)}x^a(1-x)^{b-1},$$

$$b_1 = 1,$$

$$\mu_s = \frac{b-s}{a+s},$$

$$b_{2s} = -\frac{(a+s-1)(q+s)\mu_s}{(a+2s-2)(a+2s-1)}\frac{x}{1-x},$$

$$b_{2s+1} = \frac{s(a+b+s)}{(a+2s-1)(a+2s)}\frac{x}{1-x};$$

$$I_x(a,b) = C \left[\frac{1}{1+}\frac{c_1 x}{1+}\frac{c_2 x}{1+}\cdots\right]$$

due to Aroian (1941), where

$$C = \frac{\Gamma(a+b)}{\Gamma(a+1)\Gamma(b)}x^a(1-x)^b,$$

due to Aroian (1941), where

$$C = \frac{\Gamma(a+b)}{\Gamma(a+1)\Gamma(b)} x^a (1-x)^b,$$

$$c_1 = -\frac{a+b}{a+1},$$

$$c_2 = \frac{b-1}{(a+1)(a+2)},$$

$$c_{2s} = \frac{s(b-s)}{(a+2s-1)(a+2s)},$$

$$c_{2s+1} = -\frac{(a+s)(a+b+s)}{(a+2s)(a+2s-1)};$$

$$I_x(a,b) = C \left[1 + \frac{k_1 x}{1 + l_1 x +} \frac{k_2 x^2}{1 + l_2 x +} \frac{k_3 x^2}{1 + l_3 x +} \cdots \right]$$

also due to Aroian (1941), where

$$C = \frac{\Gamma(a+b)}{\Gamma(a+1)\Gamma(b)} x^a (1-x)^b,$$

$$k_1 = \frac{a+b}{a+1},$$

$$l_1 = -\frac{a+b+1}{a+2},$$

$$k_{s+1} = \frac{s(b-s)(a+s)(a+b+s)}{(a+2s-1)(a+2s)^2(a+2s+1)},$$

$$l_{s+1} = \frac{(s+1)(b-s-1)}{(a+2s+1)(a+2s+2)} - \frac{(a+s)(a+b+s)}{(a+2s)(a+2s+1)};$$

$$I_x(a,b) = C \left[1 - \frac{\gamma_1}{1+\gamma_1 -} \frac{\gamma_2}{1+\gamma_2 -} \frac{\gamma_3}{1+\gamma_3 -} \frac{\gamma_4}{1+\gamma_4 -} \cdots \right]^{-1}$$

also due to Aroian (1941, 1959), where

$$C = \frac{\Gamma(a+b)}{\Gamma(a+1)\Gamma(b)}x^a(1-x)^b,$$

$$\gamma_r = \frac{a+b+r-1}{a+r}x;$$

and,

$$I_x(a,b) = C\left[1 + \frac{k_1 w}{1+l_1 w+}\frac{k_2 w^2}{1+l_2 w+}\frac{k_3 w^2}{1+l_3 w+}\cdots\right]$$

due to Tretter and Walster (1979), where

$$C = \frac{\Gamma(a+b)}{\Gamma(a+1)\Gamma(b)}x^a(1-x)^{b-1},$$

$$w = \frac{x}{1-x},$$

$$k_1 = \frac{b-1}{a+1},$$

$$l_1 = \frac{2-b}{a+2},$$

$$k_s = \frac{(s-1)(a+b+s-2)(a+s-1)(b-s)}{(a+2s-3)(a+2s-2)^2(a+2s-1)},$$

$$l_s = \frac{(a+s-1)(s-b)}{(a+2s-2)(a+2s-1)} + \frac{s(a+b+s-1)}{(a+2s-1)(a+2s)}.$$

E. Approximations and Inequalities

Two approximations of I_x are:

$$I_x(a,a) \approx \frac{1}{2} + \Phi(\eta)$$

due to Cadwell (1952), where

$$x = \frac{1}{2} + \sqrt{\frac{\pi}{3}}\Phi\left(\sqrt{\frac{3}{4a-1}}\eta\right),$$

$$\Phi(\eta) = \frac{1}{\sqrt{2\pi}} \int_0^\eta \exp\left(-\frac{z^2}{2}\right) dz,$$

and

$$I_x(a+b, b-s) \approx \frac{2 \sim (b-s)\pi\sqrt{\xi(\xi-1)/2\pi(a+b)}}{\xi - x}$$

$$\times \left\{ \left(\frac{x}{\xi}\right)^\xi \left(\frac{1-x}{\xi-1}\right)^{1-\xi} \right\}^{a+b},$$

$$b - s < 0 \text{ is an integer}$$

due to Aroian (1941), where $\xi = (a+s)/(a+b)$.

For $a \gg b \gg 1$, an inequality involving B_x due to Amos (1963) is:

$$\frac{1}{ba^b}\left(1 - \frac{1}{a}\right)^{a-1} < B_{1/a}(b, a) < \frac{1}{a^b}\left\{1 - \left(1 - \frac{1}{a}\right)^a\right\}.$$

Volodin (1970) derived sharp bounds for I_x for values of $a \leq 1$ and $b \leq 1$. Write

$$\epsilon = a + b,$$
$$\gamma = \frac{a}{\epsilon},$$
$$J_x(a, b) = (1 - \gamma)\left(\frac{x}{1-x}\right)^a \quad \text{if } 0 \leq x \leq 1/2,$$
$$J_x(a, b) = 1 - \gamma\left(\frac{1-x}{x}\right)^b \quad \text{if } 1/2 \leq x \leq 1,$$
$$J_x(a, b) \equiv 0 \quad \text{if } x < 0,$$
$$J_x(a, b) \equiv 1 \quad \text{if } x > 1,$$
$$A = \frac{\Gamma(1+\epsilon)}{\Gamma(1+a)\Gamma(1+b)}.$$

Volodin gave the following estimates for I_x:

$$J_x(a, b) \leq I_x(a, b) \leq J_x(a, b) + \frac{b}{a+1}\left(A^{1/\epsilon} - 1\right)^{a+1}$$

if $0 \leq x \leq 1 - A^{-1/\epsilon} \leq 1/2$;

$$J_x(a, b) - \frac{a}{b+1}\left(A^{1/\epsilon} - 1\right)^{b+1} \leq I_x(a, b) \leq J_x(a, b)$$

if $1/2 \le A^{-1/\epsilon} \le x \le 1$; and,

$$J_x(a,b) - \frac{a}{b+1}\left(A^{1/\epsilon} - 1\right)^{b+1} \le I_x(a,b)$$

$$\le J_x(a,b) + \frac{b}{a+1}\left(A^{1/\epsilon} - 1\right)^{a+1}$$

if $1 - A^{-1/\epsilon} \le x \le A^{-1/\epsilon}$. Moreover, if $0 < b \le a \le 1$ then Volodin showed that

$$\sup_{0 \le x \le 1} |I_x(a,b) - J_x(a,b)| \le \frac{a}{1+b}\left(A^{1/\epsilon} - 1\right)^{1+b}.$$

F.   Special Cases

Finally, the closed form expressions for I_x which arise as special cases are:

$$I_x(a,1) = x^a, \qquad I_x(1,b) = 1 - (1-x)^b,$$

$$I_x(a, n-a+1) = \sum_{i=a}^{n}\binom{n}{i}x^i(1-x)^{n-i}$$

if a is an integer,

$$I_x(a,b) = 1 - \sum_{i=1}^{a}\frac{\Gamma(b+i-1)}{\Gamma(b)\Gamma(i)}x^{i-1}(1-x)^b,$$

if a is an integer,

$$I_x(a,b) = \sum_{i=1}^{b}\frac{\Gamma(a+i-1)}{\Gamma(a)\Gamma(i)}x^a(1-x)^{i-1},$$

if b is an integer,

$$I_x\left(\frac{1}{2}, \frac{1}{2}\right) = \frac{2}{\pi}\arctan\sqrt{\frac{x}{1-x}},$$

$$I_x\left(k - \frac{1}{2}, \frac{1}{2}\right) = I_x\left(\frac{1}{2}, \frac{1}{2}\right) - \sqrt{x(1-x)}\sum_{l=1}^{k-1}d_l$$

and, more generally,

$$
I_x\left(k - \frac{1}{2}, j - \frac{1}{2}\right) = I_x\left(k - \frac{1}{2}, \frac{1}{2}\right) + \sum_{l=1}^{j-1} c_l,
$$

where

$$
c_l = \frac{\Gamma(k + l - 1)}{\Gamma(k - 1/2)\Gamma(l + 1/2)} x^{k-1/2}(1 - x)^{l-1/2},
$$

$$
d_l = \frac{\Gamma(l)}{\Gamma(l + 1/2)\Gamma(1/2)} x^{l-1}.
$$

References

1. Amos, D.E. Additional percentage points for the incomplete beta distribution. Biometrika **1963**, *50*, 449-457.
2. Aroian, L.A. Continued fractions for the incomplete beta function. Annals of Mathematical Statistics **1941**, *12*, 218-223.
3. Aroian, L.A. Corrections to "Continued fractions for the incomplete beta function." Annals of Mathematical Statistics **1959**, *30*, 1265.
4. Cadwell, J.H. An approximation to the symmetrical incomplete beta function. Biometrika **1952**, *39*, 204-207.
5. Didonato, A.R.; Morris, A.H. Significant digit computation of the incomplete beta function ratios, ACM Transactions on the Mathematical Software **1992**, *18*, 360-373.
6. Doman, B.G.S. An asymptotic expansion for the incomplete beta function, Mathematics of Computation **1996**, *65*, 1283-1288.
7. Gnanadesikan, R.; Hughes, L.P.; Pinkham, R.S. Numerical evaluation of partial derivatives of the incomplete beta integral, Unpublished Bell Telephone Laboratories Memo **1966**.
8. Gradshteyn, I.S.; Ryzhik, I.M. *Tables of Integrals, Series, and Products*; Academic Press: San Diego, 2000; 1163 pp.
9. Hartley, H. O. Studentization, or the elimination of the standard deviation of the parent population from the random sample-distribution of statistics. Biometrika **1944**, *33*, 173.

10. Molina, E.C. An expansion for Laplacian integrals in terms of incomplete gamma functions, and some applications. Bell Systems Technical Journal **1932**, *11*, 563-575.

11. Müller, J.H. On the application of continued fractions to the evaluation of certain integrals, with special reference to the incomplete Beta function, Biometrika **1930-31**, *22*, 284-297.

12. Pearson, K.; Pearson, M.V. On the numerical evaluation of high order incomplete Eulerian integrals. Biometrika **1935** *27*, 409-423.

13. Tretter, M.J.; Walster, G.W. Continued fractions for the incomplete beta function: additions and corrections, Annals of Statistics **1979**, *7*, 462-465.

14. Volodin, I.N. Beta distribution for small values of parameters. Theory of Probability and Its Applications **1970**, *15*, 563-566.

15. Wise, M.E. The incomplete beta function as a contour integral and a quickly converging series for its inverse, Biometrika **1950**, *37*, 208-218.

16. Wishart, J. On the approximate quadrature of certain skew curves, with an account of the researchers of Thomas Bayes. Biometrika **1927**, *19*, 1-18.

Mathematical Properties of the Beta Distribution

Arjun Gupta[1] and Saralees Nadarajah[2]

[1]Department of Mathematics and Statistics
Bowling Green State University
Bowling Green, Ohio 43403
[2]Department of Mathematics
University of South Florida
Tampa, Florida 33620

I. Definition

The pdf of the Beta distribution is:

$$f(x; a, b) = \frac{1}{B(a,b)} x^{a-1}(1-x)^{b-1}, \qquad 0 < x < 1, \qquad (I.1)$$

where $a > 0$, $b > 0$ and $B(a,b)$ is the Beta function discussed in Chapter 1. The parameters a and b are symmetrically related by

$$f(x; a, b) = f(1 - x; b, a). \qquad (I.2)$$

This implies that if X has the beta distribution with parameters a and b then $1 - X$ has the beta distribution with parameters b and a. The pdf (I.1) corresponds to type I distribution in the system of Pearson curves. The special case of (I.1) for $a = b = 1$ is the uniform distribution. The special case for $a = b = 1/2$ is the arc-sine distribution with the pdf

$$f(x) = \frac{1}{\pi\sqrt{x(1-x)}}, \qquad 0 < x < 1.$$

When $b = 1$ the beta distribution is known as the power function distribution.

II. CDF

The cumulative distribution function of (I.1) is:

$$\begin{aligned}
F(x) &= \frac{1}{B(a,b)} \int_0^x w^{a-1}(1-w)^{b-1} dw \\
&= \frac{B_x(a,b)}{B(a,b)} \\
&= I_x(a,b),
\end{aligned}$$

where B_x and I_x are the incomplete beta function and the incomplete beta function ratio, respectively (see Chapter 1). If a and b are both integers then the cdf above can be evaluated as the Binomial sum:

$$F(x) = 1 - \sum_{i=0}^{a-1} \binom{a+b-1}{i} x^i (1-x)^{a+b-1-i}.$$

For general a and b, using the relationship (II.1) in Chapter 1, we can write

$$F(x) = \frac{x^a}{aB(a,b)} \sum_{i=0}^{\infty} \frac{(a)_i (1-b)_i}{(a+1)_i} \frac{x^i}{i!}.$$

From (I.2), it follows that $F(x; a, b) = 1 - F(1-x; b, a)$.

III. Characteristic Function

The characteristic function of (I.1) is:

$$E\left[\exp(itX)\right] = {}_1F_1\left(a; a+b; it\right),$$

where $_1F_1$ is the confluent hypergeometric function defined by

$$_1F_1(\alpha, \beta; z) = \sum_{k=0}^{\infty} \frac{(\alpha)_k}{(\beta)_k} \frac{z^k}{k!};$$

see Section 9.2 of Gradshteyn and Ryzhik (2000) for detailed properties of this function. Of course, the moment generating function of (I.1) is $_1F_1(a; a + b; t)$.

IV. Moments

The rth moment about zero associated with (I.1) is:

$$\mu_r' = \frac{B(a + r, b)}{B(a, b)}$$
$$= \frac{\Gamma(a + r)\Gamma(a + b)}{\Gamma(a)\Gamma(a + b + r)}$$
$$= \frac{(a)_r}{(a + b)_r}$$

if r is an integer. In particular,

$$E(X) = \frac{a}{a + b},$$

$$Var(X) = \frac{ab}{(a + b)^2} \frac{1}{a + b + 1},$$

$$CV(X) = \sqrt{\frac{b}{a(a + b + 1)}},$$

$$E\left[X^{-k}\right] = \frac{(a + b - 1)(a + b - 2) \cdots (a + b - k)}{(a - 1)(a - 2) \cdots (a - k)},$$

$$E\left[X^{-1}\right] = \frac{a + b - 1}{a - 1}.$$

Note also that

$$E\left[X^k(1 - X)^{-k}\right] = \frac{\Gamma(a + k)\Gamma(b - k)}{\Gamma(a)\Gamma(b)},$$

$$E\left[(1 - X)^{-1}\right] = \frac{a + b - 1}{b - 1}$$

and

$$E\left(X^r\right) = \frac{a+r-1}{a+b+r-1} E\left(X^{r-1}\right), \qquad r \geq 1.$$

Pham-Gia (1994) recently established some simple bounds for $Var(X)$, showing $Var(X) < 1/4$ for any a and b; $Var(X) < 1/12$ if the pdf (I.1) is unimodal, i.e. $a > 1$ and $b > 1$; and, $Var(X) > 1/12$ if the pdf (I.1) is U-shaped, i.e. $a < 1$ and $b < 1$.

The central moments of (I.1) satisfy the following recurrence relation:

$$\mu_{s+1} = -\frac{s\lambda}{1+s\lambda} + \theta \sum_{i=1}^{s} \binom{s}{i} \frac{\lambda^i(1-\theta)^i i!}{(1+s\lambda)\cdots(1+[s-j]\lambda)} \mu_{s-i},$$

where $\lambda = 1/(a+b)$ and $\theta = a/(a+b)$ (Mühlbach, 1972). Some initial values are $\mu_0 = 1$, $\mu_1 = 0$,

$$\mu_2 = \frac{\lambda\theta(1-\theta)}{1+\lambda}$$

and

$$\mu_3 = \frac{2\lambda^2\theta(1-\theta)}{(1+\lambda)(1+2\lambda)}(1-2\theta).$$

V.   Cumulants

The cumulant generating function of (I.1) is:

$$K(t) = \log\left[\frac{\Gamma(a+b)}{\Gamma(a)}\right] - \log\left[\frac{\Gamma(a+b-t)}{\Gamma(a-t)}\right].$$

Hence the cumulants are

$$\kappa_r = (r-1)! \sum_{k=0}^{b-1} (a+k)^{-r}, \qquad r \geq 1$$

if b is an integer. In the general case

$$\kappa_r = (-1)^r \left[\psi^{(r-1)}(a) - \psi^{(r-1)}(a+b)\right],$$

where

$$\psi^{(r-1)}(x) = \frac{d^r}{dx^r} \log\Gamma(x).$$

is the rth derivative of the log-gamma function. Some values of this derivative useful for computing the cumulants are:

$$\psi^{(1)}(n+1) = -C + \sum_{k=1}^{n} \frac{1}{k},$$

$$\psi^{(1)}\left(\frac{1}{2}+n\right) = -C + 2\left\{\sum_{k=1}^{n} \frac{1}{2k-1} - \log 2\right\},$$

$$\psi^{(1)}\left(\frac{m}{n}\right) = -C - \log(2n) - \frac{\pi}{2}\cot\frac{m\pi}{n}$$
$$+2\sum_{k=1}^{[(n-1)/2]} \cos\frac{2mk\pi}{n}\log\sin\frac{k\pi}{n},$$
$$m = 1, 2, \ldots, n-1; n = 2, 3, \ldots,$$

$$\psi^{(2)}(n) = \frac{\pi^2}{6} - \sum_{k=1}^{n-1} \frac{1}{k^2},$$

$$\psi^{(2)}\left(\frac{1}{2}+n\right) = \frac{\pi^2}{2} - 4\sum_{k=1}^{n} \frac{1}{(2k-1)^2}$$

and

$$\psi^{(n)}(x) = (-1)^{n+1}n! \sum_{k=1}^{\infty} \frac{1}{(x+k)^{n+1}},$$

where $C = 0.57721566490\cdots$ is the Euler's constant.

VI. Mean Deviations

The mean deviation of (I.1) is:

$$\delta_1(X) = E\left(|X - E(X)|\right)$$
$$= \frac{2}{B(a,b)} \frac{a^a b^b}{(a+b)^{a+b+1}}.$$

For a and b large, using Stirling's approximation to the gamma function, we get the approximations

$$\delta_1(X) \approx \sqrt{\frac{2ab}{\pi(a+b)}} \frac{1}{a+b} \left(1 + \frac{1}{12(a+b)} - \frac{1}{12a} - \frac{1}{12b} \right)$$

and

$$\frac{\delta_1(X)}{SD(X)} \approx \sqrt{\frac{2}{\pi}} \left(1 + \frac{7}{12(a+b)} - \frac{1}{12a} - \frac{1}{12b} \right).$$

The mean deviation about the median (say, m) is:

$$\delta_2(X) = E\left(|X - m|\right)$$
$$= \frac{2m^a(1-m)^b}{(a+b)B(a,b)}.$$

For the symmetric beta distribution ($a = b$), both $\delta_1(X)$ and $\delta_2(X)$ reduce to the very simple expression:

$$\frac{4^{-a}}{aB(a,a)}.$$

Substituting the values of $\delta_1(X)$ and $\delta_2(X)$ into the double inequality $\delta_1(X) \leq \delta_2(X) \leq SD(X)$, several interesting inequalities involving the beta function can be obtained.

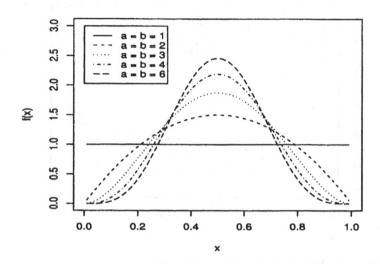

Figure 1. Unimodal, symmetric beta pdfs.

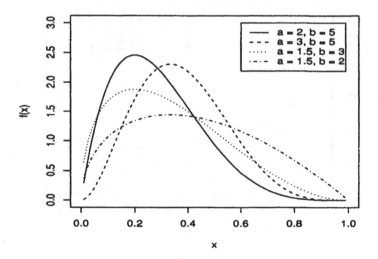

Figure 2. Unimodal, skewed beta pdfs.

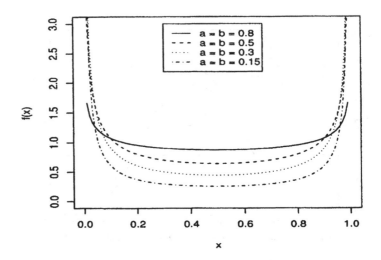

Figure 3. U-shaped, symmetric beta pdfs.

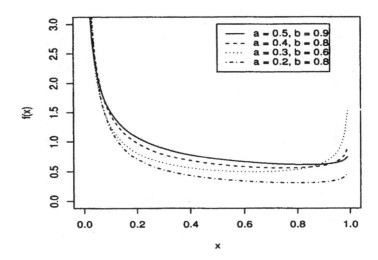

Figure 4. U-shaped, skewed beta pdfs.

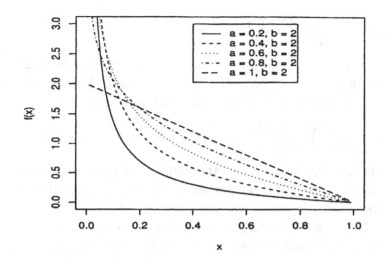

Figure 5. *J*-shaped beta pdfs.

VII. Shape

If $a > 1$ and $b > 1$ then the pdf $f(x) \to 0$ as $x \to 0, 1$; if $0 < a < 1$ then $f(x) \to \infty$ as $x \to 0$; if $0 < b < 1$ then $f(x) \to \infty$ as $x \to 1$; if $a = 1$ (respectively, $b = 1$) then $f(x)$ approaches a finite non-zero value as $x \to 0$ (respectively, $x \to 1$).

If $a > 1$ and $b > 1$ then the pdf has a single mode at $x = (a - 1)/(a + b - 2)$. If $a < 1$ and $b < 1$ then there is an anti-mode at this same value of x, and this corresponds to a *U*-shaped beta distribution. If $(a - 1)(b - 1) \leq 0$ then the pdf does not have a mode or anti-mode, and this corresponds to *J*-shaped or reverse *J*-shaped beta distributions. For all $a > 0$ and $b > 0$, there are points of inflexion at

$$\frac{a - 1}{a + b - 2} \pm \frac{1}{a + b - 2} \sqrt{\frac{(a - 1)(b - 1)}{a + b - 3}},$$

provided that there are real and lie between 0 and 1.

The skewness of (I.1) is:

$$\gamma_1 = \frac{2(b-a)}{a+b+2}\sqrt{\frac{a+b+1}{ab}}.$$

If $a = b$ then $\gamma_1 = 0$, and the pdf becomes symmetric about $x = 1/2$. If $b > a$ then $\gamma_1 > 0$ and the pdf becomes skewed to the right. Similarly, $b < a$ gives a left skewed pdf. The kurtosis of (I.1) is:

$$\gamma_2 = \frac{3(a+b+1)}{ab(a+b+2)(a+b+3)}\left\{2(a+b)^2 + ab(a+b-6)\right\}.$$

Note that both the skewness and the kurtosis are symmetrical functions of a and b; so, interchanging the parameters in the pdf yields its mirror image. In the symmetrical case $a = b = c$, $\gamma_2 \to 3$ as c becomes large, and the beta pdf approaches the normal pdf.

Karian et al. (1996) studied the contour curves of (γ_1, γ_2). They showed that the curves exhibit turning points when $a = b$, making $\gamma_1 = 0$ and

$$\gamma_2 = \frac{3(2a+1)}{2b+3}.$$

When $a = b = 0$ the turning point is at $\gamma_2 = 1$ and when $a = b = 1$ the turning point is at $\gamma_2 = 1.8$. Karian et al. also showed that the points of (γ_1, γ_2) are bounded by $1 + \gamma_1^2 < \gamma_2 < 3 + \gamma_1^2$.

Some approximations of γ_1 and γ_2 are:

$$\gamma_1^2 \approx \frac{1}{a} + \frac{1}{b} - \frac{4}{a+b},$$
$$\text{for } a, b \text{ large,}$$

$$\gamma_1^2 \approx 4\left(\frac{1}{a} + \frac{1}{b}\right) - \frac{16}{a+b},$$

$$\gamma_2 \approx 3 + \frac{2}{a} + \frac{2}{b} - \frac{6}{a+b},$$
$$\text{for } a, b \text{ large,}$$

and

$$\gamma_2 \approx 3 + 6\left(\frac{1}{a} + \frac{1}{b}\right) - \frac{30}{a+b}.$$

Ratnaparkhi and Mosimann (1990) tabulated values of γ_1 and γ_2 for all combinations of $a, b = 0.1(0.2)0.5$, 1, 34, 5(5)20 and $b = 25$.

Two concepts used to measure inequalities between two or more distributions are the Lorenz curve and the Gini index – defined, respectively, as the graph of the ratio

$$\phi(x) = \frac{F(x)E(X \mid X \leq x)}{E(X)}$$

versus $F(x)$ and

$$C(X) = \frac{Cov(X, F(X))}{E(X)}.$$

In the unit square $[0, 1] \times [0, 1]$, the Lorenz curve is convex and lies below the diagonal line $\phi = F$ at a vertical distance $d(x) = F(x) - \phi(x)$. It is known that $d(x)$ attains its maximum when $x = E(X)$. Actually,

$$d(E(X)) = \frac{\delta_1(X)}{2E(X)},$$

where $\delta_1(X)$ is the mean deviation of X about its mean. When $x = m$ (where m denotes the median of X),

$$d(m) = \frac{\delta_2(X)}{2E(X)},$$

where $\delta_1(X)$ is the mean deviation of X about its median. The Gini index defined above is equal to twice the area between the diagonal and the Lorenz curve.

Pham-Gia and Turkkan (1992) established the following properties of the Lorenz curve when X has the pdf (I.1):

1. The co-ordinates of the Lorenz curve are

   $$[\phi(x), F(x)] = [I_x(a + 1, b), I_x(a, b)]$$

2. If $a < b$ and Y is beta distributed with parameters b and a then the Lorenz curve for Y lies above that for X.

3. For a fixed and $b \to \infty$, the Lorenz curve for X approaches the

curve with the co-ordinates:

$$\left[\frac{1}{\Gamma(a)}\int_0^x z^{a-1}\exp(-z)dz - \frac{a^a\exp(-a)}{\Gamma(a+1)},\right.$$

$$\left.\frac{1}{\Gamma(a)}\int_0^x z^{a-1}\exp(-z)dz\right].$$

4. For the symmetrical beta distribution $(a = b)$, the Lorenz curve for X lies between the diagonal $\phi = F$ and the lower limit curve defined by:

$$\phi = \begin{cases} 0, & \text{if } 0 \le F \le 1/2, \\ 2F - 1, & \text{if } 1/2 < F \le 1. \end{cases}$$

5. The Gini index is

$$C(X) = \frac{2B(a, 2b)}{p\{B(a, b)\}^2}.$$

VIII.   Reliability Measures

For the beta pdf (I.1), the failure rate function and the mean residual life function defined, respectively, as

$$\lambda(t) = \frac{f(t)}{1 - F(t)}$$

and

$$\mu(t) = E\left(X - t \mid X > t\right)$$

take the forms

$$\lambda(t) = \frac{t^{a-1}(1 - t)}{B(a, b) - B_t(a, b)} \tag{VIII.1}$$

and

$$\mu(t) = \frac{\int_t^\infty \{B(a, b) - B_x(a, b)\}\, dx}{B(a, b) - B_t(a, b)}, \tag{VIII.2}$$

respectively.

Ahmed (1991) provided a characterization of the beta distribution in terms of the failure rate function. Namely, if X is a non-negative continuous random variable with cdf F, pdf f and mean m then X has the beta distribution given by (I.1) if and only if

$$E(X \mid X \geq t) = m + \frac{m}{a}t(1-t)\lambda(t),$$

$$m = \frac{a}{a+b}.$$

For the generalized beta distribution with parameters (a, b, c, d) (discussed in Chapter 5), these conditions become

$$E(X \mid X \geq t) = m + \frac{m}{bc+ad}(t-c)(d-t)\lambda(t),$$

$$m = \frac{bc+ad}{a+b}.$$

For the power function distribution (the special case of the beta distribution for $b = 1$), the conditions reduce to:

$$E(X \mid X \geq t) = m + \frac{m}{a}t(1-t)\lambda(t),$$

$$m = \frac{a}{a+1}.$$

Gupta and Gupta (2000) studied the monotonicity properties of the two measures (VIII.1) and (VIII.2), and established the following:

- if $a \geq 1$ and $b \geq 1$ then $\lambda(t)$ is increasing for all t and $\mu(t)$ is decreasing for all t.
- if $a \leq 1$ and $b \leq 1$ then $\lambda(t)$ is decreasing for all t and $\mu(t)$ is increasing for all t.
- if $a > 1$ and $0 < b < 1$ then $\lambda(t)$ is bathtub shaped and $\mu(t)$ is upside down bathtub shaped.
- if $0 < a < 1$ and $b > 1$ then $\lambda(t)$ is upside down bathtub shaped and $\mu(t)$ is bathtub shaped.

These different behaviors are an indication of the flexibility of the beta distribution.

IX.   Relationships

The beta distribution has relationships with several of the well-known univariate distributions.

A.   Gamma Distribution

Suppose that X has the gamma distribution with parameters a and r, that Y has the gamma distribution with parameters b and r, and that X and Y are independent. Then it is well known that $X/(X+Y)$ has the beta distribution with parameters a and b (Cramer, 1946). This relationship has been studied further by several authors.

Laha (1964) studied the converse of the above relationship: if X and Y are two independently and identically distributed random variables with common cdf F and if $X/(X+Y)$ has the beta distribution then the question is whether F is necessarily a gamma distribution? This question was originally posed by Mauldon (1956). Laha (1964) showed that under the conditions of the question F must have the following general properties:

- either $F(x) = 0$ for $x \leq 0$ or $F(x) = 1$ for $x \geq 0$,
- $F(x)$ is absolutely continuous and has a continuous pdf $f(x) = F'(x) > 0$.

However, if in addition to the conditions of the question we have the following conditions satisfied

- F has finite absolute moments of all orders
- and F is infinitely divisible

then F must be gamma distributed.

The above result can be generalized to more than two variables. Suppose X_1, X_2 and X_3 are three independent positive random variables, and suppose (U_1, U_2) are given by

$$U_1 = \frac{X_1}{X_1 + X_2}$$

and

$$U_2 = \frac{X_1 + X_2}{X_1 + X_2 + X_3}.$$

Then Kotlarski (1967) showed that X_k are gamma distributed with parameters p_k and a (a common – $k = 1, 2, 3$) if and only if U_1 and U_2 are independent beta distributed, U_1 with parameters (p_1, p_2), and U_2 with parameters $(p_1 + p_2, p_3)$.

Yeo and Milne (1991) provided two relationships between the gamma and beta distributions based on products of independent random variables.

1. Suppose that X and Y are independent, absolutely continuous and non-negative random variables such that X has bounded support. Then for $a > 0$ and $b > 0$, any two of the following three conditions imply the third.
   (i) XY is gamma distributed with parameters a and $1/\mu$, where $0 < \mu < \infty$;
   (ii) X is beta distributed with parameters a and b;
   (iii) Y is gamma distributed with parameters $a + b$ and $1/\mu$.

2. Suppose for a fixed positive integer m that $X_1, X_2, \ldots, X_{m+1}$ are iid non-negative random variables which are independent of another non-negative random variable X with bounded support, and that

   $$Y = X (X_1 + X_2 + \cdots + X_{m+1}).$$

   Then any two of the following three conditions imply the third.
   (i) Y has the same distribution, determined by its moments, as each of $X_1, X_2, \ldots, X_{m+1}$;
   (ii) X is beta distributed with parameters $a = 1$ and $b = m$;
   (iii) Y is exponentially distributed.
   This result also holds if (i) is replaced by
   (i)$'$ $X_1, X_2, \ldots, X_{m+1}$ are each exponentially distributed
   (i)$''$ or the weaker condition that Y has the same distribution as each of $X_1, X_2, \ldots, X_{m+1}$ and belongs to the class of distributions whose characteristic function is of the form

   $$1 - A \, |t| \, \{1 + o(t)\}, \qquad t \to 0,$$

where A is a real constant.

This result generalizes the relationship between the exponential and uniform distributions established in Kotz and Steutel (1988) and Yeo and Milne (1989).

Given these two statements, it is natural to ask the question: if the product XY has the gamma distribution then must X and Y have the beta and the gamma distributions specified in the statement 1 above? Kotlarski (1965) addressed this question and constructed examples to show that the answer is negative. If we take R_1, R_2 as mutually exclusive and exhaustive subsets of $R = \{1, 2, \ldots, n\}$ then one example is:

$$X = \prod_{k \in R_1} V_k, \qquad Y = \prod_{k \in R_2} V_k,$$

where V_k are independent random variables with the pdfs

$$f_k(v) = \frac{1}{\Gamma\left(\dfrac{q+k-1}{n}\right)} v^{q+k-2} \exp\left\{-\left(\frac{v}{n}\right)^n\right\}, \qquad v > 0.$$

If we take R_1, R_2 as mutually exclusive and exhaustive subsets of $R = \{1, 2, \ldots\}$ then another example is obtained by:

$$X = \prod_{k \in R_1} V_k, \qquad Y = \prod_{k \in R_2} V_k,$$

where V_k are independent random variables with the pdfs

$$f_0(v) = qv^{q-1}, \qquad 0 < v < 1$$

and

$$f_k(v) = (q+k)\left(1 + \frac{1}{k}\right)^{-(q+k)} v^{q+k-1},$$
$$0 < v < 1 + (1/k), \quad k \geq 1.$$

In both examples the product XY has a gamma distribution while the marginals of (X, Y) are not gamma or beta distributed.

Bailey (1992) provided the following geometric interpretation of the beta distribution based on chi-squared random variables. Let X_k, $k = 1, 2, \ldots, n$ be independent random variables such that X_k^2

has the chi-square distribution with degrees of freedom $2p_k$. Define the corresponding spherical coordinates $R \geq 0$, $\Theta_k \in [0, \pi/2]$, $j = 1, 2, \ldots, n - 1$ by

$$
\begin{aligned}
X_1 &= R \cos \Theta_1 \cos \Theta_2 \cdots \cos \Theta_{n-2} \cos \Theta_{n-1}, \\
X_2 &= R \sin \Theta_1 \cos \Theta_2 \cdots \cos \Theta_{n-2} \cos \Theta_{n-1}, \\
X_3 &= R \qquad\quad \sin \Theta_2 \cdots \cos \Theta_{n-2} \cos \Theta_{n-1}, \\
&\ \ \vdots \qquad\qquad\qquad \vdots \\
X_{n-1} &= R \qquad\qquad\qquad\qquad \sin \Theta_{n-2} \cos \Theta_{n-1}, \\
X_n &= R \qquad\qquad\qquad\qquad\qquad\ \ \sin \Theta_{n-1}.
\end{aligned}
$$

Then R, $\Theta_1, \Theta_2, \ldots, \Theta_{n-1}$ are independent. Moreover, R^2 has the chi-squared distribution with degrees of freedom $p_1 + p_2 + \cdots + p_n$ and $U_k = \cos^2 \Theta_k$ has the beta distribution with parameters $a = p_1 + p_2 + \cdots + p_k$ and $b = p_{k+1}$. Note that the product

$$
U_1 U_2 \cdots U_{n-1} = \frac{X_1^2}{R^2} =^d \frac{\chi_{2p_1}^2}{\chi_{p_1 + p_2 + \cdots + p_n}^2}.
$$

Hence a consequence of the above geometric representation is that the product of the independent beta random variables U_k also has the beta distribution with parameters $a = p_1$ and $b = p_1 + p_2 + \cdots + p_n$. Another easy consequence of the above representation is that if S is chi-squared distributed with degrees of freedom $2a + 2b$ and if W is beta distributed (with parameters a and b) independently of S then both SW and $S(1 - W)$ have the chi-square distribution with degrees of freedom $2a$ and $2b$, respectively.

Dufresne (1998) derived three of the most recent properties connecting the beta and gamma distributions. Let $X(a)$, $Y(a, b)$ and U denote random variables having the gamma distribution (with shape parameter a and scale parameter 1), the beta random distribution with parameters a & b, and an arbitrary distribution. Then the following properties hold:

1. For any $a > 0$ and $b > 0$,

$$
U X(a) + X(b) =^d X(a + b) + \{1 + (U - 1)Y(a, b)\}.
$$

2. For any $a > 0$, $b > 0$, $c > 0$ and $d > 0$,

$$Y(a,b)X(c) + X(d) =^d X(c+d)\{1 - Y(b,a)Y(c,d)\}.$$

In particular, for any $a > 0$, $b > 0$ and $c > 0$,

$$Y(a, b+c)X(b) + X(c) =^d X(b+c)Y(a+c, b)$$
$$=^d X(a+c)Y(b+c, a).$$

3. For any $a > 0$, $b > 0$, $c > 0$ and $d > 0$,

$$\frac{X(a)}{Y(c,d)} + X(b) =^d X(a+b)\left\{\frac{Y(a,b)}{Y(c,d)} + 1 - Y(a,b)\right\}$$
$$=^d X(a+b)\left\{1 + Y(a,b)\frac{X(d)}{X(c)}\right\}.$$

In particular, for any $a > 0$, $b > 0$ and $c > 0$,

$$\frac{X(a)}{Y(b, a+c)} + X(c) = \frac{X(a+c)}{Y(b,a)},$$

where it is assumed throughout that all the variables are independent.

For other relationships between the gamma and beta distributions see Huang and Chen (1989).

B. Uniform Distribution

Suppose $X_1, X_2, \ldots, X_n$ are iid random variables from a uniform distribution defined over the unit interval. Then $X_{(1)}$ has the beta distribution with parameters $a = 1$ and $b = n$. Conversely, if $X_{(1)}$ has the uniform distribution over the unit interval then each X_i has the beta distribution with parameters $a = 1$ and $b = 1/n$.

C. Power Function Distribution

Suppose $X_1, X_2, \ldots, X_n$ are iid random variables from the power function distribution $a = \alpha$ (recall that power function distribution is the particular case of the beta distribution when $b = 1$). Then $X_{(1)}^{\alpha}$ has the beta distribution with parameters $a = 1$ and $b = n$.

Conversely, if $X_{(1)}$ has the power function distribution with $a = \beta$ then each X_i^β has the beta distribution with parameters $a = 1$ and $b = 1/n$.

D. Other Distributions

If X has the F-distribution with m degrees of freedom in the numerator and n degrees of freedom in the denominator then

$$\frac{(m/n)X}{1 + (m/n)X}$$

has the beta distribution with parameters $a = m/2$ and $b = n/2$.

If X has the beta distribution with parameters a and b then

$$Y = \frac{X}{1 - X}$$

has the standard form of the Pearson type VI distribution with the pdf

$$f(y) = \frac{1}{B(a,b)} \frac{y^{a-1}}{(1 + y)^{a+b}}, \qquad y > 0.$$

If X has the power-function distribution (particular case of the beta distribution when $b = 1$) then $1/X$ has a Pareto distribution.

References

1. Ahmed, A.N. Characterization of beta, binomial, and Poisson distributions, IEEE Transactions on Reliability **1991**, *40*, 290-295.

2. Bailey, R.W. Distributional identities of beta and chi-squared variables: A geometric interpretation, The American Statistician **1992**, *46*, 117-120.

3. Cramer, H. *Mathematical Methods of Statistics*; Princeton University Press: Princeton, New Jersey, 1946.

4. Dufresne, D. Algebraic properties of beta and gamma distri-

butions, and applications, Advances in Applied Mathematics
1998, *20*, 285-299.

5. Kotlarski, I. On characterizing the gamma and the normal distribution, Pacific Journal of Mathematics **1967**, *20*, 69-76.

6. Gradshteyn, I.S.; Ryzhik, I.M. *Tables of Integrals, Series, and Products*; Academic Press: San Diego, 2000; 1163 pp.

7. Gupta, P.L.; Gupta, R.C. The monotonicity of the probability of the beta distribution, Applied Mathematics Letters **2000**, *13*, 5-9.

8. Huang, W.-J.; Chen, L.-S. Note on a characterization of gamma distributions, Statistics and Probability Letters **1989**, *8*, 485-487.

9. Karian, Z.A.; Dudewicz, E.J.; McDonald, P. The extended generalized lambda distribution system for fitting distributions to data: history, completion of theory, tables, applications, the "final word" on moment fits, Communications in Statistics–Simulation and Computation **1996**, *25*, 611-642.

10. Kotlarski, I. On pairs of independent random variables whose product follows the gamma distribution, Biometrika **1965**, *52*, 289-294.

11. Kotz, S.; Steutel, F.W. Note on a characterization of exponential distributions, Statistics and Probability Letters **1988**, *6*, 201-203.

12. Laha, R.G. On a problem connected with beta and gamma distributions, Transactions of the American Mathematical Society **1964**, *113*, 287-298.

13. Mauldon, J. G. Characterizing properties of statistical distribution, Quarterly Journal of Mathematics (Oxford, Second Series) **1956**, *7*, 155-160.

14. Mühlbach, G. von. Rekursionsformeln für die zentralen Momente der Pólyaund der Beta-Verteilung, Metrika **1972**, *19*, 173-177.

15. Norton, R.M. On properties of the arc-sine law, Sankhyā, A **1975**, *37*, 306-308.

16. Pakes, A.G. On characterizations involving mixed sums, Preprint, University of Western Australia, 1990.

17. Pham-Gia, T. Value of the beta prior information, Communications in Statistics–Theory and Methods **1994**, *23*, 2175-2195.

18. Pham-Gia, T.; Turkkan, N. Determination of the beta distribution from its Lorenz curve, Mathematical and Computer Modelling **1992**, *16*, 73-84.

19. Ratnaparkhi, M.V.; Mosimann, J.E. On the normality of transformed beta and unit-gamma random variables, Communications in Statistics–Theory and Methods **1990**, *19*, 3833-3854.

20. Yeo, G.F.; Milne, R.K. On characterizations of exponential distributions, Statistics and Probability Letters **1989**, *7*, 303-305.

21. Yeo, G.F.; Milne, R.K. On characterizations of beta and gamma distributions, Statistics and Probability Letters **1991**, *11*, 239-242.

Products and Linear Combinations

Saralees Nadarajah[1] and Arjun Gupta[2]

[1]Department of Mathematics
University of South Florida
Tampa, Florida 33620
[2]Department of Mathematics and Statistics
Bowling Green State University
Bowling Green, Ohio 43403

I. Products

A. Closed Under Products

Under certain mild conditions, the product of two or more independent Beta variables also follows the same distribution. This property has been studied in various forms by Jambunathan (1954) and Krysicki (1999). Some of the known results are:

- if $X_1, X_2, \ldots, X_n$ are independent beta random variables with parameters $a = a_i$, $b = b_i$, $i = 1, 2, \ldots, n$, and if $a_{i+1} = a_i + b_i$ for $i = 1, 2, \ldots, n - 1$, then the product $X_1 X_2 \cdots X_n$ is also a beta random variable with parameters $a = a_1$ and $b = b_1 + \cdots + b_n$.
- the pdf of a beta random variable with parameters $a = \alpha$, $b = \beta$ is equal to the pdf of the geometric mean of n beta

random variables with parameters $a = (\alpha + i)/n$, $b = \beta/n$ for $i = 0, 1, \ldots, n - 1$.

- the pdf of a beta random variable with parameters $a = \alpha$, $b = \beta$ is equal to the pdf of the infinite product

$$\prod_{i=1}^{\infty} X_i^{1/2^i},$$

where X_i is a beta random variable with parameters

$$a = \frac{1}{2} + \frac{\alpha - 1}{2^i}, \qquad b = \frac{\beta}{2^i}$$

for $i = 1, 2, \ldots$.

These results present a beta random variable as a finite or an infinite product of independent random variables of the same kind. The following question arises: if the product $X_1 X_2 \cdots X_n$ of n independent random variables has a beta distribution then must each X_k also have a beta distribution? Kotlarski (1962) showed that the answer to this question is negative. He constructed groups of independent random variables which do not have a beta distribution but their product has a beta distribution. For instance, if $R_1, R_2, \ldots, R_n$ are mutually exclusive and exhaustive subsets of the set $R = \{1, 2, \ldots, m\} (m > n)$ then define

$$X_k = \prod_{r \in R_k} Y_r, \qquad k = 1, 2, \ldots, n,$$

where Y_r are independent random variables with the pdf

$$f(y) = \frac{m}{B\left(\dfrac{a + r - 1}{m}, \dfrac{b}{m}\right)} y^{a+r-1} (1 - y^m)^{\frac{b}{m} - 1}, \qquad 0 < y < 1.$$

Then it can be shown that the product $X_1 X_2 \cdots X_n$ has the beta distribution with the parameters a and b. Another construct is to define

$$X_k = \prod_{r \in N_k} Y_r, \qquad k = 1, 2, \ldots, n,$$

where Y_r are independent random variables with the pdf

$$f(y) = \frac{2^r}{B\left(\dfrac{2^{r-1}+a-1}{2^r}, \dfrac{b}{2^r}\right)} y^{2^{r-1}-a-1} \left(1 - y^{2^r}\right)^{\frac{b}{2^r}-1}$$

for $0 < y < 1$ and $N_1, N_2, \ldots, N_n$ are mutually exclusive and exhaustive subsets of the set $N = \{1, 2, \ldots\}$.

B. Exact Distributions

In related developments, Wilks (1932), Springer and Thompson (1970), Springer (1978), Nandi (1980), Tang and Gupta (1984) and Fan (1991) studied the exact distribution of

$$Y = \prod_{k=1}^{n} X_k \tag{I.1}$$

when the X_k are independent beta random variables with parameters $a = a_k$, $b = b_k$. Wilks (1932) expressed the pdf of Y as the $(n-1)$-fold integral:

$$f(y) = K y^{a_n - 1}(1-y)^{\delta_0 - 1} \int_0^1 \cdots \int_0^1 \prod_{k=1}^{n-1} w_k^{b_k - 1}(1 - w_k)^{\delta_k - 1}$$

$$\times \left\{1 - \eta_k(1-y)\right\}^{-\alpha(k)} dw_k, \tag{I.2}$$

where

$$K = 1 \bigg/ \left\{\prod_{k=1}^{n} B(a_k, b_k)\right\},$$

$$\delta_i = \sum_{k=i+1}^{n} b_k,$$

$$\alpha(k) = a_{k+1} + b_{k+1} - a_k,$$

$$\eta_i = 1 - \prod_{k=1}^{i}(1 - w_k).$$

The above can also be written as a Meijer G-function multiplied by a constant K (Springer and Thompson, 1970). In precise terms,

$$f(y) = K G_{n0}^{n0}\left(y \,\middle|\, \begin{matrix} a_1 + b_1 - 1, \ldots, a_n + b_n - 1 \\ a_1 - 1, \qquad \ldots, a_n - 1 \end{matrix}\right), \qquad (I.3)$$

where the constant is

$$K = \prod_{k=1}^{n} \frac{\Gamma(a_k + b_k)}{\Gamma(a_k)}$$

and G_{pq}^{lm} is the Meijer G-function defined by

$$G_{pq}^{lm}\left(y \,\middle|\, \begin{matrix} a_1, a_2, \ldots, a_p \\ b_1, b_2, \ldots, b_p \end{matrix}\right)$$

$$= \frac{1}{2\pi i} \int_{c-i\infty}^{c+i\infty} y^{-z} \frac{\displaystyle\prod_{k=1}^{m} \Gamma(z + b_k) \prod_{k=1}^{m} \Gamma(1 - a_k - z)}{\displaystyle\prod_{k=m+1}^{p} \Gamma(z + a_k) \prod_{k=l+1}^{q} \Gamma(1 - b_k - z)} dz.$$

The real constant c in the integral is taken to define a Bromwich path separating the poles of $\Gamma(z + b_k)$ from those of $\Gamma(1 - a_k - z)$. For a detailed discussion of the Meijer G-function, see Bateman (1954, pp. 374-379).

For integer values of a_k and b_k, Springer and Thompson (1970) – see also Springer (1978) – refined (I.3) and gave the following explicit expression:

$$f(y) = \sum_{k=1}^{m} \sum_{j=0}^{e_k-1} \frac{K_{kj} y^{d_k-1}(-\log y)^{e_k-j-1}}{(e_k - j - 1)! j!}, \qquad (I.4)$$

where

$$K_{k0} = \sum_{q=1, q\neq k}^{m} (d_q - d_k)^{-e_q},$$

$$K_{kj} = \sum_{r=0}^{j-1} \sum_{q=1, q\neq k}^{m} (-1)^{r+1} \binom{j-1}{r} \frac{r! e_q K_{k,j-r-1}}{(d_q - d_k)^{r+1}}, \qquad j > 0$$

and d_k denotes the m different integers that occur with multiplicity e_k among the $a_i - 1, a_i, a_i + 1, \ldots, a_i + b_i - 2$ for $i = 1, 2, \ldots, n$. For an example, consider the product $Y = X_1 X_2 X_3$ with $(a_1, b_1) = (9, 3)$, $(a_2, b_2) = (8, 3)$ and $(a_3, b_3) = (4, 2)$. Applying equation (I.4), we get the pdf of Y as:

$$f(y) = \frac{3960}{7}y^3 - 1980y^4 + 99000y^7$$
$$+ (374220 + 356400 \log y)\, y^8$$
$$- (443520 - 237600 \log y)\, y^9 - \frac{198000}{7}y^{10}$$

and the corresponding cdf is

$$F(y) = \frac{990}{7}y^4 - 396y^5 + 12375y^8 + (37180 + 39600 \log y)\, y^9$$
$$- (46728 - 23760 \log y)\, y^{10} - \frac{18000}{7}y^{11}.$$

Figure 1 below graphs the pdf of Y.

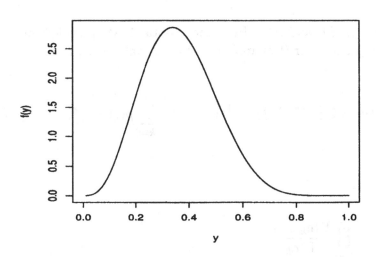

Figure 1. Pdf of the product $Y = X_1 X_2 X_3$ when $(a_1, b_1) = (9, 3)$, $(a_2, b_2) = (8, 3)$ and $(a_3, b_3) = (4, 2)$.

If we assume that all the a_k are equal to, say a, and that all the b_k are equal to, say b, then (I.4) reduces to the simpler form

$$f(y) = \left(\frac{(a+b-1)!}{(a-1)!}\right)^n \sum_{k=1}^{b} \sum_{j=0}^{n-1} K_{kj} y^{a+k-2}(-\log y)^{n-j-1},$$

where

$$K_{k0} = \sum_{q=1,q\neq k}^{b} (d_q - d_k)^{-n}$$

and

$$K_{kj} = \sum_{q=1,q\neq k}^{b} \sum_{r=0}^{j-1} (-1)^{r+1} \binom{j-1}{r} \frac{nr! K_{k,j-r-1}}{(d_q - d_k)^{r+1}}, \qquad j > 0.$$

Nandi (1980) and later Tang and Gupta (1984) provided an explicit expression for (I.2) applicable for any value of a_k and b_k:

$$f(y) = C \left\{\Gamma(\delta_0)\right\}^{-1} y^{a_n-1}(1 - y)^{\delta_0-1} \sum_{k=0}^{\infty} \rho_k(1 - y)^k, \qquad \text{(I.5)}$$

where

$$C = \prod_{k=1}^{n} \frac{\Gamma(a_k + b_k)}{\Gamma(a_k)}$$

and

$$\rho_0 = 1,$$

$$\rho_i = \sum_J \prod_{k=1}^{n-1} \left[\left\{ (\alpha(k))_{j(k)} \left(b_k + \sum_{l=1}^{k-1} (b_l + j(l)) \right)_{j(k)} \right\} \right.$$
$$\left. \Big/ \left\{ j(k)! \left(\delta_0 + \sum_{l=1}^{k-1} j(l) \right)_{j(k)} \right\} \right],$$
$$i = 1, 2, \ldots.$$

Here, $\sum_J$ denotes the sum over all sequences $J = (j_1, \ldots, j_{n-1})$ for which each $j_k \geq 0$ and $j_1 + \cdots + j_{n-1} = i$. In the particular case $n = 2$, (I.5) reduces to:

$$f(y) = C\Gamma^{-1} (b_1 + b_2) y^{a_2-1}(1 - y)^{b_1+b_2-1}$$
$$\times \; _2F_1 (b_1, a_2 + b_2 - a_1; b_1 + b_2; 1 - y).$$

Because each $\rho_i > 0$ and $0 < y < 1$, the order of integration and summation can be inter-changed in (I.5) to get the cdf of Y as

$$F(y) = C \left\{ \Gamma (\delta_0) \right\}^{-1} \sum_{k=0}^{\infty} \rho_k B (a_n, \delta_0 + k) I_y (a_n, \delta_0 + k),$$

which corresponds to a mixture of beta distributions.

Consider the following variant of (I.1):

$$Y = \prod_{k=1}^{m} X_k \prod_{k=m+1}^{n} X_k,$$

where the first m X_k's are beta distributed with parameters (a_k, b_k) while the remaining are gamma distributed with shape parameters c_k (and scale parameters set to 1). Springer and Thompson (1970) derived the pdf of Y, expressing it in a form similar to (I.3):

$$
f(y) = KG_{mn}^{n0}\left(y\,\middle|\,
\begin{matrix}
a_1 + b_1 - 1,\ a_2 + b_2 - 1, \\
a_1 - 1,\qquad a_2 - 1,
\end{matrix}
\right.
$$

$$
\left.
\begin{matrix}
\ldots,\ a_m + b_m - 1 \\
\ldots,\ a_m - 1,\qquad c_{m+1} - 1, \ldots, c_n - 1
\end{matrix}
\right),
$$

where

$$
K = \prod_{k=1}^{m} \frac{\Gamma(a_k + b_k)}{\Gamma(a_k)} \prod_{k=m+1}^{n} \frac{1}{\Gamma(c_k)}.
$$

The distribution of the product Y in (I.1) has also been studied when the X_k have the type I noncentral beta distribution. We mention work by Malik (1970) and Bhargava (1975).

Malik (1970) derived the distribution of $Y = X_1 X_2$ when the X_k have the type I noncentral beta distribution with parameters $a_k/2$, $b_k/2$ and λ_k. The pdf Y is expressed as:

$$
f(y) = \exp\{-(\lambda_1 + \lambda_2)\} \sum_{k=0}^{\infty} \sum_{m=0}^{k} \frac{\lambda_1^m \lambda_2^{k-m}}{\Gamma\left(\dfrac{2m + p_1}{2}\right)}
$$

$$
\times \frac{\Gamma\left(\dfrac{2m + p_1 + q_1}{2}\right)}{\Gamma\left(\dfrac{2k - 2m + p_2}{2}\right)}
$$

$$
\times \frac{\Gamma\left(\dfrac{2k - 2m + p_2 + q_2}{2}\right)}{m!(k - m)!\,\Gamma\left(\dfrac{q_1 + q_2}{2}\right)} y^{m + \frac{p_1}{2} - 1} (1 - y)^{\frac{q_1 + q_2}{2} - 1}
$$

$$
\times \,_2F_1\left(\frac{q_2}{2}, 2m - k + \frac{p_1 - p_2 + q_1}{2}; \frac{q_1 + q_2}{2}; 1 - y\right),
$$

$$
(I.6)
$$

where $_2F_1$ denotes the Gauss hypergeometric function. Mikhail and Tracy (1975) derived a particular case of this formula when only one of the variables has the noncentral beta distribution. Malik pointed out that (I.6) is also expressible in terms of a mixture of beta distributions, namely

$$f(y) = \exp\{-(\lambda_1 + \lambda_2)\} \sum_{k=0}^{\infty} \sum_{m=0}^{k} \sum_{r=0}^{\infty} \lambda_1^m \lambda_2^{k-m}$$

$$\times \frac{\Gamma\left(\frac{p_1 + q_1}{2} + m\right)}{\Gamma\left(2m - k + \frac{p_1 - p_2 + q_1}{2}\right)}$$

$$\times \frac{\Gamma\left(\frac{q_2}{2} + r\right)}{\Gamma\left(\frac{p_1 + q_1 + q_2}{2} + m + r\right)}$$

$$\times \frac{\Gamma\left(2m - k + \frac{p_1 - p_2 + q_1}{2} + r\right)}{B\left(\frac{p_1}{2} + m, \frac{q_1 + q_2}{2} + r\right) B\left(\frac{p_2}{2} + k - m, \frac{q_2}{2}\right)}$$

$$\times \frac{y^{m + \frac{p_1}{2} - 1}(1 - y)^{\frac{q_1 + q_2}{2} + r - 1}}{m!(k - m)!r!},$$

where the sum over r comes from the hypergeometric function. From

64 Nadarajah and Gupta

this representation it follows that the corresponding cdf is:

$$
F(z) = \exp\left\{-(\lambda_1 + \lambda_2)\right\} \sum_{k=0}^{\infty} \sum_{m=0}^{k} \sum_{r=0}^{\infty} \lambda_1^m \lambda_2^{k-m}
$$

$$
\times \frac{\Gamma\left(\dfrac{p_1 + q_1}{2} + m\right)}{\Gamma\left(2m - k + \dfrac{p_1 - p_2 + q_1}{2}\right)}
$$

$$
\times \frac{\Gamma\left(\dfrac{q_2}{2} + r\right)}{\Gamma\left(\dfrac{p_1 + q_1 + q_2}{2} + m + r\right)}
$$

$$
\times \frac{\Gamma\left(2m - k + \dfrac{p_1 - p_2 + q_1}{2} + r\right)}{B\left(\dfrac{p_2}{2} + k - m, \dfrac{q_2}{2}\right) m!(k-m)!r!},
$$

$$
\times I_y\left(\dfrac{p_1}{2} + m, \dfrac{q_1 + q_2}{2} + r\right),
$$

where I_y denotes the incomplete beta function ratio (see Chapter 1).

Bhargava (1975) studied the distribution of Y in (I.1) under a slightly different setting, assuming that X_1 and $X_n \mid X_1, \ldots, X_{n-1}$ have the type I noncentral beta distribution. Precisely, if X_1 has the type I noncentral beta distribution with parameters (a_1, b_1, λ_1) and if $X_n \mid X_1, \ldots, X_{n-1}$ has the type I noncentral beta distribution with parameters $(a_n, b_n, \lambda_n x_1 \cdots x_{n-1})$, where $a_j = a_{j+1} + b_{j+1}, j = 1, \ldots, n-1$ then it is established that Y will also have the type I noncentral beta distribution with parameters $(a_n, b_1 + \cdots + b_n, \lambda_1 + \cdots + \lambda_n)$. In this case, the unconditional pdf of $Z = X_{k+1} \cdots X_n$ is:

$$f(z) = \exp\left\{-\left(\sum_{i=1}^{n}\lambda_i\right)/2\right\}\sum_{l=0}^{\infty}\sum_{j=0}^{\infty}\sum_{m=0}^{\infty}\frac{\left(\sum_{i=s+1}^{n}\lambda_i/2\right)^{m+j}}{m!j!}$$

$$\times\frac{\left(\sum_{i=1}^{k}\lambda_i/2\right)^{l}}{l!}\frac{B\left(a_k+j,l+m+\sum_{i=1}^{k}b_i\right)}{B\left(a_k,l+\sum_{i=1}^{k}b_i\right)}$$

$$\times\frac{z^{a_n-1}(1-z)^{j-1+\sum_{i=k+1}^{n}b_i}}{B\left(a_n,j+\sum_{i=k+1}^{n}b_i\right)}.$$

In the particular case $n = 2$, this reduces to give the pdf of $Z = X_2$ as:

$$f(z) = \exp\left\{-(\lambda_1+\lambda_2)/2\right\}\sum_{l=0}^{\infty}\sum_{j=0}^{\infty}\sum_{k=0}^{\infty}\frac{(\lambda_2/2)^{k+j}}{k!j!}\frac{(\lambda_1/2)^{l}}{l!}$$

$$\times\frac{B\left(a_1+j,b_1+l+k\right)}{B\left(a_1,b_1+l\right)}\frac{z^{a_2-1}(1-z)^{b_2+j-1}}{B\left(a_2,b_2+j\right)}.$$

Pham-Gia (2000) derived the exact pdf of the quotients

$$W = \frac{X_1}{X_2},$$

$$T = \frac{X_1}{X_1+X_2}$$

when X_1 and X_2 are independent standard beta random variables with parameters (a_1, b_1) and (a_2, b_2), respectively. The pdf of W is given by:

$$
f(w) = \begin{cases}
B\left(a_1 + a_2, b_2\right) w^{a_1 - 1} \\
\quad \times\, _2F_1\left(a_1 + a_2, 1 - b_1; a_1 + a_2 + b_2; w\right)/A, \\
\quad \text{if } 0 < w \le 1, \\
B\left(a_1 + a_2, b_1\right) w^{-(1 + a_2)} \\
\quad \times\, _2F_1\left(a_1 + a_2, 1 - b_2; a_1 + a_2 + b_1; \dfrac{1}{w}\right)/A, \\
\quad \text{if } w \ge 1
\end{cases}
$$

while that of T takes the form:

$$
f(t) = \begin{cases}
B\left(a_1 + a_2, b_2\right) t^{a_1 - 1}(1 - t)^{a_1 + 1} \\
\quad \times\, _2F_1\left(a_1 + a_2, 1 - b_1; a_1 + a_2 + b_2; \dfrac{t}{1 - t}\right)/A, \\
\quad \text{if } 0 < w \le 1/2, \\
B\left(a_1 + a_2, b_1\right) t^{-(1 + a_2)}(1 - t)^{a_2 - 1} \\
\quad \times\, _2F_1\left(a_1 + a_2, 1 - b_2; a_1 + a_2 + b_1; \dfrac{1 - t}{t}\right)/A, \\
\quad \text{if } 1/2 \le t \le 1,
\end{cases}
$$

where $A = B(a_1, b_1)B(a_2, b_2)$. The figures below illustrate the pdfs of W and T when $(a_1, b_1) = (2.5, 3.75)$ and $(a_2, b_2) = (1.25, 4.06)$.

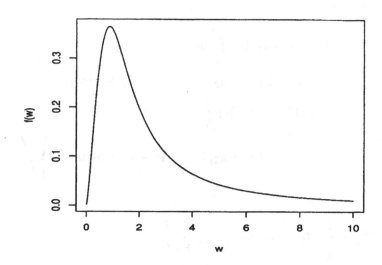

Figure 2. Pdf of W when $(a_1, b_1) = (2.5, 3.75)$ and $(a_2, b_2) = (1.25, 4.0$

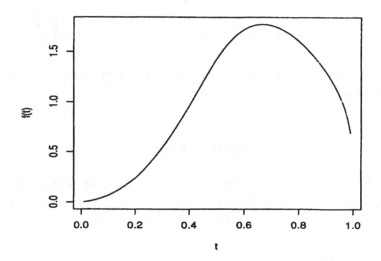

Figure 3. Pdf of T when $(a_1, b_1) = (2.5, 3.75)$ and $(a_2, b_2) = (1.25, 4.0$

Pham-Gia (2000) also derived the moments of W and T (about zero), expressing them as:

$$
E(W^n) = \left\{ B(a_1 + a_2, b_2) \int_0^1 w^{n+a_1-1} \right.
$$
$$
\times {}_2F_1(a_1 + a_2, 1 - b_1; a_1 + a_2 + b_2; w)\, dw
$$
$$
+ B(a_1 + a_2, b_1) \int_0^1 w^{a_2-n-1}
$$
$$
\left. \times {}_2F_1(a_1 + a_2, 1 - b_2; a_1 + a_2 + b_1; w)\, dw \right\} \Big/ A
$$

and

$$
E(T^n) = \left\{ B(a_1 + a_2, b_2) \int_0^1 \frac{t^{n+a_1-1}}{(1+t)^n} \right.
$$
$$
\times {}_2F_1(a_1 + a_2, 1 - b_1; a_1 + a_2 + b_2; t)\, dt
$$
$$
+ B(a_1 + a_2, b_1) \int_0^1 \frac{t^{a_2-1}}{(1+t)^n}
$$
$$
\left. \times {}_2F_1(a_1 + a_2, 1 - b_2; a_1 + a_2 + b_1; t)\, dt \right\} \Big/ A.
$$

Gupta and Nadarajah (2002) pointed out these expressions can be simplified further.

C. Approximation

Fan (1991) approximated the product Y in (I.1) by a standard beta random variable Z with parameters a and b. By equating the expectation and variance, a and b are determined as

$$
a = \frac{S}{T - S^2}
$$

and

$$
b = \frac{(1 - S)(S - T)}{T - S^2},
$$

where

$$S = \prod_{k=1}^{n} \left(\frac{p_k}{p_k + q_k} \right)$$

and

$$T = \prod_{k=1}^{n} \left(\frac{p_k \left(p_k + 1 \right)}{\left(p_k + q_k \right) \left(p_k + q_k + 1 \right)} \right).$$

Comparison of the first ten moments showed that the approximation is quite good. For example, when $n = 3$, $(p_1, p_2, p_3) = (778, 43, 23)$ and $(q_1, q_2, q_3) = (567, 57, 12)$, the approximate eighth moment and its exact value are 0.10554×10^{-5} and 0.103925×10^{-5}, respectively.

II. Linear Combinations

A. Sums and Differences

The pdf of the sum or the difference of two random variables can be obtained by using the Fourier transform (Springer, 1978, page 52). In practice, however, the mathematics can be intractable. Pham-Gia and Turkkan (1993) and Pham and Turkkan (1994) derived the exact distribution of the sum and the difference of two standard beta random variables in terms of the Appell function. The Appell function is the hypergeometric function in two variables defined by:

$$F_1 \left(a, b_1, b_2; c; x, y \right) = \sum_{n=0}^{\infty} \sum_{m=0}^{\infty} \left(b_1 \right)_n \left(b_2 \right)_n \frac{\left(a \right)_{m+n}}{\left(c \right)_{m+n}} \frac{x^m \, y^n}{m! \, n!}.$$

If X_1 is a beta random variable with parameters (a_1, b_1) and if X_2 is an independent beta random variable with parameters (a_2, b_2) then Pham-Gia and Turkkan (1993) showed that the pdf of the difference

$D = X_1 - X_2$ is:

$$
f(d) = \begin{cases}
\dfrac{B(a_2, b_1)}{A} d^{b_1+b_2-1}(1-d)^{a_2+b_1-1} \\
\quad \times F_1\left(b_1, a_1 + a_2 + b_1 + b_2 - 2, 1 - a_1; \right.\\
\quad \left. \times a_2 + b_1; 1 - d, 1 - d^2\right), \\
\qquad \text{if } 0 \le d \le 1, \\
\dfrac{B(a_1, b_2)}{A}(-d)^{b_1+b_2-1}(1+d)^{a_1+b_2-1} \\
\quad \times F_1\left(b_2, 1 - a_2, a_1 + a_2 + b_1 + b_2 - 2, 1 - a_1;\right.\\
\quad \left. \times a_1 + b_2; 1 - d^2, 1 + d^2\right), \\
\qquad \text{if } -1 \le d \le 0,
\end{cases}
$$

where $A = B(a_1, b_1)B(a_2, b_2)$. Pham and Turkkan (1994) derived the pdf of the sum $Y = X_1 + X_2$ as:

$$
f(y) = \begin{cases}
B^*(a_1, b_1, a_2, b_2)\, y^{a_1+a_2-1}(1-y)^{b_1-1} \\
\quad \times F_1\left(a_2, 1 - b_1, 1 - b_2; a_1 + a_2; \dfrac{y}{y-1}, y\right), \\
\qquad \text{if } 0 \le y < 1, \\
B(a_1 + b_2 - 1, a_2 + b_1 - 1)\,/\,\{B(a_1, b_1)\,B(a_2, b_2)\}, \\
\qquad \text{if } y = 1, \\
B^*(b_2, a_2, b_1, a_1)(y-1)^{a_2-1}(2-y)^{b_1+b_2-1} \\
\quad \times F_1\left(b_1, 1 - a_1, 1 - a_2; b_1 + b_2; 2 - y, \dfrac{2-y}{1-y}\right), \\
\qquad \text{if } 1 < y \le 2,
\end{cases}
$$

where

$$
B^*(j_1, k_1; j_2, k_2) = \frac{\Gamma(j_1 + k_1)\,\Gamma(j_2 + k_2)}{\Gamma(k_1)\,\Gamma(k_2)\,\Gamma(j_1 + j_2)}.
$$

If X_1 and X_2 follow the generalized beta distribution (equation (VII.2) in Chapter 5) on (c_1, d_1) and (c_2, d_2), respectively, such that $d_1 - c_1 = d_2 - c_2 = \delta$ then the pdf of the sum $Y = X_1 + X_2$ generalizes to a location-scale transform of the above:

$$
\frac{1}{\delta} f\left(\frac{y - c_1 - c_2}{\delta}\right). \tag{II.1}
$$

The figures below illustrate the great variety of shapes of the distribution of D.

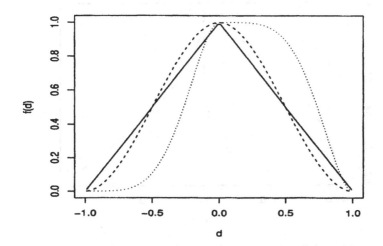

Figure 4. Pdfs of D: ———— for $(a_1, b_1, a_2, b_2) = (1, 1, 1, 1)$; – – – – for $(a_1, b_1, a_2, b_2) = (1, 1, 2, 2)$; ············ for $(a_1, b_1, a_2, b_2) = (1, 1, 2, 5)$.

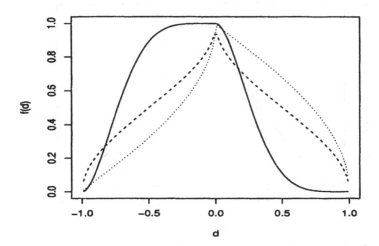

Figure 5. Pdfs of D: ———— for $(a_1, b_1, a_2, b_2) = (1, 1, 5, 2)$; – – – – – for $(a_1, b_1, a_2, b_2) = (1, 1, 0.5, 0.5)$; ············ for $(a_1, b_1, a_2, b_2) = (1, 1, 0.5, 0.9)$.

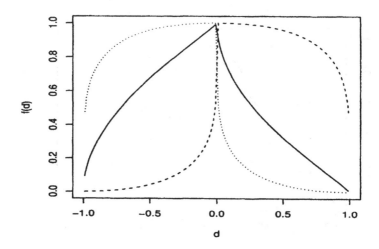

Figure 6. Pdfs of D: ———— for $(a_1, b_1, a_2, b_2) = (1, 1, 0.9, 0.5)$; – – – – for $(a_1, b_1, a_2, b_2) = (1, 1, 0.2, 2)$; .... for $(a_1, b_1, a_2, b_2) = (1, 1, 2, 0.2)$.

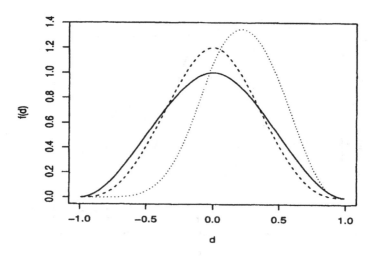

Figure 7. Pdfs of D: ———— for $(a_1, b_1, a_2, b_2) = (2, 2, 1, 1)$; – – – – for $(a_1, b_1, a_2, b_2) = (2, 2, 2, 2)$; .... for $(a_1, b_1, a_2, b_2) = (2, 2, 2, 5)$.

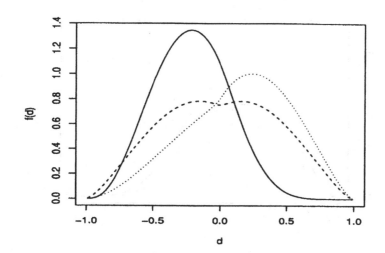

Figure 8. Pdfs of D: ——— for $(a_1, b_1, a_2, b_2) = (2, 2, 5, 2)$; – – – – for $(a_1, b_1, a_2, b_2) = (2, 2, 0.5, 0.5)$; .... for $(a_1, b_1, a_2, b_2) = (2, 2, 0.5, 0.9)$.

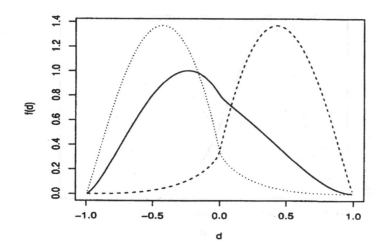

Figure 9. Pdfs of D: ——— for $(a_1, b_1, a_2, b_2) = (2, 2, 0.9, 0.5)$; – – – – for $(a_1, b_1, a_2, b_2) = (2, 2, 0.2, 2)$; .... for $(a_1, b_1, a_2, b_2) = (2, 2, 2, 0.2)$.

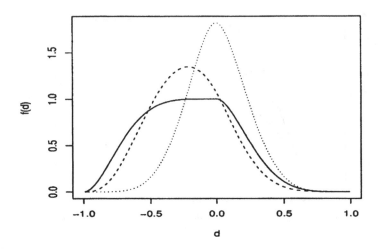

Figure 10. Pdfs of D: ———— for $(a_1, b_1, a_2, b_2) = (2, 5, 1, 1)$; – – – – for $(a_1, b_1, a_2, b_2) = (2, 5, 2, 2)$; .... for $(a_1, b_1, a_2, b_2) = (2, 5, 2, 5)$.

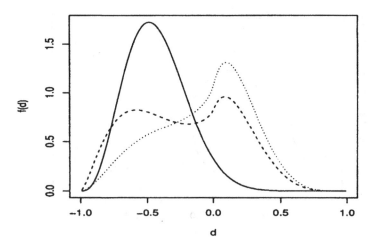

Figure 11. Pdfs of D: ———— for $(a_1, b_1, a_2, b_2) = (2, 5, 5, 2)$; – – – – for $(a_1, b_1, a_2, b_2) = (2, 5, 0.5, 0.5)$; .... for $(a_1, b_1, a_2, b_2) = (2, 5, 0.5, 0.9)$.

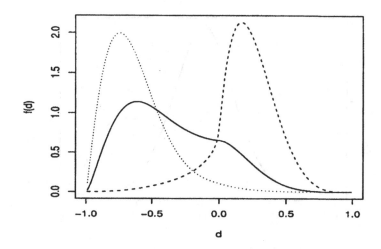

Figure 12. Pdfs of D: —————— for $(a_1, b_1, a_2, b_2) = (2, 5, 0.9, 0.5)$;
– – – – – for $(a_1, b_1, a_2, b_2) = (2, 5, 0.2, 2)$; .... for $(a_1, b_1, a_2, b_2) = (2, 5, 2, 0.2)$.

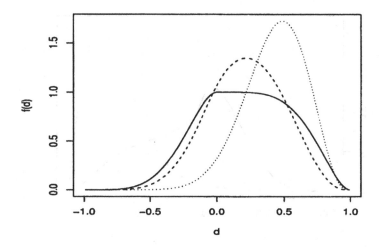

Figure 13. Pdfs of D: —————— for $(a_1, b_1, a_2, b_2) = (5, 2, 1, 1)$; –
– – – – for $(a_1, b_1, a_2, b_2) = (5, 2, 2, 2)$; .... for $(a_1, b_1, a_2, b_2) = (5, 2, 2, 5)$.

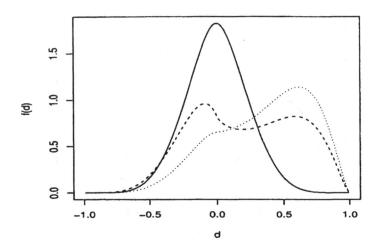

Figure 14. Pdfs of D: ——— for $(a_1, b_1, a_2, b_2) = (5, 2, 5, 2)$; $- - - - -$ for $(a_1, b_1, a_2, b_2) = (5, 2, 0.5, 0.5)$; .... for $(a_1, b_1, a_2, b_2) = (5, 2, 0.5, 0.9)$.

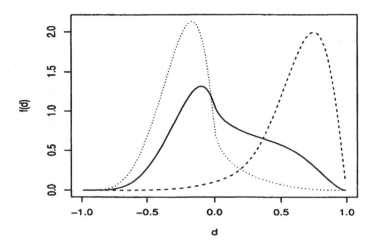

Figure 15. Pdfs of D: ——— for $(a_1, b_1, a_2, b_2) = (5, 2, 0.9, 0.5)$; $- - - - -$ for $(a_1, b_1, a_2, b_2) = (5, 2, 0.2, 2)$; .... for $(a_1, b_1, a_2, b_2) = (5, 2, 2, 0.2)$.

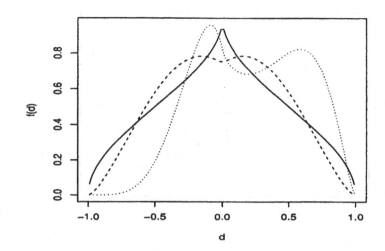

Figure 16. Pdfs of D: ———— for $(a_1, b_1, a_2, b_2) = (0.5, 0.5, 1, 1)$; $-----$ for $(a_1, b_1, a_2, b_2) = (0.5, 0.5, 2, 2)$; .... for $(a_1, b_1, a_2, b_2) = (0.5, 0.5, 2, 5)$.

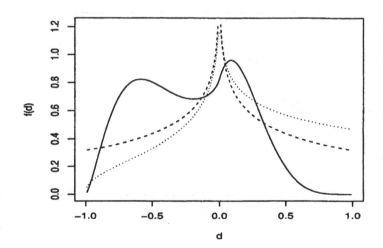

Figure 17. Pdfs of D: ———— for $(a_1, b_1, a_2, b_2) = (0.5, 0.5, 5, 2)$; $-----$ for $(a_1, b_1, a_2, b_2) = (0.5, 0.5, 0.5, 0.5)$; .... for $(a_1, b_1, a_2, b_2) = (0.5, 0.5, 0.5, 0.9)$.

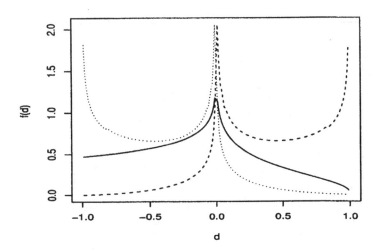

Figure 18. Pdfs of D: ——— for $(a_1, b_1, a_2, b_2) = (0.5, 0.5, 0.9, 0.5)$; $-----$ for $(a_1, b_1, a_2, b_2) = (0.5, 0.5, 0.2, 2)$; .... for $(a_1, b_1, a_2, b_2) = (0.5, 0.5, 2, 0.2)$.

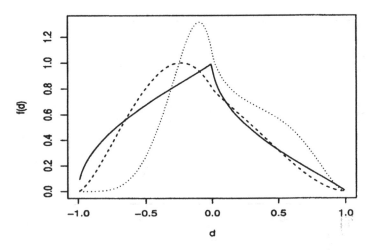

Figure 19. Pdfs of D: ——— for $(a_1, b_1, a_2, b_2) = (0.5, 0.9, 1, 1)$; $-----$ for $(a_1, b_1, a_2, b_2) = (0.5, 0.9, 2, 2)$; .... for $(a_1, b_1, a_2, b_2) = (0.5, 0.9, 2, 5)$.

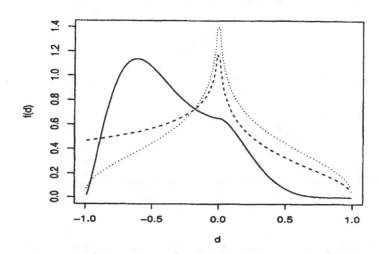

Figure 20. Pdfs of D: ———— for $(a_1, b_1, a_2, b_2) = (0.5, 0.9, 5, 2)$;
$----$ for $(a_1, b_1, a_2, b_2) = (0.5, 0.9, 0.5, 0.5)$; .... for $(a_1, b_1, a_2, b_2) = (0.5, 0.9, 0.5, 0.9)$.

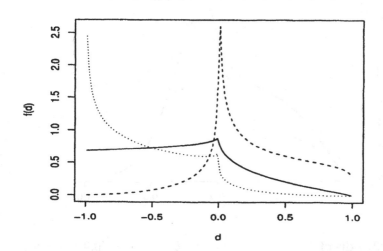

Figure 21. Pdfs of D: ———— for $(a_1, b_1, a_2, b_2) = (0.5, 0.9, 0.9, 0.5)$;
$----$ for $(a_1, b_1, a_2, b_2) = (0.5, 0.9, 0.2, 2)$; .... for $(a_1, b_1, a_2, b_2) = (0.5, 0.9, 2, 0.2)$.

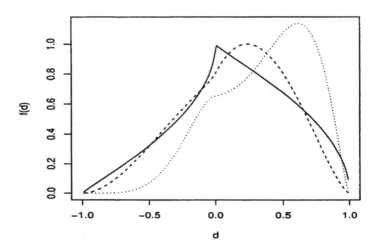

Figure 22. Pdfs of D: ———— for $(a_1, b_1, a_2, b_2) = (0.9, 0.5, 1, 1)$; $----$ for $(a_1, b_1, a_2, b_2) = (0.9, 0.5, 2, 2)$; .... for $(a_1, b_1, a_2, b_2) = (0.9, 0.5, 2, 5)$.

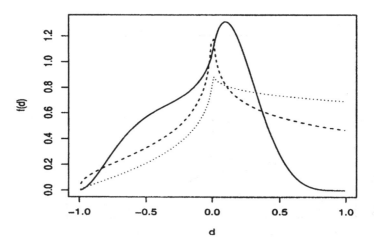

Figure 23. Pdfs of D: ———— for $(a_1, b_1, a_2, b_2) = (0.9, 0.5, 5, 2)$; $----$ for $(a_1, b_1, a_2, b_2) = (0.9, 0.5, 0.5, 0.5)$; .... for $(a_1, b_1, a_2, b_2) =$ $(0.9, 0.5, 0.5, 0.9)$.

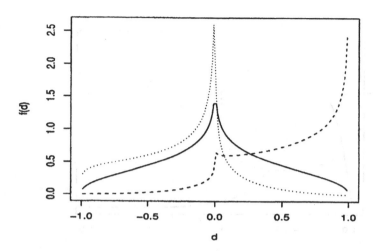

Figure 24. Pdfs of D: ———— for $(a_1, b_1, a_2, b_2) = (0.9, 0.5, 0.9, 0.5)$;
– – – – – for $(a_1, b_1, a_2, b_2) = (0.9, 0.5, 0.2, 2)$; .... for $(a_1, b_1, a_2, b_2) = (0.9, 0.5, 2, 0.2)$.

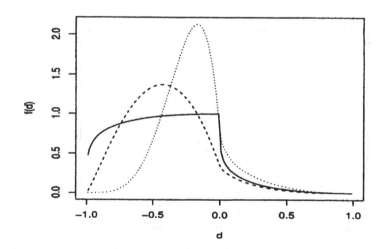

Figure 25. Pdfs of D: ———— for $(a_1, b_1, a_2, b_2) = (0.2, 2, 1, 1)$; –
– – – – for $(a_1, b_1, a_2, b_2) = (0.2, 2, 2, 2)$; .... for $(a_1, b_1, a_2, b_2) = (0.2, 2, 2, 5)$.

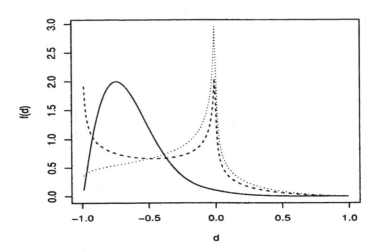

Figure 26. Pdfs of D: ———— for $(a_1, b_1, a_2, b_2) = (0.2, 2, 5, 2)$; −−−−− for $(a_1, b_1, a_2, b_2) = (0.2, 2, 0.5, 0.5)$; .... for $(a_1, b_1, a_2, b_2) = (0.2, 2, 0.5, 0.9)$.

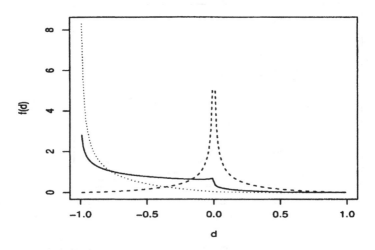

Figure 27. Pdfs of D: ———— for $(a_1, b_1, a_2, b_2) = (0.2, 2, 0.9, 0.5)$; −−−−− for $(a_1, b_1, a_2, b_2) = (0.2, 2, 0.2, 2)$; .... for $(a_1, b_1, a_2, b_2) = (0.2, 2, 2, 0.2)$.

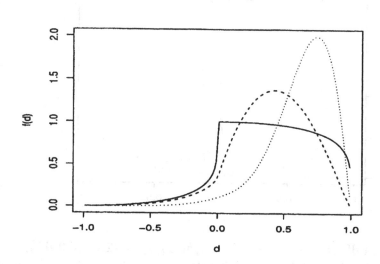

Figure 28. Pdfs of D: ———— for $(a_1, b_1, a_2, b_2) = (2, 0.2, 1, 1)$; – – – – for $(a_1, b_1, a_2, b_2) = (2, 0.2, 2, 2)$; .... for $(a_1, b_1, a_2, b_2) = (2, 0.2, 2, 5)$.

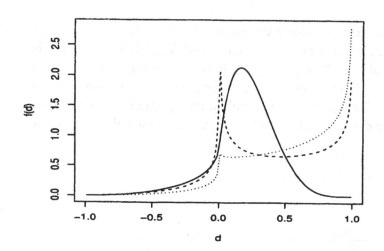

Figure 29. Pdfs of D: ———— for $(a_1, b_1, a_2, b_2) = (2, 0.2, 5, 2)$; – – – – for $(a_1, b_1, a_2, b_2) = (2, 0.2, 0.5, 0.5)$; .... for $(a_1, b_1, a_2, b_2) = (2, 0.2, 0.5, 0.9)$.

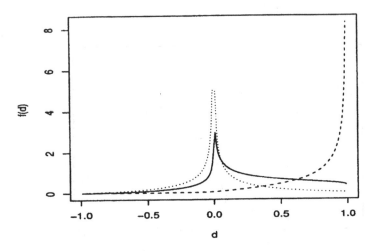

Figure 30. Pdfs of D: ——————— for $(a_1, b_1, a_2, b_2) = (2, 0.2, 0.9, 0.5)$; — — — — for $(a_1, b_1, a_2, b_2) = (2, 0.2, 0.2, 2)$; .... for $(a_1, b_1, a_2, b_2) = (2, 0.2, 2, 0.2)$.

B.   Approximations

When X_1 and X_2 follow the generalized beta distribution (equation (VII.2) in Chapter 5) on (c_1, d_1) and (c_2, d_2), respectively, the exact pdf of $Y = X_1 + X_2$ is given in (II.1). An approximation is to assume that Y also has the generalized beta distribution on $(c_1 + c_2, d_1 + d_2)$ with parameters a_3 and b_3. By equating the first two moments, one can see that a_3 and b_3 are the solutions of the equations

$$\frac{a_3}{a_3 + b_3} = \frac{E(Z) - a_1 - a_2}{b_1 + b_2 - a_1 - a_2}$$

and

$$\frac{a_3 b_3}{(a_3 + b_3)^2 (a_3 + b_3 + 1)} = \frac{Var(Z)}{(b_1 + b_2 - a_1 - a_2)^2}$$

(Sculli and Wong, 1985).

Now consider the linear combination in its general form:

$$Y = \sum_{k=1}^{n} d_k X_k$$

when the X_k are independent beta random variables with parameters (a_k, b_k). Johannesson and Giri (1995) suggested two approximations for the distribution of Y. The first is based on approximating $Y = \rho Z$, where Z is a standard beta random variable with parameters e and f. By equating the expectation and variance and solving them, e and f are determined as:

$$e = Ff$$

and

$$f = \frac{F}{S(1+F)^3} - \frac{1}{1+F},$$

where

$$E = \sum_{k=1}^{n} d_k E(X_k),$$

$$F = \frac{E}{1-E}$$

and

$$S = \frac{1}{\rho^2} \sum_{k=1}^{n} d_k^2 Var(X_k).$$

The second approximation is based on the equality $Y = \rho Z/\gamma$, where Z is a standard beta random variable with parameters g and h. By equating the first three central moments and solving them, γ, g and h are determined as:

$$\gamma = -\frac{B + \sqrt{B^2 - 4AC}}{2A},$$

$$g = \frac{(\gamma k_3 - 1)(\gamma k_2 - 1)}{\gamma(k_3 - k_2)}$$

and

$$h = \frac{\gamma k_3 g + 2\gamma k_3 - 2}{1 - \gamma k_3},$$

where

$$A = k_3^2 k_2 + k_3 k_2 k_1 - 2k_3^2 k_1,$$

$$B = k_3^2 - 3k_3 k_2 + 3k_3 k_1 - k_2 k_1,$$

$$C = 2k_2 - k_1 - k_3,$$

$$k_1 = E(Y),$$

$$k_2 = \frac{E(Y^2)}{E(Y)}$$

and

$$k_3 = \frac{E(Y^3)}{E(Y^2)}.$$

References

1. Bateman, H. *Tables of Integral Transforms* (volume 1); McGraw-Hill: New York, 1954.
2. Bhargava, R.P. Some results on beta distributions with application to multivariate problem, Annals of the Institute of Statistical Mathematics **1975**, *27*, 109-116.
3. Fan, D.-Y. The distribution of the product of independent beta variables, Communications in Statistics–Theory and Methods **1991**, *20*, 4043-4052.

4. Gupta, A.K.; Nadarajah, S. Letter to the editor, Communications in Statistics–Theory and Methods **2002**.

5. Jambunathan, M.V. Some properties of beta and gamma distributions, Annals of Mathematical Statistics **1954**, *25*, 401-405.

6. Johannesson, B.; Giri, N. On approximations involving the beta distribution, Communications in Statistics–Simulation and Computation **1995**, *24*, 489-503.

7. Kotlarski, I. On groups of n independent random variables whose product follows the beta distribution, Colloquium Mathematicum **1962**, *2*, 325-332.

8. Krysicki, W. On some new properties of the beta distribution, Statistics and Probability Letters **1999**, *42*, 131-137.

9. Malik, H.J. The distribution of the product of two noncentral beta variates, Naval Research Logistics Quarterly **1970**, *17*, 327-330.

10. Mikhail, N.N.; Tracy, D.S. The exact non-null distribution of Wilk's Λ criterion in the bivariate collinear case, Canadian Mathematical Bulletin **1975**, *17*, 757-758.

11. Nandi, S.B. On the exact distribution of a normalized ratio of the weighted geometric mean to the unweighted arithmetic mean in samples from gamma distributions, Journal of the American Statistical Association **1980**, *75*, 217-220.

12. Pham-Gia, T. Distributions of the ratios of independent beta variables and applications, Communications in Statistics–Theory and Methods **2000**, *29*, 2693-2715.

13. Pham-Gia, T.; Turkkan, N. Bayesian analysis of the difference of two proportions, Communications in Statistics–Theory and Methods **1993**, *22*, 1755-1771.

14. Pham-Gia, T.; Turkkan, N. Reliability of a standby system with beta-distributed component lives, IEEE Transactions on Reliability **1994**, *43*, 71-75.

15. Sculli, D.; Wong, K.L. The maximum and sum of two beta variables in the analysis of PERT networks, Omega International Journal of Management Science **1985**, *13*, 233-240.

16. Springer, M.D. *Algebra of Random Variables*; John Wiley and Sons: New York, 1978.

17. Springer, M.D.; Thompson, W.E. The distribution of products of beta, gamma and Gaussian random variables, SIAM Journal on Applied Mathematics **1970**, *18*, 721-737.

18. Tang, J.; Gupta, A.K. On the distribution of the product of independent beta random variables, Statistics and Probability Letters **1984**, *2*, 165-168.

19. Wilks, S.S. Certain generalizations in the analysis of variance, Biometrika **1932**, *24*, 471-494.

Order Statistics and Records

Arjun Gupta[1] and Saralees Nadarajah[2]

[1]Department of Mathematics and Statistics
Bowling Green State University
Bowling Green, Ohio 43403
[2]Department of Mathematics
University of South Florida
Tampa, Florida 33620

I. Order Statistics

A. Uniform Distribution

The beta distribution arises naturally in the theory of uniform order statistics. Suppose $X_1, X_2, \ldots, X_n$ is a random sample from the uniform distribution on $(0,1)$. Let $X_{1:n} \leq X_{2:n} \leq \cdots \leq X_{n:n}$ be the corresponding order statistics. Then the following facts are well-known:

- $W_1 = X_{i:n}/X_{j:n}$ and $W_2 = X_{j:n}$ are statistically independent. Moreover, W_1 and W_2 are beta distributed with parameters $(i, j-i)$ and $(j, n-j+1)$, respectively.
- $W_3 = X_{i:n}/(1 + X_{i:n} - X_{j:n})$ and $W_4 = X_{j:n} - X_{i:n}$ are statistically independent. Moreover, W_3 and W_4 are beta distributed with parameters $(i, n-j+1)$ and $(j-r, n-j+$

$i + 1$), respectively.

- For $1 \leq i_1 < i_2 < \cdots < i_j \leq n$, the random variables defined by $W_k = X_{i_k:n}/X_{i_{k+1}:n}$, $k = 1, 2, \ldots, j - 1$ and $W_j = X_{i_j:n}$ are statistically independent. Moreover, W_k and W_j are beta distributed with parameters $(i_k, i_{k+1} - i_k)$ and $(i_j, n - i_j + 1)$, respectively.

Using the fact that the product of beta random variables can be also beta (see Chapter 2), Malmquist (1971) established some interesting properties concerning products of independent uniform order statistics. To be specific, let $U^1_{n_1:n_2-1}$ be the n_1th uniform order statistic from a random sample $\{U^1_i\}$ of size $n_2 - 1$. Let $U^2_{n_2:n_3}$ be the n_2th uniform order statistic from a random sample $\{U^2_i\}$ of size n_3 independent of $\{U^1_i\}$. Then $U^1_{n_1:n_2-1}$ and $U^2_{n_2:n_3}$ are independently beta distributed with parameters $(n_1, n_2 - n_1)$ and $(n_2, n_3 - n_2 + 1)$, respectively. So using properties in Chapter 2, note that the product

$$Z = U^1_{n_1:n_2-1} U^2_{n_2:n_3}$$

is also beta distributed with parameters $(n_1, n_3 - n_1 + 1)$. Hence, Z is distributed as the n_1th uniform order statistic from a random sample of size n_3. By repeating this process $\nu - 1$ times, note that

$$Z = U^1_{n_1:n_2-1} U^2_{n_2:n_3-1} \cdots U^{\nu-1}_{n_{\nu-1}:n_\nu},$$
$$n_1 < n_2 - 1 < \cdots < n_\nu$$

would be distributed as the n_1th uniform order statistic from a random sample of size n_ν. More generally, Malmquist (1971) established that if $\{Y_{n_k}\}$ are defined in terms of the above by

$$Y_{n_1} = \prod_{k=1}^{\nu-1} U^k,$$

$$Y_{n_2} = \prod_{k=2}^{\nu-1} U^k$$

$$\vdots \quad \vdots$$

$$Y_{n_{\nu-1}} = U^{\nu-1}$$

and if $\{X_{n_k:n_\nu}\}$ are a subsequence of the uniform order statistics

$$X_{1:n_\nu} < \cdots < X_{n_1:n_\nu} < \cdots < X_{n_2:n_\nu}$$
$$< \cdots < X_{n_{\nu-1}:n_\nu} < \cdots < X_{n_\nu:n_\nu}$$

from a random sample of size n_ν then $\{Y_{n_k}\}$ and $\{X_{n_k:n_\nu}\}$ have the same simultaneous distribution.

The beta distribution also arises in connection with fractional uniform order statistics (Stigler, 1977). For a random sample of size n from the uniform distribution on $(0,1)$, Stigler defined the fractional uniform order statistic (denoted by $U_{a:n}$, $0 < a < n+1$) as a quantity following the beta distribution with parameters a and $n+1-a$. If $U_{1:n} \le U_{2:n} \le \cdots \le U_{n:n}$ are the uniform order statistics from the sample then Jones (2002a) showed that

$$U_{a:n} = (1 - C)U_{[a]:n} + CU_{[a]+1:n},$$

where C is beta distributed – independently of the U's – with parameters $a - [a]$ and $1 - a + [a]$.

B. Beta Distribution

Suppose now that $X_{1:n} \le X_{2:n} \le \cdots \le X_{n:n}$ are order statistics for a random sample of size n from the beta distribution with parameters a and b. Van Zwet (1964) established the following bounds for the moments of the order statistics:

$$\frac{i-1}{n} \le F(E(X_{i:n})) \le \frac{i}{n}, \qquad \text{if } a > 1, b > 1,$$

$$\frac{i-1}{n} \le F(E(X_{i:n})) \le \frac{i}{n+1}, \qquad \text{if } a > 1, b = 1,$$

$$\frac{i}{n+1} \le F(E(X_{i:n})) \le \frac{i}{n}, \qquad \text{if } a = 1, b > 1,$$

$$F(E(X_{i:n})) \le \frac{i}{n+1}, \qquad \text{if } a \ge 1, b < 1$$

and

$$\frac{i}{n+1} \leq F\left(E\left(X_{i:n}\right)\right), \qquad \text{if } a < 1, b \geq 1,$$

where F denotes the cdf of the beta distribution.

C. Power Function Distribution

Rider (1964) studied the behavior of the largest order statistics for two independent samples from the power function distribution. If L denotes the largest order statistic for a random sample of size n from the power function distribution then it is easy to show that its pdf is:

$$f(l) = nl^{na-1}.$$

If there are two random samples of size m and n and if L_1 and L_2 denote the corresponding largest order statistics then the product $U = L_1 L_2$ and the quotient $V = L_2/L_1$ can be shown to have the pdfs:

$$f(u) = \begin{cases} \dfrac{mna\left(u^{na} - u^{ma}\right)}{(m-n)u}, & \text{if } m \neq n, \\ -n^2 a^2 u^{na-1} \log u, & \text{if } m = n \end{cases}$$

and

$$f(v) = \begin{cases} \dfrac{mna}{m+n} v^{na-1}, & \text{if } 0 \leq v \leq 1, \\ \dfrac{mna}{m+n} v^{-ma-1}, & \text{if } 1 \leq v < \infty. \end{cases}$$

Rider (1964) also showed that the geometric mean (say G) for a random sample of size n from the power function distribution has the pdf:

$$f(g) = \frac{a^n n^n g^{na-1}}{(n-1)!} \left(-\log g\right)^{n-1}.$$

Likes (1967) considered the distribution of the products and quotients within one sample. Let $X_{1:n} \leq X_{2:n} \leq \cdots \leq X_{n:n}$ be the order statistics for a random sample of size n from the power function distribution. Consider the product and the quotient defined by

$$U = X_{n-i+1:n}X_{n-i+2:n}\cdots X_{n:n}$$

and

$$V = \frac{X_{n-j+1:n}}{X_{n-i+1:n}}.$$

Likes established that the corresponding cdfs are:

$$F(u) = \frac{a^n}{\Gamma(n)}\int_0^x x^{a-1}\left[\log\left(\frac{1}{x}\right)\right]^{n-1}dx,$$
$$0 < v \le 1, \qquad \text{if } i = n,$$

$$F(u) = \binom{n}{i}\sum_{k=1}^i (-1)^{k-1}\binom{i}{k-1}\left(\frac{i-k+1}{n-i}\right)^{i-1}\frac{i-k+1}{n-k+1},$$
$$0 < v \le 1, \qquad \text{if } 1 \le i < n$$

and

$$F(v) = (j-i)\binom{n-i}{j-i}\int_0^{v^a} x^{n-j}(1-x)^{j-i-1}dx, \quad 0 < v \le 1.$$

In particular, if $j = i+1$ then $V = X_{n-i:n}/X_{n-i+1:n}$ has the cdf

$$F(v) = v^{a(n-i)}, \qquad 0 < v \le 1.$$

D. Complementary Beta Distribution

This distribution is discussed in Chapter 5 and is due to Jones (2002b). Jones established some elementary properties of order statistics $X_{1:n} \le X_{2:n} \le \cdots \le X_{n:n}$ for a random sample of size n from this distribution. For instance, the pdf and the expectation of $X_{i:n}$ are:

$$f(x) = \frac{B(a,b)}{B(i,n-i+1)}\left\{I_x^{-1}(a,b)\right\}^{i-a}\left\{1 - I_x^{-1}(a,b)\right\}^{n-i+1-b}$$

and

$$E(X_{i:n}) = \frac{1}{B(a,b)}\sum_{k=0}^{i-1}\binom{n}{k}B(a+k,b+n-k).$$

Thus the expectation of the spacing $S_{i:n} = X_{i+1:n} - X_{i:n}$ is

$$E(S_{i:n}) = \binom{n}{i} \frac{B(a+i, b+n-i)}{B(a,b)}.$$

II. Records

Given an infinite sequence $X_1, X_2, \ldots$ of iid random variables, an observation X_j is called an upper record value (or simply a record) if $X_j > X_i$ for every $i < j$. The times at which records appear $\{T_n, n \geq 0\}$ (known as the record time sequence) is defined by:

$T_0 = 1$ with probability 1

and, for $n \geq 1$,

$$T_n = \min\{j : X_j \geq X_{T_{n-1}}\}.$$

The record value sequence $\{R_n\}$ is then defined by

$$R_n = X_{T_n}, \qquad n = 0, 1, 2, \ldots.$$

An analogous definition deals with lower record value sequence $\{\tilde{R}_n\}$.

A. Power Function Distribution

If the iid sequence $X_1, X_2, \ldots$ is assumed to come from the power function distribution (the special case of the beta distribution for $b = 1$) then it can be shown that

$$R_n =^d \left[1 - \prod_{k=0}^{n} (1 - U_k)\right]^{1/a},$$

where $\{U_k, k \geq 0\}$ is a sequence of iid uniform random variables on $(0,1)$. The lower record sequence, $\{\tilde{R}_n\}$, admits a slightly simpler representation:

$$\tilde{R}_n =^d \left[\prod_{k=0}^{n} U_k\right]^{1/a}.$$

In the particular case $a = 1$, one notes the properties:

(i) $(1 - R_n)/(1 - R_{n-1})$ and $1 - R_{n-1}$ are independent.

(ii) $E(R_n \mid R_{n-1} = r_{n-1}) = (1/2) + (1/2)r_{n-1}$.

(iii) $(1 - R_n)/(1 - R_{n-1}) =^d 1 - R_0$.

All of these are characteristic properties of the uniform distribution. The rth moment of the nth record is given by

$$E(R_n^r) = \int_0^1 z^{r/a} \frac{(-\log(1-z))^n}{n!} dz$$

$$= \int_0^\infty \{1 - \exp(-z)\}^{r/a} \frac{z^n}{n!} \exp(-z) dz$$

$$= \sum_{k=0}^\infty \binom{r/a}{k} (-1)^k (1+k)^{-(n+1)}.$$

In the particular case $a = 1$, the expectation and the variance of R_n are:

$$E(R_n) = 1 - 2^{-(n+1)}$$

and

$$Var(R_n) = 3^{-(n+1)} - 4^{-(n+1)}.$$

References

1. Jones, M.C. On fractional uniform order statistics, Statistics and Probability Letters 2002a, 58, 93-96.

2. Jones, M.C. The complementary beta distribution, Journal of Statistical Planning and Inference 2002b, 104, 329-337.

3. Likes, J. Distributions of some statistics in samples from exponential and power-function populations, Journal of the American Statistical Association 1967, 62, 259-271.

4. Malmquist, S. A note on a property of beta-distributed variables with applications to ordered observations, Skand. AktuarTidskr 1971, 97-101.

5. Rider, P.R. Distribution of product and of quotient of maximum values in samples from a power-function population, Journal of the American Statistical Association 1964, 59, 877-880.

6. Stigler, S.M. Fractional order statistics, with applications, Journal of the American Statistical Association **1977**, *72*, 544-550.

7. van Zwet, W.R. Convex transformations of random variables, Mathematical Center Tracts **1964**, *7*, Mathematisch Centrum, Amsterdam.

Generalizations and Related Univariate Distributions

Saralees Nadarajah[1] and Arjun Gupta[2]

[1]Department of Mathematics
University of South Florida
Tampa, Florida 33620
[2]Department of Mathematics and Statistics
Bowling Green State University
Bowling Green, Ohio 43403

I. Arc-Sine Distribution

As mentioned in Chapter 2, the arc-sine distribution is the particular case of the Beta distribution when $a = b = 1/2$. The general form of the arc-sine pdf is:

$$f(x) = \begin{cases} \dfrac{1}{\pi\sqrt{\left(\frac{2}{c}\right)^2 - x^2}}, & \mid x \mid < \left|\frac{2}{c}\right|, \\ 0, & \mid x \mid \geq \left|\frac{2}{c}\right|, \end{cases} \qquad (I.1)$$

where $c \neq 0$. When $c = 2$ the distribution is said to be in its standard form with the pdf

$$f(x) = \frac{1}{\pi\sqrt{1 - x^2}}, \qquad -1 < x < 1. \qquad (I.2)$$

The arc-sine distribution arises naturally in statistical communication theory; see Lee (1960, Chapter 6) and Middleton (1965, Chapter 14), where (I.2) is used as a model for the amplitude of a periodic signal in thermal noise and the limiting spectral density function of a high-index-angle modulated carrier, respectively. The arc-sine distribution arises also in the study of the simple random walk. Feller (1967, Chapter 3) showed that the limiting distribution of the proportion of time spent on the positive side of the x axis by a simple symmetric random walk has the pdf:

$$f(x) = \frac{1}{\pi\sqrt{x(1-x)}}, \qquad 0 < x < 1.$$

More generally, Erdös and Kac (1947) proved that if $X_1, X_2, \ldots$ are independent random variables each having mean zero and variance one then

$$N_n = \sum_{k=1}^{n} I\left\{\sum_{i=1}^{k} X_i > 0\right\}$$

has the limiting distribution

$$\lim_{n\to\infty} \Pr\left\{\frac{N_n}{n} < y\right\} = \frac{2}{\pi} \arcsin\sqrt{y}, \qquad (I.3)$$

which is the cdf of (I.2). The proof of this limiting result given by Erdös and Kac is complicated. Recently Hoffmann-Jorgensen (1999) provided a simple and elementary proof of a more general version of (I.3). For other work on how the arc-sine distribution arises in stochastic processes and their applications, see Chung and Feller (1949), Andersen (1953), Spitzer (1964), Chen et al. (1981), Karlin and Taylor (1981), Pitman and Yor (1992), Getoor and Sharpe (1994), Breitung and Gourieroux (1997), Neuts et al. (1999) and Breitung (2001).

The central moments of (I.1) are

$$\mu_{2j} = c^{-2j} \binom{2j}{j}$$

and these moments characterize the distribution. Norton (1975) showe that if X_1 and X_2 are iid and non-degenerate with a symmetric and moment-determined distribution then $X_1 X_2$ and $(X_1 + X_2)/2$ have the same distribution if and only if the common pdf is (I.2). Arnold and Groeneveld (1980) provided three alternative characterization based on the trigonometric relationships:

$$\sin^2 U = (1 - \cos 2U)/2,$$

$$\sin 2U = 2 \sin U \cos U$$

and

$$\sin^2 U - \sin^2 V = \sin(U + V) \sin(U - V).$$

If U and V are taken to be iid uniform $(-\pi, \pi)$ then $\sin U$, $\sin 2U$, $-\cos 2U$, $\sin(U + V)$ and $\sin(U - V)$ have arc-sine distributions. This suggests the three characterizations:

1. If X is a symmetric random variable then X^2 and $(1 + X)/2$ are identically distributed if and only if X has the pdf (I.2).
2. If X is a symmetric random variable with the property that X^2 and $1 - X^2$ are identically distributed then the random variables X and $2X\sqrt{1 - X^2}$ are identically distributed if and only if X has the pdf (I.2).
3. If X_1, X_2 are symmetric iid random variables with the property that X_i^2 and $1 - X_i^2$ are identically distributed then the random variables $X_1^2 - X_2^2$ and $X_1 X_2$ are identically distributed if and only if the common pdf of X_1, X_2 is (I.2).

These characterizations are useful in quantifying the heavy-tailed behavior of (I.2).

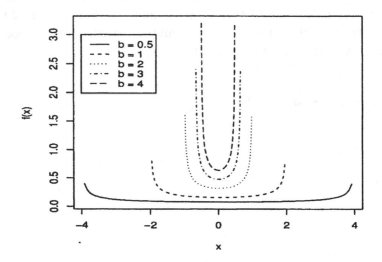

Figure 1. Arcsine pdfs.

II.   Log Beta Distribution

As the name indicates, X is said to have the log beta distribution if $\log X$ has a beta distribution. Clearly the support of X must be a finite positive interval, say $0 < c \leq X \leq d$. Then $Z = (\log X - \log c)/(\log d - \log c)$ has the standard beta distribution (equation (I.1) of Chapter 2) for some parameters a and b. The moments of X are

$$
\begin{aligned}
\mu_r' &= \frac{\exp(ra)}{B(a,b)} \int_0^1 z^{a-1}(1-z)^{b-1} \exp\left\{r(b-a)z\right\} dz \\
&= \exp(ra)\ _1F_1\left(a; a+b; r(b-a)\right),
\end{aligned}
$$

where $_1F_1$ is the confluent hypergeometric function. If both a and b are integers then we can write

$$\mu_r' = \{r(b-a)\}^{1-a-b} \exp(ra) \frac{(a+b-2)!(1-a-b)_a}{(a-1)!}$$

$$\times \left[\sum_{k=0}^{b-1} \frac{(1-b)_k}{k!(2-a-b)_k} \{r(b-a)\}^k \right.$$

$$\left. - \exp\{r(b-a)\} \sum_{k=0}^{a-1} \frac{(-1)^k(1-a)_k}{k!(2-a-b)_k} \{r(b-a)\}^k \right].$$

If in addition $a = 1$ then we can reduce the above to the simple form:

$$\mu_r' = b! \{r(b-1)\}^{-b} \left[\exp\{r(b-1)\} - \sum_{k=0}^{b-1} \frac{r^k(b-1)^k}{k!} \right].$$

As for applications, the log beta distribution has been used to model the evolution of the size distribution for aerosols (Bunz et al., 1987; Chang et al., 1988; Runyan et al., 1988; Han et al., 1989) and the frequency of high microbial counts (e.g. standard plate count, coliforms, yeasts) in commercial food products (Corradini et al., 2001). Barrett et al. (1991) provided a comparison of the log-beta distribution with the log-normal distribution for particle size populations. They demonstrated that populations with a wide but finite size range can be described by the log beta distribution irrespective of the direction or magnitude of their skewness. They also suggested that the log beta distribution should be preferred for populations in which the size distribution is skewed to the left.

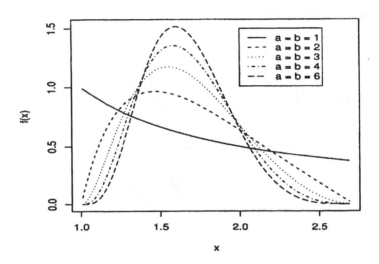

Figure 2. Left skewed log Beta pdfs.

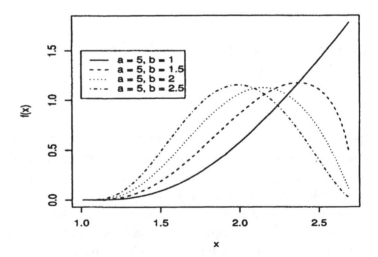

Figure 3. Right skewed log Beta pdfs.

III. Type I Noncentral Beta Distribution

If S is a non-central chi-squared random variable with degrees of freedom $2a$ and noncentrality parameter λ and if T is a central chi-squared random variable with degrees of freedom $2b$ (distributed independently of S) then

$$X = \frac{S}{S+T} \qquad\qquad (\text{III.1})$$

is said to have the type I noncentral beta distribution with shape parameters a, b and noncentrality parameter λ. Tang (1938) showed that the pdf and the cdf of X can be represented as Poisson mixtures:

$$f(x; a, b, \lambda) = \frac{x^{a-1}(1-x)^{b-1}}{\Gamma(b)} \sum_{k=0}^{\infty} \frac{\Gamma(a+b+k)u_k x^k}{\Gamma(a+k)} \qquad\qquad (\text{III.2})$$

and

$$F(x; a, b, \lambda) = \sum_{k=0}^{\infty} u_k I_x(a+k, b), \qquad\qquad (\text{III.3})$$

respectively, where

$$u_k = \frac{(\lambda/2)^k \exp(-\lambda/2)}{k!} \qquad\qquad (\text{III.4})$$

is the pmf of the Poisson distribution with parameter $\lambda/2$. When b is an integer, Nicholson (1954) and Hodges (1955) showed that (III.3) can be reduced to the equivalent forms:

$$F(x; a, b, \lambda) = \exp\{-\lambda(1-x)\} \sum_{k=0}^{b-1} \frac{\lambda^k (1-x)^k}{k!} I_x(a+k, b-k)$$

or

$$F(x; a, b, \lambda) = \exp\{-\lambda(1-x)\} \left\{ I_x(a, b) \right.$$
$$\left. + x^a \sum_{k=1}^{b-1} x^k (1-x)^k \lambda^k \frac{P_k}{k!} \right\},$$

where

$$P_k = \sum_{l=0}^{b-k-1} \binom{a+k+l-1}{a+k-1}(1-x)^l.$$

Using properties of the incomplete beta function (see Chapter 1), Nicholson also derived closed form expressions for F when $b = 1, \ldots, 5$ giving:

$$F(x; a, 1, \lambda) = (1-\alpha)\exp\left[-\lambda\left\{1 - (1-\alpha)^{1/a}\right\}\right],$$

$$F(x; a, 2, \lambda) = \exp\left\{-\lambda(1-x)\right\}\left[1 - \alpha + x^{a+1}(1-x)\lambda\right],$$

$$F(x; a, 3, \lambda) = \exp\left\{-\lambda(1-x)\right\}\left[1 - \alpha \right.$$
$$+ x^{a+1}(1-x)\left\{a + 2 - (a+1)x\right\}\lambda$$
$$\left. + \frac{1}{2}x^{a+2}(1-x)^2\lambda^2\right],$$

$$F(x; a, 4, \lambda) = \exp\left\{-\lambda(1-x)\right\}\left[1 - \alpha \right.$$
$$+ \frac{1}{2}x^{a+1}(1-x)\left\{(a+3)(a+2)\right.$$
$$- 2(a+3)(a+1)x + (a+2)(a+1)x^2\right\}\lambda$$
$$+ \frac{1}{2}x^{a+2}(1-x)^2\left\{a + 3 - (a+2)x\right\}\lambda^2$$
$$\left. + \frac{1}{6}x^{a+3}(1-x)^3\lambda^3\right]$$

and

$$F(x; a, 5, \lambda) = \exp\{-\lambda(1 - x)\} \Bigg[1 - \alpha$$
$$+ \frac{1}{6} x^{a+1}(1 - x) \Big\{ (a + 4)(a + 3)(a + 2)$$
$$- 3(a + 4)(a + 3)(a + 1)x$$
$$+ 3(a + 4)(a + 2)(a + 1)x^2$$
$$- (a + 3)(a + 2)(a + 1)x^3 \Big\} \lambda$$
$$+ \frac{1}{4} x^{a+2}(1 - x)^2 \Big\{ (a + 4)(a + 3)$$
$$- 2(a + 4)(a + 2)x + (a + 3)(a + 2)x^2 \Big\} \lambda^2$$
$$+ \frac{1}{6} x^{a+3}(1 - x)^3 \Big\{ a + 4 - (a + 3)x \Big\} \lambda^3$$
$$+ \frac{1}{24} x^{a+4}(1 - x)^4 \lambda^4 \Bigg],$$

where $\alpha = 1 - x^a$.

Ding (1994) and later Chattamvelli (1995a) provided the following alternative representations of (III.2) and (III.3):

$$f(x; a, b, \lambda) = \sum_{k=0}^{\infty} u_k s_k$$

and

$$F(x; a, b, \lambda) = \sum_{k=0}^{\infty} v_k t_k,$$

where s_k, t_k, u_k and v_k are given the recurrence relations:

$$s_0 = \frac{at_0(1-x)}{x},$$

$$s_k = \frac{t_{k-1}(a+b+k-1)}{1-x}, \quad k \geq 1,$$

$$t_0 = \frac{\Gamma(a+b)}{\Gamma(a)\Gamma(b)}x^a(1-x)^b,$$

$$t_k = \frac{t_{k-1}x(a+b+k-1)}{a+k}, \quad k \geq 1,$$

$$u_0 = \exp\left(-\frac{\lambda}{2}\right),$$

$$u_k = \frac{u_{k-1}\lambda}{2k}, \quad k \geq 1,$$

$$v_0 = \exp\left(-\frac{\lambda}{2}\right),$$

$$v_k = v_{k-1} + u_k, \quad k \geq 1.$$

These are similar to the representations given in Ding (1992) for the non-central chi-squared distribution. If b is an integer then the above reduce to a recursion formula due to Tang (1938):

$$F(x; a, b, \lambda) = x^{a+b-1}\exp\{-\lambda(1-x)\}\sum_{k=0}^{b-1}t_k,$$

where

$$t_k = \frac{1-x}{kx}\{(a+b-k+\lambda x)t_{k-1} + \lambda(1-x)t_{k-2}\},$$

$$t_1 = \frac{1-x}{x}(x\lambda + a + b - 1),$$

$$t_0 = 1.$$

Chen and Chou (2000) provided a simple representation for the pdf:

$$f(x; a, b, \lambda) = \psi(x)\exp(-\lambda/2)f(x; a, b, 0),$$

where

$$\psi(x) = \sum_{k=0}^{\infty}\frac{(\lambda x)^k}{2^k k!}\kappa(k)$$

with

$$\kappa(k) = \prod_{l=1}^{k} \frac{a+b+l-1}{a+l-1}.$$

They showed further that $\psi(\cdot)$ satisfies the differential equation

$$2x\psi''(x) = (\lambda x - 2a)\psi'(x) + \lambda(a+b)\psi(x)$$

and this enables computation of the pdf using the S-system.

Starting with the relationship (8) in Chapter 1, Seber (1963) derived a finite expansion of the cdf F when b is an even integer:

$$F(x; a, b, \lambda) = x^a \exp\{-\lambda(1-x)\} \sum_{k=0}^{b-1} (1-x)^k L_k^{a-1}(-\lambda x) \quad \text{(III.5)}$$

where

$$L_k^{a-1}(-\lambda x) = \sum_{l=0}^{k} \binom{k+a-1}{k-l} \frac{(\lambda x)^l}{l!}$$

are the generalized Laguerre polynomials. These polynomials are readily obtained by the recurrence relations:

$$kL_k^{a-1}(-\lambda x) = (2k - 2 + a + \lambda x)L_{k-1}^{a-1}(-\lambda x)$$
$$-(k + a - 2)L_{k-2}^{a-1}(-\lambda x),$$
$$L_1^{a-1}(-\lambda x) = a + \lambda x,$$
$$L_0^{a-1}(-\lambda x) = 1.$$

Chattamvelli (1995b) suggested the following representation of (III.5) for large values of x:

$$F(x; a, b, \lambda) = x^{a+b-1} \exp\{-\lambda(1-x)\}$$
$$\times \sum_{k=0}^{b-1} \left(\frac{1-x}{x}\right)^k L_k^{a+b-k-1}(-\lambda x).$$

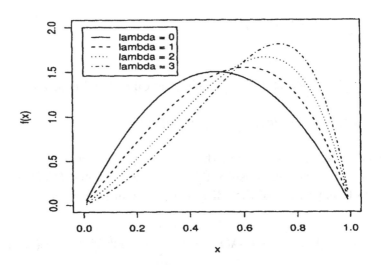

Figure 4. Type I noncentral beta pdfs for $a = 2$, $b = 2$ and $\lambda = 0$, 1, 2, 3.

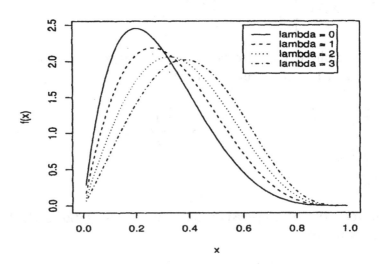

Figure 5. Type I noncentral beta pdfs for $a = 2$, $b = 5$ and $\lambda = 0$, 1, 2, 3.

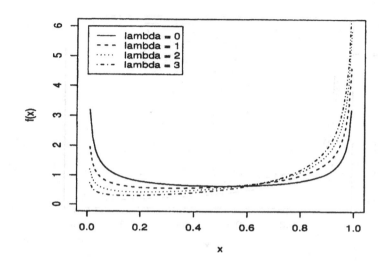

Figure 6. Type I noncentral beta pdfs for $a = 0.5$, $b = 0.5$ and $\lambda = 0$, 1, 2, 3.

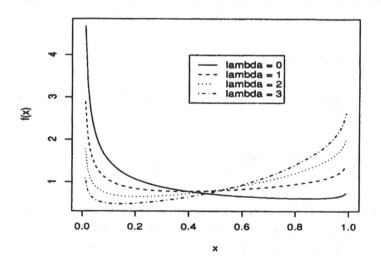

Figure 7. Type I noncentral beta pdfs for $a = 0.5$, $b = 0.9$ and $\lambda = 0$, 1, 2, 3.

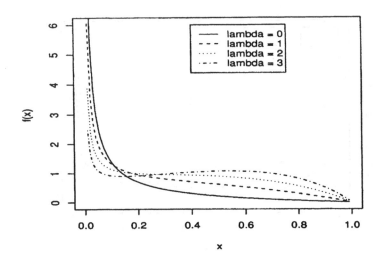

Figure 8. Type I noncentral beta pdfs for $a = 0.2$, $b = 2$ and $\lambda = 0$, 1, 2, 3.

The noncentral beta distribution given by (III.2) and (III.3) can be evaluated also through the noncentral F since the cdf of noncentral F, say G, is related to F in (III.3) by:

$$F\left(x; a, b, \lambda\right) = G\left(y; 2a, 2b, \lambda\right),$$

where $y = bx/\{a(1 - x)\}$. The corresponding relationship in terms of the pdfs is:

$$f\left(x; a, b, \lambda\right) = \frac{\left(2b + 2ay\right)^2}{4ab} g\left(y; 2a, 2b, \lambda\right).$$

The moments of the noncentral beta distribution can be easily obtained using the representations (III.2) and (III.3). For instance, the first two moments of X are:

$$\mu_1' = \exp\left(-\frac{\lambda}{2}\right) \sum_{k=0}^{\infty} \frac{\lambda^k}{2^k k!} \frac{a + k}{a + b + k}$$

and

$$\mu_2' = \exp\left(-\frac{\lambda}{2}\right) \sum_{k=0}^{\infty} \frac{\lambda^k}{2^k k!} \frac{(a+k)(a+k+1)}{(a+b+k)(a+b+k+1)}.$$

Using the definition of hypergeometric functions, the above can be rewritten as:

$$\mu_1' = \frac{a\exp(-\lambda/2)}{a+b} \, {}_2F_2\left(a+1, a+b; a, a+b+1; \frac{\lambda}{2}\right)$$

and

$$\mu_2' = \frac{a(a+1)\exp(-\lambda/2)}{(a+b)(a+b+1)} \, {}_2F_2\left(a+2, a+b; a, a+b+2; \frac{\lambda}{2}\right).$$

A more-straightforward closed form expression for the first two moments was obtained by Marchand (1997). For $c > 1$, define

$$g(c,\delta) = \sum_{k=0}^{[c]-1} (-1)^k \delta^{-(k+1)} \frac{\Gamma(c)}{\Gamma(c-k)}$$

$$+ (-1)^{[c]} \delta^{-c} D_{c-[c]}\left(\delta^{c-[c]}\right) \frac{\Gamma(c)}{\Gamma(c-[c]+1)}, \qquad \text{(III.6)}$$

where

$$D_d(x) = \exp\left\{-x^{1/d}\right\} \int_0^x \exp\left\{z^{1/d}\right\} dz.$$

Then the first two moments can be expressed as:

$$\mu_1' = 1 - bg\left(a+b, \frac{\lambda}{2}\right)$$

and

$$\mu_2' = 1 - \frac{2b(b+1)}{\delta}$$

$$+ \left\{b(b-1)\frac{2b(b+1)(a+b)}{\delta}\right\} g\left(a+b, \frac{\lambda}{2}\right).$$

Note that if $c = 1$ in (III.6) then

$$g(1,\delta) = \frac{1 - \exp(-\delta)}{\delta}.$$

If $c \geq 2$ is an integer then (III.6) reduces to:

$$g(c, \delta) = \sum_{k=0}^{c-2} (-1)^k \delta^{-(k+1)} \frac{(c-1)!}{(c-k-1)!}$$

$$+ \frac{(-1)^{c-1}}{\delta^c} (c-1)! \left\{ 1 - \exp(-\delta) \right\}.$$

Some other properties of the g function (which may be useful in the computation of the moments) are:

$$g(c, \delta) = \exp(-\delta) \int_0^1 z^{c-1} \exp(\delta z) dz,$$

$$g(c+1, \delta) = \frac{1}{\delta} - \frac{c}{\delta} g(c, \delta)$$

and

$$g(c, \delta) = \frac{D_c(\delta^c)}{c \delta^c}.$$

Das Gupta (1968) suggested two approximations for the non-central distribution. The first is based directly on the definition $X = S/(S+T)$ in (III.1). Applying Patnaik (1949)'s approximation for the chi-squared variables S and T, one gets

$$X \approx \frac{4a + 2\lambda}{4a + \lambda} Z,$$

where Z is a standard beta random variable with parameters $(4a + \lambda)^2/\{8(2a + \lambda)\}$ and b. The second approximation is based on rewriting (III.1) as:

$$X = \frac{S/T}{1 + (S/T)}$$

and approximating S by cS', where $c = 2(a + \lambda)/(2a + \lambda)$ and S' is a standard chi-squared random variable with degrees of freedom $(2a + \lambda)^2/\{2(a + \lambda)\}$. Since the ratio of two chi-squared variables relates to the F distribution, it follows that one can approximate

$$X \approx \frac{\gamma F}{1 + \gamma F},$$

where $\gamma = (2a + \lambda)/(2b)$ and F has the standard F distribution with degrees of freedom $(2a + \lambda)^2/\{2(a + \lambda)\}$ and $2b$.

A third approximation for the noncentral distribution was suggested by Johannesson and Giri (1995). This is based on approximating $X \approx Z/\gamma$, where Z is a standard beta variable. The parameters of Z and γ can be determined by equating the first three moments.

IV. Type II Noncentral Beta Distribution

If S is a central chi-squared random variable with degrees of freedom $2a$ and if T is a non-central chi-squared random variable with degrees of freedom $2b$ and noncentrality parameter λ (distributed independently of S) then $X = S/(S + T)$ is said to have the type II noncentral beta distribution with shape parameters a, b and noncentrality parameter λ. The Poisson mixture representation of the cdf of X is:

$$F(x; a, b, \lambda) = \sum_{k=0}^{\infty} u_k I_x(a, b + k),$$

where u_k are as defined in (III.4). Note that the pdf of $1 - X$ takes the same form as (III.2) with the shape parameters reversed. Thus the above cdf can also be expressed as:

$$F(x; a, b, \lambda) = 1 - \sum_{k=0}^{\infty} u_k I_{1-x}(b + k, a).$$

Hence the various representations of the type I noncentral beta distribution can be applied to compute this distribution also.

V. Doubly Noncentral Beta Distribution

If S is a non-central chi-squared random variable with degrees of freedom $2a$ and noncentrality parameter λ_1 and if T is a non-central chi-squared random variable with degrees of freedom $2b$ and noncentrality parameter λ_2 (distributed independently of S) then $X = S/(S + T)$ is said to have the doubly noncentral beta distribution with shape parameters a, b and noncentrality parameters λ_1, λ_2. The Poisson mixture representation of the cdf of X is:

$$F(x; a, b, \lambda_1, \lambda_2) = \sum_{k=0}^{\infty} \sum_{l=0}^{\infty} u_k v_l I_x(a + k, b + l), \qquad (V.1)$$

where u_k and v_l are the Poisson weights with parameters λ_1 and λ_2, respectively. Note from the definition that $1 - X$ also has the doubly noncentral beta distribution with both the shape and the noncentrality parameters reversed. Thus the above cdf can also be expressed as:

$$F(x; a, b, \lambda_1, \lambda_2) = 1 - \sum_{k=0}^{\infty} \sum_{l=0}^{\infty} v_k u_l I_{1-x}(b + k, a + l).$$

For double infinite series representations of (V.1), see Bulgren (1971), Tiku (1972, 1974, 1975), Chattamvelli (1994) and Johnson et al. (1994). For Laguerre series representations (single series) of (V.1), see Chattamvelli (1995b).

VI.   Park's Noncentral Beta Distribution

Let S be a normal random variable with mean ξ and unit variance and let T be an independent noncentral chi-squared variable with degrees of freedom N and noncentrality parameter κ. Park (1964) studied the distribution of $X = S/\sqrt{S^2 + T^2}$. It enjoys certain engineering applications, see Park and Glaser (1957) and Middleton (1959). Since the numerator has only one degree of freedom and the denominator has been generalized to noncentral, X can be regarded as a variation of the noncentral beta variable. Park established the following properties:

(i)   the pdf of X admits the behavior

$$f(x) \sim K U^{-(N+1)} \left(1 - x^2\right)^{(N-2)/2}, \qquad |u| \to 1,$$

where

$$K = \frac{2^{1-N/2} \exp\left(-\kappa^2/2\right)}{\sqrt{\pi}} \int_0^{\infty} z^N \exp\left\{-(z - \xi)^2/2\right\} dz.$$

(ii) The mode of X, say x_0, is

$$
x_0 = \begin{cases}
\dfrac{\xi}{\sqrt{\xi + \kappa^2}} \left[1 - \dfrac{1}{2} \left\{ 1 + \dfrac{\xi^2}{\kappa^2} + \dfrac{\xi^2 + \kappa^2}{N + 3} \right\}^{-1} \right], \\
\quad \text{if } \kappa \gg 1, \\[2mm]
\dfrac{\xi}{\sqrt{\xi^2 + N + 1}}, \\
\quad \text{if } \kappa \ll 1 \ll \mid \xi \mid, \\[2mm]
\xi \Gamma \left(\tfrac{N}{2} + 1 \right) \left\{ \xi^2 \Gamma^2 \left(\tfrac{N}{2} + 1 \right) + 4 \Gamma^2 \left(\dfrac{N + 3}{2} \right) \right\}^{-1}, \\
\quad \text{if } \kappa, \mid \xi \mid \ll 1.
\end{cases}
$$

(iii) the expectation is:

$$
E(X) = \frac{\xi}{\sqrt{\xi^2 + \kappa^2 + N - 1}} \left\{ 1 - \frac{1}{2 \left(\xi^2 + \kappa^2 + N - 1 \right)} \right\}
$$
$$
+ O\left(\kappa^4 \right) + O\left(\xi^4 \right) + O\left(\kappa \right) O\left(\mid \xi \mid^3 \right)
$$
$$
+ O\left(\kappa^2 \right) O\left(\xi^2 \right).
$$

(iv) the variance is:

$$
Var(X) = \frac{\kappa^2 + N - 1}{\left(\xi^2 + \kappa^2 + N - 1 \right)^2} + O\left(\kappa^6 \right)
$$
$$
+ O\left(\kappa^4 \right) O\left(\xi^2 \right) + O\left(\kappa^2 \right) O\left(\xi^4 \right).
$$

The cdf of X can be expressed in terms of Ruben's W function (Ruben, 1964). Ruben's W is the probability content of an infinite sector – under a circular normal distribution – bounded by an arbitrary line through the center of the distribution and a ray drawn from a point C_0 on the x-axis at an angle θ. Denoting this probability by $W(c_0, \theta)$, where c_0 is the distance between C_0 and the center of the distribution, the cdf of X can be written as:

$$1 - F(x/\xi) = \begin{cases} W\left[\sqrt{\kappa^2 + \xi^2}, \arccos x - \arctan\left(-\dfrac{\kappa}{\xi}\right)\right] \\ \quad + W\left[\sqrt{\kappa^2 + \xi^2}, \arccos x + \arctan\left(-\dfrac{\kappa}{\xi}\right)\right], \\ \quad \text{if } \arccos x \geq \arctan(-\kappa/\xi), \\ W\left[\sqrt{\kappa^2 + \xi^2}, \arccos x + \arctan\left(-\dfrac{\kappa}{\xi}\right)\right] \\ \quad - W\left[\sqrt{\kappa^2 + \xi^2}, \arctan\left(-\dfrac{\kappa}{\xi} - \arccos x\right)\right], \\ \quad \text{if } \arccos x \leq \arctan(-\kappa/\xi). \end{cases}$$

In the particular case $\kappa = 0$,

$$1 - F\left(-\frac{x}{c_0}\right) = 2W\left(c_0, \arccos x\right),$$

or equivalently

$$F\left(-\frac{\cos\theta}{c_0}\right) = 1 - 2W\left(c_0, \theta\right).$$

VII. Generalized Beta Distribution

A four-parameter generalization of the standard beta pdf

$$f(x; a, b) = \frac{1}{B(a,b)} x^{a-1}(1-x)^{b-1}, \qquad 0 < x < 1, \qquad \text{(VII.1)}$$

which accommodates different finite sample spaces is:

$$f(y) = \frac{1}{(d-c)B(a,b)} \left(\frac{y-c}{d-c}\right)^{a-1} \left(1 - \frac{y-c}{d-c}\right)^{b-1},$$
$$c \leq y \leq d. \qquad \text{(VII.2)}$$

We can see c and d as location parameters and $d - c$ as a scale parameter. Since the transformation $X = (Y - c)/(d - c)$ reduces (VII.2) to (VII.1), the mathematical properties of (VII.2) can be deduced easily from Chapter 2. For instance,

$$E(Y) = c + (d - c)\frac{a}{a + b}$$

and

$$Var(Y) = (d - c)^2 \frac{ab}{(a + b)^2(a + b + 1)}.$$

The kurtosis and skewness for (VII.2) are the same as the expressions for (VII.1) given in Chapter 2.

VIII.   McDonald and Richards's Generalized Beta Distribution

McDonald and Richards (1987b, 1987c) presented a four-parameter generalized beta distribution given by the pdf

$$f(x) = \frac{|p| \, x^{ap-1} (bq^p - x^p)^{b-1}}{(bq^p)^{a+b-1} B(a, b)}. \tag{VIII.1}$$

This includes as special cases three and two-parameter beta, generalized gamma, Weibull, power function, Pareto, lognormal, half-normal, uniform and others. The cdf and the rth moment (about zero) are:

$$F(x) = \frac{x^{ap}}{aq^{ap}B(a, b)} \, {}_2F_1\left(a, 1 - b; a + 1; \frac{x^p}{bq^p}\right)$$

and

$$\mu_r' = \frac{q^r b^{r/p} B\left(a + b, r/p\right)}{B\left(a, r/p\right)}.$$

For the generalized gamma distribution which arises as the special case for $b \to \infty$, we get the simpler expressions:

$$F(x) = \frac{(x/b)^{ap} \exp\left\{-(x/q)^p\right\}}{\Gamma(a + 1)} \, {}_1F_1\left(1; a + 1; \left(-\frac{x}{q}\right)^p\right)$$

and

$$\mu_r' = \frac{q^r \Gamma\left(a + r/p\right)}{\Gamma(a)}.$$

In this case the corresponding hazard rate function exhibits the behavior summarized in the table below (where D represents a strictly decreasing shape, I represents a strictly increasing shape, $\cup$ represents a bathtub shape and $\cap$ represents an upside down bathtub shape).

Sign of $p(p-1)$	Sign of $ap-1$	Shape of Hazard Rate
$-$	$-$	D
$-$	$+$	$\cap$
$+$	$+$	I
$+$	$-$	$\cup$

The shape of hazard rate for the general pdf (VIII.1) is more involved, see McDonald and Richards (1987a).

A re-scaled version of the pdf (VIII.1) which has been studied more widely is:

$$f(x; p, q, a, b) = \frac{|p| \, x^{ap-1} \left\{ 1 - (x/q)^p \right\}^{b-1}}{q^{ap} B(a, b)}, \ 0 \leq x \leq q, \text{(VIII.2)}$$

where $a > 0$, $b > 0$, $p > 0$ and $q > 0$. This distribution is described from a probabilistic point of view in McDonald (1984). It can be reduced to the form of (VII.2) by the change of variable $Y = X^p$; hence, (VIII.2) can be thought of as a "Weibullized" generalized beta distribution. As in the case of (VIII.1), three and two-parameter beta, generalized gamma, Weibull, power function, Pareto, lognormal, half-normal and uniform are contained as special cases of (VIII.2). A special case that is not so obvious is the unit gamma distribution. It arises as the limit of (VIII.2) when $a = \delta/p$ and $p \to 0$:

$$f(x) = \frac{\delta^p x^{\delta-1}}{a^\delta \Gamma(b)} \left\{ \ln \left(\frac{q}{x} \right) \right\}^{b-1}, \qquad 0 < x < q.$$

The unit gamma with $q = 1$ is mentioned in Patil et al. (1984) and has been used as a mixing distribution for the parameter a in the binomial (Grassia, 1977). An additional special case of (VIII.2) for $p = -1$ is the invert beta distribution.

The cdf, the rth moment about zero and the Gini index (see Chapter 2) of (VIII.2) are:

$$F(x) = \frac{(x/q)^{ap}}{a B(a, b)} \, {}_2F_1 \left(a, 1 - b; a + 1; \left(\frac{x}{q} \right)^p \right),$$

$$\mu_r' = \frac{q^r B(a+b, r/p)}{B(a, r/p)}$$

and

$$C(X) = \frac{B(2a+1/p, b)}{a(ap+1)B(a,b)B(a+1/p, b)} \, {}_4F_3\left(2a+\frac{1}{p}, a, a+\frac{1}{p},\right.$$

$$\left. 1-b; 2a+b+\frac{1}{p}, a+1, a+\frac{1}{p}+1; 1\right).$$

In particular, the first moment is:

$$\mu_1' = \frac{qB(a+b, 1/p)}{B(a, 1/p)}.$$

If we consider this as a function $\mu(b) = \mu_1'$ of the parameter b then it can be shown that the limit of $\mu(b)$ as $b \to 0$ is q, the limit of $\mu(b)$ as $b \to \infty$ is 0, and that $\mu(b)$ is strictly decreasing in b (Wilfling, 1996, Lemma 2).

A random variable X is said to exhibit less inequality in the Lorenz sense than a random variable Y (written in symbols as $X \leq_L Y$) if the Lorenz curve of X is greater than or equal to the Lorenz curve of Y (see Chapter 2 for the meaning of Lorenz curve). Wilfling (1996) provided the following sufficient conditions for $X \leq_L Y$ when X and Y are distributed according to (VIII.2) with parameters $(p_1, 1, a_1, b_1)$ and $(p_2, 1, a_2, b_2)$, respectively:

(a) $p_1 = p_2 \geq 1$ and $a_1 = b_1$ and $a_2 = b_2 = a_1 - 1$.
(b) $p_1 = p_2$ and $a_1 > a_2$ and $b_1 = b_2 \geq 1$.
(c) $p_1 = p_2 = p$ and $a_1 > a_2$ and $b_2 > \xi > 0$, with ξ satisfying $E(X)$ $E(Z)$ for a random variable Z having (VIII.2) as its pdf with $q = 1$, $a = a_2$ and $b = \xi$.
(d) $p_1 \geq p_2$ and $a_1 = a_2$ and $b_1 = b_2$.

Since the Lorenz ordering is scale-invariant, we have set the parameter q in (VIII.2) to be equal to 1 without loss of generality.

IX. Libby and Novick's Generalized Beta
Distribution

Libby and Novick (1982) proposed a three-parameter generalization of (VII.1) with the pdf

$$f(x; a, b, \lambda) = \frac{\lambda^a x^{a-1}(1-x)^{b-1}}{B(a,b)\left\{1 - (1-\lambda)x\right\}^{a+b}}, \qquad 0 \leq x \leq 1, \text{ (IX.1)}$$

where $a > 0$, $b > 0$ and $\lambda > 0$. When $\lambda = 1$ this reduces to the standard beta pdf (VII.1). If a random variable X has the pdf (IX.1) then we shall write $X \sim G3B(a, b, \lambda)$. Note that if $X \sim G3B(a, b, \lambda)$ then $1 - X \sim G3B(a, b, 1/\lambda)$, a property similar to the one enjoyed by the standard beta pdf (VII.1). Also if $X \sim G3B(a, b, \lambda)$ then $Y = \lambda X/(1 + \lambda X - X)$ has the standard beta distribution with parameters a and b. Using this property, we get the cdf of X as

$$F(x) = \Pr\left[Y \leq \frac{\lambda x}{1 + \lambda x - x}\right]$$
$$= I_{\lambda x/(1+\lambda x - x)}(a, b),$$

where $I_x(a, b)$ is the incomplete beta function discussed in Chapter 1.

The rth moment of $X \sim G3B(a, b, \lambda)$ about zero is:

$$\mu_r' = \frac{\lambda^a \Gamma(a+b)\Gamma(a+r)}{\Gamma(a)\Gamma(a+b+r)} \; {}_2F_1\left(a+r, a+b; a+b+r; 1-\lambda\right).$$

Using properties of the Gauss hypergeometric function, we can rewrite the above as

$$\mu_r' = \frac{\Gamma(a+b)\Gamma(a+r)}{\lambda^r \Gamma(a)\Gamma(a+b+r)} \; {}_2F_1\left(a+r, r; a+b+r; 1-\frac{1}{\lambda}\right)$$

when $\lambda \geq 1$ and as

$$\mu_r' = \frac{\Gamma(a+b)\Gamma(a+r)}{\lambda^r \Gamma(a)\Gamma(a+b+r)} \; {}_2F_1\left(b, r; a+b+r; 1-\lambda\right)$$

when $\lambda < 1$. Further, if a and b are large then we can approximate

$$\mu_r' \approx \frac{\Gamma(a+b)\Gamma(a+r)}{\Gamma(a)\Gamma(a+b+r)} \left(\frac{a+b}{a+\lambda b}\right)^r.$$

Depending on the relative values of a, b and λ, (IX.1) can be unimodal, U-shaped, J-shaped, reversed J-shaped or a strictly increasing or decreasing function of x. The mode or anti-mode of (IX.1) is the root in $(0,1)$ of the quadratic equation:

$$2(\lambda - 1)x^2 + \{3 - a - \lambda(b+1)\}\, x + a - 1 = 0.$$

For $a > 1$ and $b > 1$, the mode can be shown to be:

$$\frac{\lambda(b+1) + a - 3}{\lambda - 1} - \frac{\sqrt{\{\lambda(b+1) + a + 1\}^2 - 8\lambda(a+b)}}{4(\lambda - 1)}$$

provided that $\lambda \neq 1$.

The parameter λ allows (IX.1) to take a much wider variety of shapes than the the standard beta pdf. For example, when $a = b$, the standard beta is symmetric with mean at $1/2$. But (IX.1) can be positively or negatively skewed, depending on λ, because the mode, skewness and kurtosis of (IX.1) depend on λ as well.

To compare the analytical shapes of (VII.1) and (IX.1), write

$$g(x) = \frac{f(x; a, b, \lambda)}{f(x; a, b)} = \frac{\lambda^a}{\{1 - (1 - \lambda)x\}^{a+b}}.$$

Its derivative

$$g'(x) = \frac{\lambda^a(a+b)(1-\lambda)}{\{1 - (1-\lambda)x\}^{a+b+1}}$$

is positive for $0 < \lambda < 1$ and negative for $\lambda > 1$. So, for $0 < \lambda < 1$, $g(x)$ is strictly increasing starting at $g(0) = \lambda^a$, which means that (IX.1) is below the standard beta pdf near zero but crosses the latter to rise above it at

$$x_0 = \frac{1 - \lambda^{a/(a+b)}}{1 - \lambda}.$$

For $\lambda > 1$ the reverse holds with the same starting point.

The figure below shows the pdf (IX.1) for selected values of a, b and λ. It can be seen that $G3B(a, b, \lambda)$ and $G3B(a, b, 1/\lambda)$ are symmetrical with respect to $x = 1/2$. For $a = b = 1/2$ and $\lambda = 5/2$, (IX.1) is U-shaped with antimode at $x_0 = 2/3$. Also, for $a = b = 1$ and $\lambda = 5/2$, (IX.1) is strictly increasing while (VII.1) is the uniform pdf on $(0, 1)$. Finally, for $a = 3$, $b = 1/2$ and $\lambda = 4/5$, (IX.1) is J-shaped like (VII.1) but crosses the latter from below near $x_0 = 0.8704$.

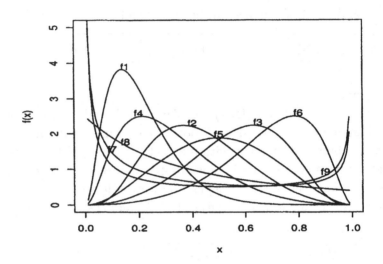

Figure 9. Generalized pdf (IX.1): f1 for $a = 3$, $b = 5$, $\lambda = 2.5$; f2 for $a = 5$, $b = 3$, $\lambda = 2.5$; f3 for $a = 3$, $b = 5$, $\lambda = 0.4$; f4 for $a = 3$, $b = 3$, $\lambda = 2.5$; f5 for $a = 3$, $b = 3$, $\lambda = 1$; f6 for $a = 3$, $b = 3$, $\lambda = 0.4$; f7 for $a = 0.5$, $b = 0.5$, $\lambda = 2.5$; f8 for $a = 1$, $b = 1$, $lambda = 2.5$; and, f9 $a = 0.3$, $b = 0.5$, $\lambda = 0.8$.

The parameter λ in (IX.1) has the following interpretation: if T has the pdf (VII.1) and $\lambda > 0$ then $X = T/\{\lambda + (1 - \lambda)T\}$ is a $G3B(a, b, \lambda)$ variate and hence λ is the intercept of the line which passes through $(1, 1)$ and X is the ratio AB/AC with A having a standard beta distribution (see Figure 10).

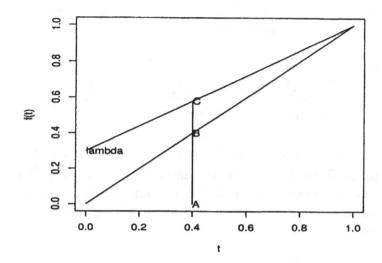

Figure 10. Meaning of λ in (IX.1).

The pdf (IX.1) was first used by Libby and Novick (1982) for utility function fitting and by Chen and Novick (1984) as a prior in some binomial sampling model. Two other applications are to the problem of Bayesian estimation of the ratio of two variances (Gelfand, 1987) and to model the proportion of time devoted to a specific work function in Bayesian work sampling (Pham-Gia, 1989).

X. Exton's Generalized Beta Distribution

Exton (1976) introduced a generalized beta distribution in $(2n + 2)$ parameters with the pdf

$$
f(x) = \frac{x^{a-1}(1-x)^{b-1}}{B(a,b)}
$$
$$
\times \frac{(1-\alpha_1 x)^{a_1-1}\cdots(1-\alpha_n x)^{a_n-1}}{F_D^{(n)}\left(a; 1-a_1,\ldots,1-a_n; a+b; \alpha_1,\ldots,\alpha_n\right)} \quad (X.1)
$$

where $a > 0$, $b > 0$, $-1 < \alpha_i < 1$, $i = 1, \ldots, n$ and $F_D^{(n)}$ is one of the four Lauricella functions defined by

$$
F_D^{(n)} (p; \beta_1, \ldots, \beta_n; q; x_1, \ldots, x_n)
$$
$$
= \sum_{k_n=0}^{\infty} \cdots \sum_{k_1=0}^{\infty} \frac{(p)_{k_1+\cdots+k_n} (\beta_1)_{k_1} \cdots (\beta_n)_{k_n}}{(q)_{k_1+\cdots+k_n}} \frac{x_1^{k_1}}{k_1!} \cdots \frac{x_n^{k_n}}{k_n!}.
$$

This series converges for $|x_i| < 1$, $i = 1, \ldots, n$. When $n = 1$, $F_D^{(n)}$ reduces to the Gauss hypergeometric function. When $n = 2$, $F_D^{(n)}$ is the Appell's hypergeometric function in two variables. When $n = 1$, $a_1 < 1$ and $\lambda < 1$, setting $\gamma = 1 - a_1$ and $1 - \lambda = \alpha_1$, we see that (X.1) reduces to:

$$
f(x) = \frac{B(a, b)}{{}_2F_1(a, \gamma, a + b, 1 - \lambda)} \{1 - (1 - \lambda)x\}^{-\gamma}, \qquad (X.2)
$$

a four-parameter generalized beta distribution. When $\lambda = 1$, (X.2) reduces to the standard beta pdf. Since

$$
{}_2F_1(a, a + b, a + b; 1 - \lambda) = \sum_{k=0}^{\infty} (a)_j \frac{(1 - \lambda)^k}{k!} = \frac{1}{\lambda^a},
$$

it follows that the pdf (IX.1) is a particular case of (X.2) for $\gamma = a + b$. Hence both (IX.1) and (X.2) are particular cases of (X.1).

Chen and Novick (1984) derived expressions for the cdf and the rth moment (about zero) of (X.2), giving the forms:

$$
F(x) = \frac{\Gamma(a + b)x^a}{a\Gamma(a)\Gamma(b)} \frac{F_D^{(2)} (a; \gamma, 1 - \beta; a + 1; (1 - \lambda)x, x)}{{}_2F_1 (a, \gamma, a + b; 1 - \lambda)}.
$$

and

$$
\mu_r' = \frac{\Gamma(a + b)\Gamma(a + r) \, {}_2F_1 (a + r, \gamma, a + b + r; 1 - \lambda)}{\Gamma(a)\Gamma(a + b + r) \, {}_2F_1 (a, \gamma, a + b; 1 - \lambda)}.
$$

The parameters λ and γ affect not just the moments but also the skewness and kurtosis.

XI. McDonald and Xu's Generalized Beta Distribution

McDonald and Xu (1995) introduced a five-parameter generalization of both (VIII.2) and (IX.1) with the pdf:

$$f(x; p, q, a, b, c) = \frac{|p| \, x^{ap-1} \{1 - (1-c)(x/q)^p\}^{b-1}}{q^{ap} B(a, b) \{1 + c(x/q)^p\}^{a+b}},$$
$$0 \le x^p \le q^p/(1-c), \qquad (\text{XI.1})$$

where $a > 0$, $b > 0$, $0 \le c \le 1$, $p > 0$ and $q > 0$. Setting $c = 0$, we see that this reduces to (VIII.2). When $q = 1/\lambda$ and $c = 1 - 1/\lambda$, (XI.1) reduces to (IX.1). The generalized gamma is the limiting case $b \to \infty$ of (XI.1) when $q = \beta b^{1/p}$. Another special case is obtained by taking the limit $c \to 1$ when $q = \beta(1 - c)^{1/p}$ and $a = 1/\delta(1 - c)$, which yields

$$f(x) = \frac{|p|}{\delta^b x \Gamma(x)} \left(\frac{\beta}{x}\right)^{b(p-1)+1} \exp\left\{\frac{1}{\delta}\left(\frac{\beta}{x}\right)^{a-1}\right\},$$
$$0 < x < \beta.$$

This distribution might be thought of as a translated inverse generalized gamma since $(\beta/X)^p - 1$ has a gamma distribution. Other special cases of (XI.1) include Burr types 3 and 12, beta-κ, beta-P, lognormal, Weibull, gamma, Lomax, F, Fisk or Rayleigh, chi-squared, half-normal, half-Student's t, exponential and log-logistic.

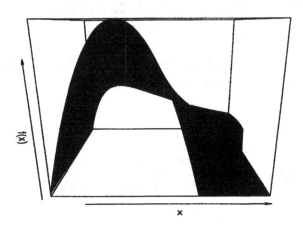

Figure 11. Pdfs of McDonald and Xu's generalized beta distribution for $p = 2$, $q = 20$, $a = 2$ and $b = 2$.

Figure 11 depicts the shape of the pdf (XI.1) as the parameter c changes for representative values for a, b, p and q. For $c = 0$, we observe that the relative frequency for values of $x > 20$ is zero. As the value of c increases, the upper limit, $q/(1 - c)^{1/p}$, increases. The parameter p impacts the peakedness of the pdf, whereas q is basically a scale parameter, and a and b control the shape and skewness.

The rth moment of (XI.1) about zero is:

$$\mu'_r = \frac{q^r B\left(a + r/p, b\right)}{B(a, b)} \; {}_2F_1\left(a + \frac{r}{p}, \frac{r}{p}; a + b + \frac{r}{p}; c\right),$$

where the hypergeometric term converges for all r if $c < 1$ or for $r/p < b$ if $c = 1$.

XII. Exponential Generalized Beta Distributions .

If a random variable Y has the pdf (XI.1) then $X = \log Y$ is said to have the exponential generalized beta distribution with the pdf defined by

$$
f(x) = \frac{\exp\left(\dfrac{a(x-\delta)}{\sigma}\right)\left\{1 - (1-c)\exp\left(\dfrac{x-\delta}{\sigma}\right)\right\}^{b-1}}{|\sigma|\, B(a,b)\left\{1 + c\exp\left(\dfrac{x-\delta}{\sigma}\right)\right\}^{a+b}},
$$
$$
-\infty < (x-\delta)/\sigma < -\log(1-c). \qquad \text{(XII.1)}
$$

This distribution contains as special cases generalized exponential, generalized Gompertz, generalized logistic, generalized Gumbel, Burr type 2 and others. The moment generating function of (XII.1) is:

$$
M(t) = \frac{B(a + t\sigma, b)\exp(\delta t)}{B(a,b)}\, {}_2F_1\left(a + t\sigma, t\sigma; a + b + t\sigma; c\right).
$$

In particular, the first moment of (XII.1) can be shown to be

$$
\mu_1' = \delta + \sigma\left\{\psi(a) - \psi(a+b)\right\}
$$
$$
+\frac{c\sigma a}{a+b}\, {}_3F_2\left(1, 1, a+1; 2, a+b+1; c\right),
$$

where $\psi(z) = d\log\Gamma(z)/dz$.

XIII. Mauldon's Generalized Beta Distribution

Given a random variable X if there is a set of constants $p_k > 0$, $k = 1, \ldots, r$ and c_k, $k = 1, \ldots, r$ such that

$$
E\left\{(t - aX)^{-p}\right\} = \prod_{k=1}^{r}(t - ac_k)^{-p_k}, \qquad p > 0
$$

then X is said to have the β-distribution with vertices $c_1, \ldots, c_r$, indices $p_1, \ldots, p_r$ and exponent $p = p_1 + \cdots + p_r$. This distribution was defined and studied by Mauldon (1959). It contains as special cases the standard beta distribution, the triangular distribution and the uniform distribution among others. Some of its properties are:

(i) if $r \geq 2$ and $c_1 < \cdots < c_r$ then for each $k = 1, \ldots, r - 1$ there exists a unique function $G_k(z)$ such that $G_k(x) = F(x)$ for $c_k < x < c_{k+1}$ and $G_k(z)$ is a regular function of z throughout the complex plane, cut along the real axis from $-\infty$ to c_k and from c_{k+1} to $+\infty$.

(ii) if $r = 2$ then for $c_1 < c_2$,

$$G'_1(z) = \frac{\Gamma(p)}{\Gamma(p_1)\Gamma(p_2)} \left\{ \frac{(z - c_1)(c_2 - z)}{c_2 - c_1} \right\}^{p-1}$$
$$\times (z - c_1)^{-p_1} (c_2 - z)^{-p_2}.$$

(iii) if p is an integer then under the conditions of (i),

$$G_k^{(p)}(z) = (-1)^{p-1}(p-1)! \frac{\sin \alpha \pi}{\pi} \prod_{l=1}^{k} (z - c_l)^{-p_l}$$
$$\times \prod_{l=k+1}^{r} (c_l - z)^{-p_l},$$

where $\alpha = p_1 + \cdots + p_k$.

(iv) under the conditions of (i),

$$G'_k(z) \sim A_k (z - c_k)^{Q-1} (c_{k+1} - z)^{P-1}, \qquad |z| \to \infty,$$

where $A_k > 0$, $P = p_1 + \cdots + p_k$ and $Q = p_{k+1} + \cdots + p_r$ (the branch of the function which is positive for $c_k < x < c_{k+1}$ is used).

(v) under the conditions of (i), $G'_k(z) \sim B_k z^{p-2}$ as $|z| \to \infty$, where $B_k \neq 0$.

(vi) any β-distribution which is not concentrated at a single point admits only one values for its exponent.

(vii) if F_1, F_2 are the cdfs of two β-distributions and if, at infinity of distinct points x, either $F'_1(x) = AF'_2(x) \neq 0$ or $F_1(x) = AF_2(x) + B \neq 0$ then the two β-distributions have the same exponent.

Mauldon (1959) also gave detailed mathematical theory of an n-dimensional generalization of the β-distribution.

XIV.   Volodin's Generalized Beta Distribution

Volodin (1994) considered some generalizations of the standard beta distribution. One of his generalizations has the pdf, the cdf and the moments given by

$$f(x) = \frac{\beta}{B(a, b + \gamma)} \int_0^x z^{a-1}(x - z)^{b-1}(1 - z)^{\gamma-1}dz, \quad \text{(XIV.1)}$$

$$F(x) = \frac{1}{B(a, b + \gamma)} \int_0^x z^{a-1}(x - z)^b(1 - z)^{\gamma-1}dz$$

and

$$\mu'_r = \frac{b}{B(a, b + \gamma)} \sum_{k=0}^r \binom{r}{k} \frac{B(a + r - k, b + \gamma + k)}{b + k}.$$

Another generalization is obtained by setting $Y = 1 - X$ (for a random variable X with the pdf (XIV.1)), then the pdf, the cdf and the moments of Y are:

$$f(x) = \frac{\beta}{B(a + b, \gamma)} \int_y^1 z^{\gamma-1}(z - y)^{b-1}(1 - z)^{a-1}dz, \quad \text{(XIV.2)}$$

$$F(x) = 1 - \frac{1}{B(a + b, \gamma)} \int_y^1 z^{\gamma-1}(z - y)^b(1 - z)^{a-1}dz$$

and

$$\mu'_r = \frac{b}{B(a, b + \gamma)} B(a + b + r, \gamma)B(r + 1, b).$$

Note that in the particular case $b = 0$, both (XIV.1) and (XIV.2) reduce to the standard beta distribution.

XV.   Gauss Hypergeometric Distribution

A random variable X is said to have the Gauss hypergeometric distribution with parameters $a > 0$, $b > 0$, γ and z if its pdf is:

$$f(x) = C\frac{x^{a-1}(1 - x)^{b-1}}{(1 + zx)^\gamma}, \qquad 0 < x < 1, \quad \text{(XV.1)}$$

where the proportionality constant C is given by

$$\frac{1}{C} = B(a, b) \, {}_2F_1(\gamma, a; a + b; -z)$$

and ${}_2F_1$ is the Gauss hypergeometric function. This distribution was suggested by Armero and Bayarri (1994) in connection with marginal prior/posterior distribution for parameter ρ $(0 < \rho < 1)$ representing traffic intensity in a M/M/1 queue. It reduces to the standard beta distribution when either γ or z equals 0. The rth moment of X about zero is:

$$\mu_r' = \frac{B(r + a, b)}{B(a, b)} \frac{{}_2F_1(\gamma, a + r; a + b + r; -z)}{{}_2F_1(\gamma, a; a + b; -z)}.$$

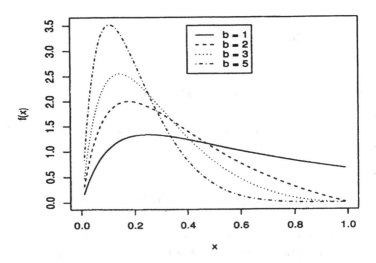

Figure 12. Gauss hypergeometric pdfs for $a = 2$, $\gamma = 3$, $z = 2$ and $b = 1, 2, 3, 5$.

XVI.   Confluent Hypergeometric Distribution

A random variable X is said to have the confluent hypergeometric distribution with parameters $a > 0$, $b > 0$ and γ if its pdf is:

$$f(x) = Cx^{a-1}(1-x)^{b-1}\exp(-\gamma x), \qquad 0 < x < 1, \qquad \text{(XVI.1)}$$

where the proportionality constant C is given by

$$\frac{1}{C} = B(a,b) \; {}_1F_1(a; a+b; -\gamma)$$

and ${}_1F_1$ is the confluent hypergeometric function. This distribution was introduced by Gordy (1998) who applied it to auction theory. It generalizes the standard beta distribution using the confluent hypergeometric function in the way Armero and Bayarri (1994) use ${}_2F_1$ to generalize the beta to the Gauss hypergeometric distribution.

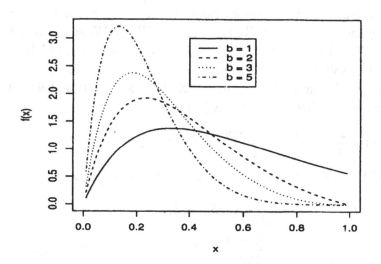

Figure 13. Confluent hypergeometric pdfs for $a = 2$, $\gamma = 3$ and $b = 1, 2, 3, 5$.

XVII.   Compound Confluent Hypergeometric Distribution

A random variable X is said to have the compound confluent hypergeometric distribution with parameters $a > 0$, $b > 0$, $-\infty < r < \infty$, $-\infty < s < \infty$, $0 \leq \nu \leq 1$ and $\theta > 0$ if its pdf is:

$$f(x) = \frac{x^{a-1}(1 - \nu x)^{b-1}\left\{\theta + (1 - \theta)\nu x\right\}^{-r}\exp\left(-sx\right)}{B(a,b)H(a,b,r,s,\nu,\theta)},$$
$$0 < x < 1/\nu. \qquad\qquad\qquad (XVII.1)$$

The function H is given by

$$H(a,b,r,s,\nu,\theta) = \nu^{-a}\exp\left(-\frac{s}{\nu}\right)\Phi_1\left(b,r,a+b,\frac{s}{\nu},1-\theta\right),$$

where Φ_1 is the confluent hypergeometric function of two variables defined by

$$\Phi_1\left(\alpha,\beta,\gamma,x,y\right) = \sum_{m=0}^{\infty}\sum_{n=0}^{\infty}\frac{(\alpha)_{m+n}(\beta)_n}{(\gamma)_{m+n}m!n!}x^m y^n.$$

This distribution was introduced by Gordy (2000). It contains McDonald and Xu's generalized beta, Gauss hypergeometric and confluent hypergeometric as special cases. When $s = 0$, $r = a + b$, $\nu = (1 - c)/q$ and $\theta = 1 - c$, (XVII.1) reduces to (XI.1). When $s = 0$ and $\nu = 1$, (XVII.1) reduces to (XV.1). When $\nu = 1$ and $\theta = 1$, (XVII.1) reduces to (XVI.1). When $\nu = 0$ and $s > 0$, (XVII.1) is the gamma distribution. An interesting limit of (XVII.1) as $\nu \to 0$ and $\theta \to 0$ such that $\nu(1 - \theta)/\theta \to \lambda$ and $s/\lambda \geq 0$ is:

$$f(x) = \frac{x^{a-1}(1 + \lambda x)^{-r}\exp(-sx)}{\Gamma(a)U\left(a,a+1-r,s/\lambda\right)}, \qquad x > 0,$$

where U is the degenerate hypergeometric function defined by

$$U\left(\alpha,\beta,z\right) = \frac{\Gamma(1-\beta)}{\Gamma(\alpha-\beta+1)}\,{}_1F_1\left(\alpha;\beta;z\right)$$
$$+ \frac{\Gamma(\beta-1)}{\Gamma(\alpha)}z^{1-\beta}\,{}_1F_1\left(\alpha-\beta+1;2-\beta;z\right).$$

Like the standard beta distribution, (XVII.1) enjoys a simple reflective property: if a random variable X has the pdf (XVII.1) then

$1/\nu - X$ also has the same pdf with s replaced by $-s$, θ replaced by $1/\theta$ and a & b interchanged.

It is straightforward to check that the moment generating function of (XVII.1) is:

$$M(t) = \frac{H(a, b, r, s - t, \nu, \theta)}{H(a, b, r, s, \nu, \theta)}.$$

Thus the kth moment about zero is:

$$\mu'_k = \frac{(a)_k}{(a+b)_k} \frac{H(a + k, b, r, s, \nu, \theta)}{H(a, b, r, s, \nu, \theta)}.$$

Since $H(a, b, r, s, \nu, \theta)$ is a finite positive real number for all admissible values of the parameters (Gordy, 2000, Theorem 1), all the moments above must exist. The first moment changes monotonically with s; actually, it can be shown that

$$\frac{d\mu'_1}{ds} = -\mu_2 < 0.$$

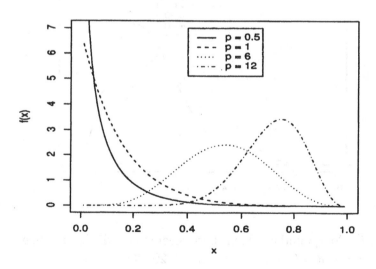

Figure 14. Compound confluent hypergeometric pdfs for $b = 4$, $r = 10$, $s = 0$, $\nu = 1$, $\theta = 0.75$ and $a = 0.5, 1, 6, 12$.

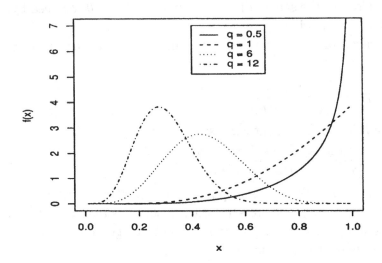

Figure 15. Compound confluent hypergeometric pdfs for $a = 6$, $r = 10$, $s = 0$, $\nu = 1$, $\theta = 0.75$ and $b = 0.5, 1, 6, 12$.

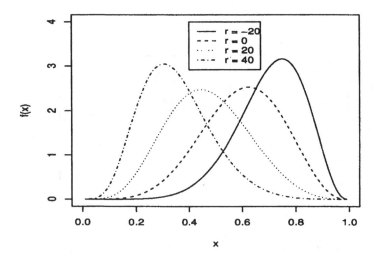

Figure 16. Compound confluent hypergeometric pdfs for $a = 6$, $b = 4$, $s = 0$, $\nu = 1$, $\theta = 0.75$ and $r = -20, 0, 20, 40$.

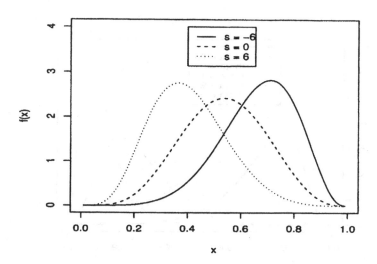

Figure 17. Compound confluent hypergeometric pdfs for $a = 6$, $b = 4$, $r = 10$, $\nu = 1$, $\theta = 0.75$ and $s = -6$, 0, 6.

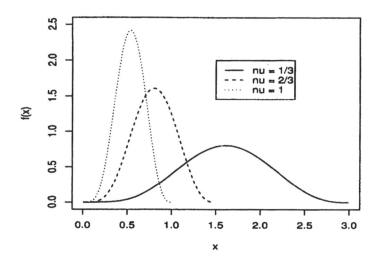

Figure 18. Compound confluent hypergeometric pdfs for $a = 6$, $b = 4$, $r = 10$, $s = 0$, $\theta = 0.75$ and $\nu = 1/3, 2/3, 1$.

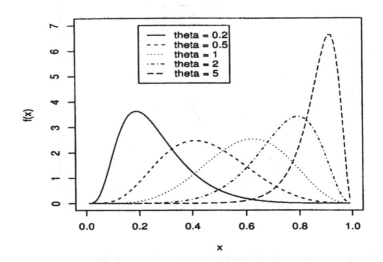

Figure 19. Compound confluent hypergeometric pdfs for $a = 6$, $b = 4$, $r = 10$, $s = 0$, $\nu = 1$ and $\theta = 0.2, 0.5, 1, 2, 5$.

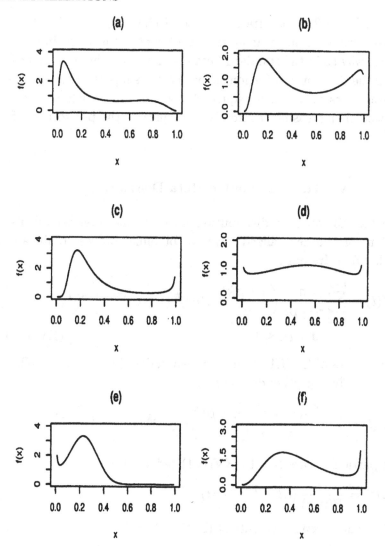

Figure 20. Multi-modal and long-tailed compound confluent hypergeometric pdfs. (a): $a = 6$, $b = 3$, $r = 15$, $s = -20$, $\nu = 1$, $\theta = 0.25$; (b): $a = 5$, $b = 1.2$, $r = 37$, $s = -26$, $\nu = 1$, $\theta = 0.36$; (c): $a = 12$, $b = 0.4$, $r = 25$, $s = -10$, $\nu = 1$, $\theta = 0.14$; (d): $a = 0.8$, $b = 0.7$, $r = -15$, $s = -10$, $\nu = 1$, $\theta = 2$; (e): $a = 0.2$, $b = 25$, $r = -40$, $s = 10$, $\nu = 1$, $\theta = 0.4$; (f): $a = 4$, $b = 0.2$, $r = 0$, $s = 10$, $\nu = 1$, $\theta = 2$.

Figures 14 to 20 illustrate the shape of (XVII.1) for a variety of parameter values. The parameters a and b control the shape in much the same way as in the standard beta distribution. The parameter ν rescales the distribution for longer or shorter support. The remaining parameters r, s and θ "squeeze" the pdf to the left or right. In particular, increasing (decreasing) s squeezes the pdf to the left (right).

XVIII. Lagrangian-Beta Distribution

The Lagrangian beta distribution arises as the inter-arrival distribution for the generalized negative binomial process (Jain and Consul, 1971). Its pdf is:

$$f(t) = \sum_{k=0}^{r-1} \frac{n}{n+\beta k} \binom{n+\beta k}{k} \alpha \left(\alpha t\right)^k \left(1 - \alpha t\right)^{n+\beta k-k-1},$$

$$0 \leq \alpha t \leq 1. \qquad \qquad \text{(XVIII.1)}$$

The properties of (XVIII.1) were studied in detail by Patel and Khatri (1994). The corresponding cdf is:

$$F(t) = 1 - \sum_{k=0}^{r-1} \frac{n}{n+\beta k} \binom{n+\beta k}{k} \alpha \left(\alpha t\right)^k \left(1 - \alpha t\right)^{n+\beta k-k}.$$

In the particular case $\beta = 1$, (XVIII.1) reduces to:

$$f(t) = \frac{\alpha}{B(n,r)} \left(\alpha t\right)^{r-1} \left(1 - \alpha t\right)^{n-1}.$$

The first and second moments of (XVIII.1) are:

$$\mu_1' = \frac{n}{\alpha} \sum_{k=0}^{r-1} \frac{1}{(n+\beta k)(n+\beta k+1)}$$

and

$$\mu_2' = \frac{2}{n\alpha} \mu_1' + \frac{2}{\alpha^2} \sum_{k=0}^{r-1} \frac{k(n-\beta)-2}{(n+\beta k)(n+\beta k+1)(n+\beta k+2)}.$$

Note that the first moment decreases as β increases.

XIX. Binomial Mixtures

Roy et al. (1993) introduced the binomial mixture of the beta distributions given by the pdf

$$f(x) = \sum_{k=0}^{n} \binom{n}{k} p^k (1-p)^{n-k} \frac{x^{(a/2)+k-1}(1-x)^{(b/2)-1}}{B\left((a/2)+k, b/2\right)},$$
$$0 < x < 1.$$

The rth moment about zero is:

$$\mu_r' = \sum_{k=0}^{n} \binom{n}{k} p^k (1-p)^{n-k} \frac{\Gamma\left((a/2)+r+k\right)\Gamma\left((a+b)/2+k\right)}{\Gamma\left((a+b)/2+r+k\right)\Gamma\left((a/2)+k\right)}.$$

In particular, the first and the second moments are:

$$\mu_1' = \sum_{k=0}^{n} \binom{n}{k} p^k (1-p)^{n-k} \frac{a+2k}{a+b+2k}$$

and

$$\mu_2' = \sum_{k=0}^{n} \binom{n}{k} p^k (1-p)^{n-k} \frac{(a+2k)(a+2k+2)}{(a+b+2k)(a+b+2k+2)}.$$

XX. Compound Beta Distributions

The compounding of distributions has been known since the 1940s. Suppose that a random variable X has cdf $F(x \mid \Theta)$ parameterized by Θ. Suppose too that $\Theta = cY$, where Y is a random variable with cdf G and c is an arbitrary constant. Then a distribution with cdf defined as:

$$H(x) = \int_{-\infty}^{\infty} F(x \mid cy)\,dG(y) \qquad (XX.1)$$

will be called a compound distribution.

Take G to be a four-parameter generalized beta distribution with the pdf

$$g(y) = \frac{ay^{p-1}}{(bw)^{p/a} B\left(p/a, w\right)} \left(1 - \frac{y^a}{bw}\right)^{w-1}$$

and let F be the binomial cdf with parameters n and Θ. Compounding F and G using (XX.1), Gerstenkorn (1982) obtained the compound binomial-generalized beta distribution with the cdf

$$H(x) = \binom{n}{x} \sum_{k=0}^{n-x} \binom{n-x}{k} (-1)^k (bwc^a)^{(x+k)/a}$$

$$\times B\left(\frac{p+x+k}{a}, w\right) \bigg/ B\left(p/a, w\right). \qquad \text{(XX.2)}$$

The compound binomial-beta distribution arises as the special case for $a = 1$, $b = 1/w$ and $c = 1$ with the cdf

$$H(x) = \binom{n}{x} \sum_{k=0}^{n-x} \binom{n-x}{k} (-1)^k \frac{B(p+x+k, w)}{B(p, w)}. \qquad \text{(XX.3)}$$

Further setting $p = 1/2$ and $w = 1/2$, we get the compound binomial-arcsine distribution with the cdf

$$H(x) = \binom{n}{x} \frac{(2x-1)!!(2n-2x-1)!!}{2^n n!}.$$

Other special cases of (XX.2) include compound binomial-generalized gamma, compound binomial-gamma, compound binomial-exponential, compound binomial-Erlang, compound binomial-chisquare, compound binomial-Rayleigh, compound binomial-Maxwell, compound binomial-truncated normal and compound binomial-Weibull.

Perhaps the most well-known compound distribution is the binomia beta distribution (popularly known as the beta-binomial distribution) with the cdf (XX.3). Sometimes this distribution is also refered to as the negative hypergeometric or Polya distribution. It has attracted applications in many areas. For the most recent applications, see Bi et al. (2000), Shiyomi et al. (2000), Smith et al. (2000), Turechek and Madden (2000), Dominici and Parmigiani (2001), Shah et al. (2001), Turechek et al. (2001), Hughes and Madden (2002), Shah and Bergstrom (2002) and Venette et al. (2002). The probability mass function (pmf) corresponding to (XX.3) is:

$$p(x) = \binom{n}{x} \frac{B(p+x, n+w-x)}{B(p, w)}, \qquad x = 0, 1, 2, \ldots. \quad \text{(XX.4)}$$

The properties of this are discussed in Moran (1968) and Ishii and Hayakawa (1960). The mean and the variance are given by

$$E(X) = \frac{np}{p + w}$$

and

$$Var(X) = \frac{npw(n + p + w)}{(p + w)^2(1 + p + w)}.$$

If the mode of (XX.4) is zero then $p < 1$. In addition, if there is also a mode at n, then also $w < 1$.

Morrison and Brockway (1979) proposed a modification of (XX.4) with the pmf

$$p(x) = \binom{n}{x} \sum_{k=0}^{x} \binom{x}{k} \frac{(m - 1)^{n-x+k} B(p + k, n + w - x)}{m^n B(p, w)},$$
$$x = 0, 1, \ldots, n.$$

The probability generating function of this is:

$$\phi(t) = \frac{1}{m^n B(p, w)} \int_0^1 \{m + t - 1 + (t - 1)(m - 1)z\}^n$$
$$\times z^{p-1}(1 - z)^{w-1} dz$$

from which it is easily seen that

$$E(X) = \frac{n}{p + w} \left(p + \frac{w}{m}\right)$$

and

$$Var(X) = \frac{n}{m^2} \left\{ \frac{m^2 pw + mw(w - p + 1) - w(w + 1)}{(p + w)(p + w + 1)} \right\}$$
$$+ \frac{n^2 pw}{m^2} \left\{ \frac{(m - 1)^2}{(p + w)^2(p + w + 1)} \right\}.$$

Another well-known compound distribution (not a special case of (XX.2)) obtained by mixing the beta and geometric distributions is the beta-geometric distribution. For applications of this distribution, see Weinberg and Gladen (1986), Hershlag et al. (1991), Ridout and Morgan (1991), Crouchley and Dassios (1998), Ecochard and Clayton (2000) and Fader and Hardie (2001). The pmf of the beta-geometric distribution is:

$$p(x) = \mu \prod_{k=1}^{x-1} \{1 - \mu + (k-1)\theta\} \bigg/ \prod_{k=1}^{x} \{1 + (k-1)\theta\}. \quad (\text{XX.5})$$

The pdf, mean and the variance are:

$$\phi(t) = \mu \; {}_2F_1\left(1, \frac{1-\mu}{\theta}; \frac{1+\theta}{\theta}; t\right),$$

$$E(X) = \frac{1-\theta}{\mu - \theta}$$

and

$$Var(X) = \frac{\mu(1-\mu)(1-\theta)}{(\mu-\theta)^2(\mu-2\theta)}.$$

Also the expected residual lifetime takes the simple form:

$$E(X - j \mid X > j) = \frac{1-\theta}{\mu - \theta} + \frac{j\theta}{\mu - \theta}.$$

Kemp (2001) suggested an extension of (XX.5) – referred to as the q-beta-geometric distribution – with the pmf:

$$p(x) = \frac{(1-\rho)(1/u; m)_x \rho^x}{(\rho/u; m)_x}, \qquad x = 1, 2, \ldots,$$

$$p(0) = \frac{1-\rho}{1-\rho/u},$$

where

$$(a; q)_n = (1-a)(1-aq) \cdots \left(1 - aq^{n-1}\right), \qquad n = 1, 2, \ldots,$$
$$(a; q)_0 = 1$$

is the hypergeometric notation of Gasper and Rahman (1990). The probabilities satisfy $0 \le p(x) \le 1$ provided that either (i) $0 < m < 1$, $0 < \rho < 1$, $u > 1$, or (ii) $m > 1$, $\rho > 1$, $0 < u < 1$. The corresponding pgf involves complicated hypergeometric functions.

XXI.   Complementary Beta Distribution

A random variable X is said to have the complementary beta distribution with parameters a and b if its cdf is:

$$F(x) = I_x^{-1}(a, b), \qquad 0 \leq x \leq 1, \tag{XXI.1}$$

where $I_x^{-1}(a, b)$ denotes the inverse of the incomplete beta function ratio. This distribution was suggested by Jones (2002). It was obtained by reversing the roles of the cdf and the quantile function for the standard beta distribution. The pdf corresponding to (XXI.1) is:

$$f(x; a, b) = \frac{B(a, b)}{\left\{I_x^{-1}(a, b)\right\}^{a-1} \left\{1 - I_x^{-1}(a, b)\right\}^{b-1}},$$
$$0 \leq x \leq 1. \tag{XXI.2}$$

Like the standard beta distribution, (XXI.2) is symmetric if and only if $a = b$. Also the parameters a and b are symmetrically related by:

$$f(x; a, b) = f(1 - x; b, a).$$

The particular case for $b = 1$ is the power function distribution with parameter $1/a$. The particular case for $a = 1$ is the standard beta distribution with parameters 1 and $1/b$. The uniform distribution on $(0, 1)$ is the particular case for $a = b = 1$. When $a = b = 1/2$ and $a = b = 2$, (XXI.2) reduces to:

$$f(x) = \frac{\pi}{2} \sin(\pi x)$$

and

$$f(x) = \frac{1}{3\sqrt{x(1 - x)}} \cos\left(\frac{\arcsin(2x - 1)}{3}\right),$$

respectively.

The figures below show the similarities between the complementary and the standard beta distributions. The solid lines are pdfs of the complementary beta distribution with parameters a and b; the dashed lines are pdfs of the standard beta distribution with parameters $1/a$ and $1/b$.

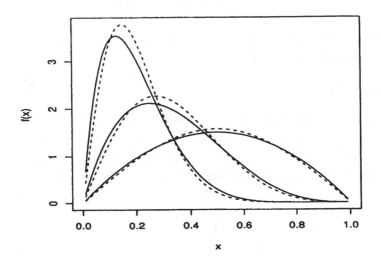

Figure 21. Complementary and standard beta pdfs for $a = 1/2$ and, from the right, $b = 1/2$, $1/4$, $1/8$.

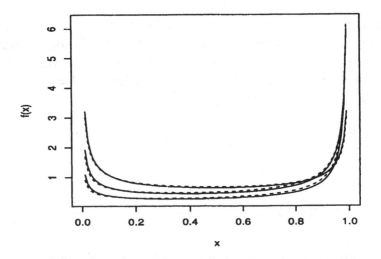

Figure 22. Complementary and standard beta pdfs for $a = 2$ and, reading down at the left hand end, $b = 2$, 4, 8.

The mode of (XXI.2) is the solution of:

$$F(x_0) = \frac{a-1}{a+b-2}, \qquad a+b \neq 2.$$

Note $0 < x_0 < 1$ if and only if the right-hand side also lies in $(0,1)$. If $a < 1$ and $b < 1$, (XXI.2) is unimodal with mode at x_0, provided that $0 < x_0 < 1$. If $a > 1$ and $b > 1$, (XXI.2) is U-shaped with antimode at x_0, provided that $0 < x_0 < 1$. If $a < 1$ and $b > 1$, (XXI.2) is J-shaped, and if $a > 1$ and $b < 1$, (XXI.2) is reverse J-shaped.

The first two moments of the complementary beta distribution are:

$$E(X) = \frac{b}{a+b}$$

and

$$E(X^2) = \frac{2}{a} \frac{B(2a, 2b+1)}{B^2(a,b)} \; {}_3F_2\left(a+b, 1, 2a; a+1, 2(a+b)+1; 1\right)$$

The rth L-moment (Hosking, 1990) of the complementary beta distribution is:

$$\lambda_r = \frac{1}{rB(a,b)} \sum_{k=0}^{r-2} (-1)^k \binom{r-2}{k} \binom{r}{k+1}$$
$$\times B(a+r-k-1, b+k+1). \qquad (\text{XXI.3})$$

The second L-moment λ_2 is a measure of spread and is actually one-half of Gini's index (Hosking, 1990). The third and fourth L-moment ratios $\tau_3 = \lambda_3/\lambda_2$ and $\tau_4 = \lambda_4/\lambda_2$ are very good measures pf the skewness and kurtosis of a distribution, respectively (Hosking, 1990, 1992). From (XXI.3), we note that

$$\lambda_2 = \frac{ab}{(a+b)(a+b+1)},$$

$$\tau_3 = \frac{a-b}{a+b+2}$$

and

$$\tau_4 = \frac{(a-b)^2 - ab + 1}{(a+b+2)(a+b+3)}.$$

Evidently the measure of spread λ_2 increases with a and b. The skewness measure $\tau_3 = 0$ when $a = b$, > 0 when $a > b$ and < 0 when $a < b$. Furthermore, $\tau_3 \to 1$ when $a \to \infty$ for b fixed and $\tau_3 \to -1$ when $b \to \infty$ for a fixed; these are the maximum and minimum permissible skewnesses (Hosking, 1990).

XXII.   Beta Normal Distribution

If G denote the cdf of a random variable then a generalized class of distributions can be defined by

$$F(x) = \frac{1}{B(a,b)} \int_0^{G(x)} w^{a-1}(1-w)^{b-1} dw, \qquad a > 0, \quad b > 0.$$
(XXII.1)

If G is taken as the cdf of the normal distribution with parameters μ and σ then X is said to have beta normal distribution. Substituting the normal cdf into (XXII.1) and differentiating, we get the corresponding pdf as:

$$f(x) = \frac{1}{\sigma B(a,b)} \left\{ \Phi\left(\frac{x-\mu}{\sigma}\right) \right\}^{a-1} \left\{ 1 - \Phi\left(\frac{x-\mu}{\sigma}\right) \right\}^{b-1}$$
$$\times \phi\left(\frac{x-\mu}{\sigma}\right),$$
(XXII.2)

where $\phi(\cdot)$ and $\Phi(\cdot)$, respectively, denote the pdf and the cdf of the standard normal distribution. This distribution was introduced Eugene et al. (2002). The parameters a and b are shape parameters, which characterize the skewness, kurtosis and bimodality of the distribution. The parameter μ is a location parameter and the parameter σ is a scale parameter that stretches out or shrinks the distribution. The distribution can be both unimodal and bimodal. Some properties of unimodal shapes include:

- the distribution is skewed to the right when $a > b$. As a increases the degree of right skewness increases.
- the distribution is skewed to the left when $a < b$. As b decreases the degree of left skewness increases.
- the distribution is symmetric and platykurtic when $a < 1$, $b < 1$ and $a = b$. As both a and b decrease until bimodality occurs, the heavier the tail.
- the distribution is leptokurtic when $a > 1$ and $b > 1$. As a and b increase the higher the peak of the distribution.

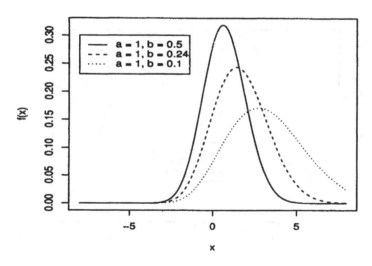

Figure 23. Beta normal pdfs for $\mu = 0$, $\sigma = 1$ and $(a, b) = (1, 0.5)$, $(1, 0.24)$, $(1, 0.1)$.

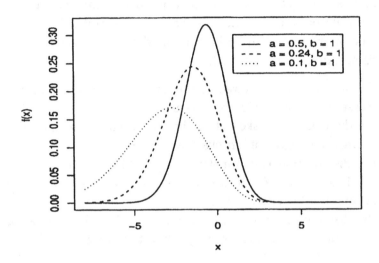

Figure 24. Beta normal pdfs for $\mu = 0$, $\sigma = 1$ and $(a, b) = (0.5, 1)$, $(0.24, 1)$, $(0.1, 1)$.

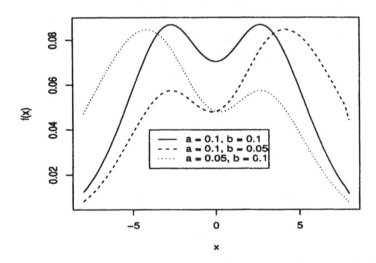

Figure 25. Beta normal pdfs for $\mu = 0$, $\sigma = 1$ and $(a, b) = (0.1, 0.1)$, $(0.1, 0.05)$, $(0.05, 0.1)$.

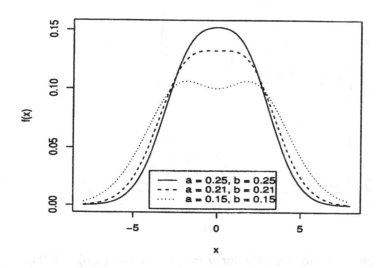

Figure 26. Beta normal pdfs for $\mu = 0$, $\sigma = 1$ and $(a, b) = (0.25, 0.25)$, $(0.21, 0.21)$, $(0.15, 0.15)$.

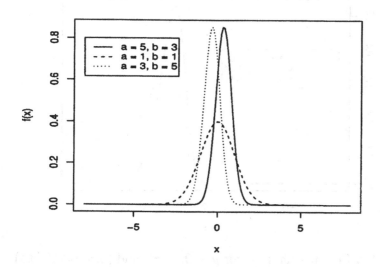

Figure 27. Beta normal pdfs for $\mu = 0$, $\sigma = 1$ and $(a, b) = (5, 3)$, $(1, 1)$, $(3, 5)$.

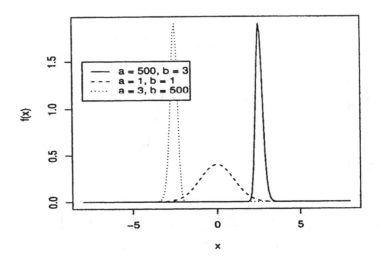

Figure 28. Beta normal pdfs for $\mu = 0$, $\sigma = 1$ and $(a, b) = (500, 3)$, $(1, 1)$, $(3, 500)$.

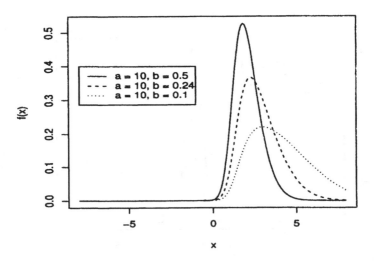

Figure 29. Beta normal pdfs for $\mu = 0$, $\sigma = 1$ and $(a, b) = (10, 0.5)$, $(10, 0.24)$, $(10, 0.1)$.

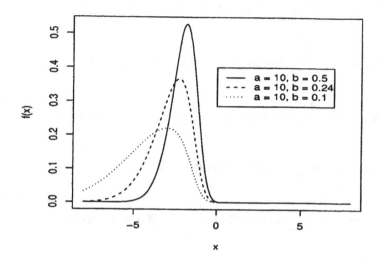

Figure 30. Beta normal pdfs for $\mu = 0$, $\sigma = 1$ and $(a, b) = (0.5, 10)$, $(0.24, 10)$, $(0.1, 10)$.

In general, exact moments of the beta normal distribution cannot be evaluated. However, the following relationships between the first two moments can be noted.

- $E(X)$ is an increasing function of a and a decreasing function of b.
- $SD(X)$ decreases as either a or b increases.
- for $b < a$, the kurtosis of X is an increasing function of a. Similarly, for $a < b$, the kurtosis of X is an decreasing function of b.
- $E(X) > \mu$ when $a > b$ and $E(X) < \mu$ when $a < b$.

Eugene et al. (2002) also showed that closed form expressions for $E(X)$ can be derived for certain values of a and b. For instance,

$$E(X) = \mu + \frac{\sigma}{\sqrt{\pi}} \quad \text{if } a = 2 \text{ and } b = 1,$$

$$E(X) = \mu + \frac{3\sigma}{2\sqrt{\pi}} \qquad \text{if } a = 3 \text{ and } b = 1,$$

$$E(X) = \mu + \frac{6\sigma}{\pi\sqrt{\pi}} \arctan\sqrt{2} \qquad \text{if } a = 4 \text{ and } b = 1,$$

$$E(X) = \mu + \frac{15\sigma}{\pi\sqrt{\pi}} \arctan\sqrt{2} - \frac{5\sigma}{2\sqrt{\pi}} \qquad \text{if } a = 5 \text{ and } b = 1,$$

$$E(X) = \mu - \frac{\sigma}{\sqrt{\pi}} \qquad \text{if } a = 1 \text{ and } b = 2,$$

$$E(X) = \mu - \frac{18\sigma}{\pi\sqrt{\pi}} \arctan\sqrt{2} + \frac{6\sigma}{\sqrt{\pi}} \qquad \text{if } a = 3 \text{ and } b = 2$$

and

$$E(X) = \mu - \frac{30\sigma}{\pi\sqrt{\pi}} \arctan\sqrt{2} + \frac{10\sigma}{\sqrt{\pi}} \qquad \text{if } a = 4 \text{ and } b = 2.$$

XXIII. Triangular Distributions

The main weakness of the beta distribution is that it suffers from difficulties involved in its maximum likelihood parameter estimation and its parameters do not have a clear-cut meaning. Recently there has some work on using more intuitively obvious distributions as alternatives for the beta distribution. Johnson (1997) used the triangular distribution with the pdf

$$f(x) = \begin{cases} \dfrac{2}{b-a}\dfrac{x-a}{m-a}, & \text{if } a \leq x \leq m, \\ \dfrac{2}{b-a}\dfrac{b-x}{m-a}, & \text{if } m \leq x \leq b \end{cases} \qquad \text{(XXIII.1)}$$

as a proxy to the beta distribution, specifically in problems of assessment of risk and uncertainty, such as the projection evaluation and review technique (PERT). The parameters of this distribution has intuitive appeal (see, for example, Williams (1992)). Like the beta distribution, (XXIII.1) can be positively or negatively skewed but must remain unimodal. Johnson (1997) pointed out, however, that there is no (XXIII.1) which would reasonably approximate uniform, J-shaped or U-shaped distributions.

Van Dorp and Kotz (2002a, 2002b) provided an extension of (XXIII.1) with the pdf

$$f(x) = \begin{cases} \dfrac{n}{b-a} \left(\dfrac{x-a}{m-a} \right)^{n-1}, & \text{if } a \leq x \leq m, \\ \dfrac{n}{b-a} \left(\dfrac{b-x}{b-m} \right)^{n-1}, & \text{if } m \leq x \leq b \end{cases} \qquad \text{(XXIII.2)}$$

as a meaningful alternative of the beta distribution. This is called the two-sided power (TSP) distribution. Unlike (XXIII.1), this new form allows for J-shaped and U-shaped pdfs. It has several attractive properties (not shared by the beta distribution) relating to the meaning of the parameters, the structure of its expected value as a function of parameters, a closed for expression for its cdf, a maximum likelihood estimation procedure involving only elementary functions and a transparent form of its entropy function. For instance, the expected value and the cdf of (XXIII.2) take the simple forms:

$$E(X) = \frac{a + (n-1)m + b}{n+1}$$

and

$$F(x) = \begin{cases} \dfrac{m-a}{b-a} \left(\dfrac{x-a}{m-a} \right)^{n}, & \text{if } a \leq x \leq m, \\ 1 - \dfrac{b-m}{b-a} \left(\dfrac{b-x}{b-m} \right)^{n}, & \text{if } m \leq x \leq b. \end{cases}$$

Although (XXIII.2) is more flexible compared to (XXIII.1), Van Dorp and Kotz pointed out that the beta distribution enjoys greater flexibility than (XXIII.2) among the J-shaped distributions. Nadarajah (2002) has provided an extension of (XXIII.2) which overcomes this weakness.

XXIV.   Tukey Lambda Distributions

If X is a uniform random variable on $(0, 1)$ then

$$Y = \frac{aX^\lambda - (1 - X)^\lambda}{\lambda}, \qquad a > 0 \qquad\qquad (XXIV.1)$$

is said to have the Tukey lambda distribution. The transformation (XXIV.1) is monotonic increasing for all values of λ ($\lambda \neq 0$). If $a = 1$, the distribution is symmetrical. The Tukey lambda distribution has been studied by various authors, see Hastings et al. (1947), Tukey (1962), Shapiro and Wilk (1965) and Joiner and Rosenblatt (1971).

Johnson and Kotz (1973) provided an extension of (XXIV.1) by defining

$$Y = \frac{X^\lambda - (1 - X)^\lambda}{\lambda},$$

where X is assumed to have the standard beta distribution with parameters a and b. If $\lambda > 0$, the range of Y is from $-1/\lambda$ to $1/\lambda$. If $\lambda < 0$, the range is unlimited. When $\lambda = 0$, take

$$Y = \log\left(\frac{X}{1 - X}\right).$$

An explicit expression for the pdf of Y is not available, but it can be expressed as

$$f(y) = \frac{1}{B(a, b)} \frac{x^{a-1}(1 - x)^{b-1}}{x^{\lambda-1} + (1 - x)^{\lambda-1}}, \qquad\qquad (XXIV.2)$$

where each x has to be expressed in terms of y to satisfy the relation

$$y = \frac{x^\lambda - (1 - x)^\lambda}{\lambda}.$$

Clearly $f(y) \to 0$ or ∞ at the extremes of the range of Y, according as $f(x) \to 0$ or ∞, for $\lambda > 1$. The mode of (XXIV.2) is given by $x = u/(1 + u)$, where u is the solution of the equation:

$$u^{\lambda-1} = \frac{(b - \lambda)u - (a - 1)}{a - \lambda - (b - 1)u} = \frac{1 - \lambda u + bu - a}{u - \lambda - bu + a}. \qquad\qquad (XXIV.3)$$

For $\lambda < 1 < \min(a, b)$, this equation has just one root and so the distribution of Y is unimodal. A similar situation arises when $1 <$

$\lambda < \min(a, b)$. If $b < \min(1, \lambda)$ and $a > \max(1, \lambda)$ then (XXIV.3) has no solution. In this case the pdf of Y is J-shaped. Similar results hold if $b > \max(1, \lambda)$ and $a < \min(1, \lambda)$. In the particular case $\lambda = a + b - 1$, (XXIV.3) becomes

$$u = \left(\frac{a-1}{b-1}\right)^{1/(a+b-2)}$$

In the symmetrical case $a = b$, (XXIV.3) is satisfied by $u^{\lambda-1} = 1$; hence, the corresponding mode of Y is 0.

The rth moment of Y about zero is:

$$\mu'_r = \frac{1}{\lambda^r B(a, b)} \sum_{k=0}^{r} (-1)^k \binom{r}{k} B\left(a + (r-k)\lambda, b + k\lambda\right).$$

If $\lambda = 0$, the rth cumulant of Y is:

$$\kappa_r = \psi^{(r-1)}(a) + (-1)^r \psi^{(r-1)}(b),$$

where

$$\psi^{(r-1)}(x) = \frac{d^r}{dx^r} \log \Gamma(x)$$

is the rth derivative of the log-gamma function (see Chapter 2 for properties).

References

1. Andersen, E.S. On the fluctuations of sums of random variables, Mathematica Scandinavica **1953**, *1*, 262-285.

2. Armero, C.; Bayarri, M.J. Prior assessments for prediction in queues, The Statistician **1994**, *43*, 139-153.

3. Arnold, B.C.; Groeneveld, R.A. Some properties of the arcsine distribution, Journal of the American Statistical Association **1980**, *75*, 173-175.

4. Barrett, A.M.; Normand, M.D.; Peleg, M. A 'log-beta' vs. the log-normal distribution for particle populations with a wide finite size range, Power Technology **1991**, *66*, 195-199.

5. Bi, J.; Templeton-Janik, L.; Ennis, J.M.; Ennis, D.M. Repli-

cated difference and preference tests: how to account for inter-
trial variation, Food Quality and Preference **2000**, *11*, 269-273.

6. Breitung, J. Rank tests for nonlinear cointegration, Journal of
   Business and Economic Statistics **2001**, *19*, 331-340.

7. Breitung, J.; Gourieroux, C. Rank tests for unit roots, Journal
   of Econometrics **1997**, *81*, 7-28.

8. Bulgren, W.G. On representations of the doubly noncentral F
   distribution, Journal of the American Statistical Association
   1971, *66*, 184-186.

9. Bunz, C.; Schock, W.; Koyro, M.; Gentry, J.; Plunkett, S.; Runyan, M.; Pearson, C.; Wang, C. Application of the log beta-
   distribution to aerosol size distributions, Journal of Aerosol Science **1987**, *18*, 663-666.

10. Chang, Y.; Han, R.J.; Pearson, C.L.; Runyan, M.R.; Ranade,
    M.B.; Gentry, J.W. Applications of the log beta-distribution
    to the evolution of aerosol growth, Journal of Aerosol Science
    1988, *19*, 879-882.

11. Chattamvelli, R. Another derivation of two algorithms for the
    noncentral χ^2 and F distributions, Journal of Statistical Com-
    putation and Simulation **1994**, *49*, 207-214.

12. Chattamvelli, R. A note on the noncentral beta distribution
    function, The American Statistician **1995**, *49*, 231-234.

13. Chattamvelli, R. On the doubly noncentral F distribution,
    Computational Statistics and Data Analysis **1995**, *20*, 481-489.

14. Chen, J.C.; Novick, M.R. Bayesian analysis for binomial mod-
    els with generalized beta prior distributions, Journal of Educa-
    tional Statistics **1984**, *9*, 163-175.

15. Chen, R.; Lin, E.; Zame, A. Another arc sine law, Sankhyā, A
    1981, *43*, 371-373.

16. Chen, Z.Y.; Zhou, Y.C. Computing the noncentral beta distri-
    bution with S-system, Computational Statistics and Data Anal-
    ysis **2000**, *33*, 343-360.

17. Chung, K.L.; Feller, W. Fluctuations in coin tossing, Proceed-
    ings of the National Academy of Sciences, USA **1949**, **35**, 605-
    608.

18. Corradini, M.G.; Normand, M.D.; Nussinovitch, A.; Horowitz,

J.; Peleg, M. Estimating the frequency of high microbial counts in commercial food products using various distribution functions, Journal of Food Protection **2001**, *64*, 674-681.

19. Crouchley, R.; Dassios, A. Interpreting the beta geometric in comparative fecundability studies, Biometrics **1998**, *54*, 161-

20. Das Gupta, P. Two approximations for the distribution of double non-central beta, Sankhyā, B **1968**, *30*, 83-88.

21. Ding, C.G. Algorithm AS 275: Computing the noncentral χ^2 distribution function, Applied Statistics **1992**, *41*, 478-482.

22. Ding, C.G. On the computation of the noncentral beta distribution, Computational Statistics and Data Analysis **1994**, *18*, 449-455.

23. Dominici, F.; Parmigiani, G. Bayesian semiparametric analysis of developmental toxicology data, Biometrics **2001**, *57*, 150-157.

24. Ecochard, R.; Clayton, D.G. Multivariate parametric random effect regression models for fecundability studies, Biometrics **2000**, *56*, 1023-1029.

25. Erdös, P.; Kac, M. On the number of positive sums of independent random variables, Bulletin of the American Mathematical Society **1947**, *53*, 1011-1020.

26. Eugene, N.; Lee, C.; Famoye, F. Beta-normal distribution and its applications, Communications in Statistics–Theory and Methods **2002**, *31*, 497-512.

27. Exton, H. *Multiple Hypergeometric Functions and Applications*; Halstead Press: New York, 1976.

28. Fader, P.S.; Hardie, B.G.S. Forecasting repeat sales at CD-NOW: A case study, Interfaces **2001**, *31*, S94-S107.

29. Feller, W. *An Introduction to Probability Theory and Its Applications* (volume 1, third edition); John Wiley and Sons: New York, 1967.

30. Gasper, G.; Rahman, M. *Basic Hypergeometric Series*; Cambridge University Press: Cambridge, 1990.

31. Gelfand, A. Estimation of a restricted variance ratio, in *Proceedings of the 2nd International Tampere Conference in Statistics*, Tampere, Finland (edited by T. Pukkila and S. Puntanen) **1987**, 457-466.

32. Gerstenkorn, T. The compounding of the binomial and general-
    ized beta distributions, in *Probability and Statistical Inference*
    (edited by W. Grossmann *et al.*) **1982**, 87-99, D. Reidel Pub-
    lishing Company.

33. Getoor, R.K.; Sharpe, M.J. Local times on rays for a class of
    planar Lévy processes, Journal of Theoretical Probability **1994**,
    7, 799-811.

34. Gordy, M.B. A generalization of generalized beta distributions.

35. Gordy, M.B. Computationally convenient distributional as-
    sumptions for common-value auctions, Computational Eco-
    nomics **1998**, *12*, 61-78.

36. Grassia, A. On a family of distributions with argument between
    0 and 1 obtained by transformation of the gamma and derived
    compound distributions, Australian Journal of Statistics **1977**,
    19, 108-114.

37. Han, R.J.; Chang, Y.; Pao, J.R.; Ranade, M.B.; Gentry, J.W.
    Applications of probit analysis with the log beta-distribution,
    Staub Reinhaltung Der Luft **1989**, *49*, 125-130.

38. Hastings, C.; Mosteller, F.; Tukey, J.W.; Winsor, C.P. Low mo-
    ments for small samples: a comparative study of order statistics,
    Annals of Mathematical Statistics **1947**, *18*, 413-426.

39. Hershlag, A.; Kaplan, E.H.; Loy, R.A.; Decherney, A.H.; Lavy,
    G. Heterogeneity in patient populations explains differences in
    invitro fertilization programs, Fertility and Sterility **1991**, *56*,
    913-917.

40. Hodges, J.L. On the noncentral beta-distribution, Annals of
    Mathematical Statistics **1955**, *26*, 648-653.

41. Hoffmann-Jorgensen, J. The arcsine law, Journal of Theoretical
    Probability **1999**, *12*, 131-145.

42. Hosking, J.R.M. *L*-moments: analysis and estimation of distri-
    butions using linear combinations of order statistics, Journal of
    the Royal Statistical Society B **1990**, *52*, 105-124.

43. Hosking, J.R.M. Moments or *L*-moments? An example compar-
    ing two measures of distributional shape, American Statistician
    1992, *46*, 186-189.

44. Hughes, G.; Madden, L.V. Some methods for eliciting expert

knowledge of plant disease epidemics and their application in cluster sampling for disease incidence, Crop Protection **2002**, *21*, 203-215.

45. Ishii, G.; Hayakawa, R. On the compound binomial distribution, Annals of the Institute of Statistical Mathematics **1960**, *12*, 69-80.

46. Jain, G.C.; Consul, P.C. A generalized negative binomial distribution, SIAM Journal on Applied Mathematics **1971**, *21*, 501-513.

47. Johannesson, B.; Giri, N. On approximations involving the beta distribution, Communications in Statistics–Simulation and Computation **1995**, *24*, 489-503.

48. Johnson, D. The triangular distribution as a proxy for the beta distribution in risk analysis, The Statistician **1997**, *46*, 387-398.

49. Johnson, N.L.; Kotz, S. Extended and multivariate Tukey lambda distributions, Biometrika **1973**, *60*, 655-661.

50. Johnson, N.L.; Kotz, S.; Balakrishnan, N. *Continuous Univariate Distributions* (volume 2); John Wiley and Sons: New York, 1995.

51. Joiner, B.L.; Rosenblatt, J.R. Some properties of the range in samples from Tukey's symmetric lambda distributions, Journal of the American Statistical Association **1971**, *66*, 394-399.

52. Jones, M.C. The complementary beta distribution, Journal of Statistical Planning and Inference **2002**, *104*, 329-337.

53. Karlin, S.; Taylor, H.M. *A Second Course in Stochastic Processes*; Academic Press: New York, 1981.

54. Kemp, A.W. The q-beta-geometric distribution as a model for fecundability, Communications in Statistics–Theory and Methods **2001**, *30*, 2373-2384.

55. Lee, Y.W. *Statistical Theory of Communications*; John Wiley and Sons: New York, 1960.

56. Libby, D.L.; Novick, M.R. Multivariate generalized beta-distributions with applications to utility assessment, Journal of Educational Statistics **1982**, *7*, 271-294.

57. Marchand, E. On moments of beta mixtures, the noncentral beta distribution, and the coefficient of determination, Journal

of Statistical Computation and Simulation 1997, *59*, 161-178.

58. Mauldon, J.G. A generalization of the beta-distribution, Annals of Mathematical Statistics 1959, *30*, 509-520.

59. McDonald, J.B. Some generalized functions for the size distribution of income, Econometrica 1984, *52*, 647-664.

60. McDonald, J.B.; Richards, D.O. Hazard rates and generalized beta distributions, IEEE Transactions on Reliability 1987, *36*, 463-466.

61. McDonald, J.B.; Richards, D.O. Model selection: some generalized distributions, Communications in Statistics (volume A) 1987, *16*, 1049-1074.

62. McDonald, J.B.; Richards, D.O. Some generalized models with application to reliability, Journal of Statistical Planning and Inference 1987, *16*, 365-376.

63. McDonald, J.B.; Xu, Y.J. A generalization of the beta distribution with applications, Journal of Econometrics 1995, *66*, 133-152.

64. Middleton, D. A note on the estimation of signal waveform, IRE Transactions of Information Theory 1959, *5*, 86-89.

65. Middleton, D. *Introduction to Statistical Communications Theory*; McGraw-Hill Book Company: New York, 1960.

66. Moran, P.A.P. *An Introduction to Probability Theory*; Oxford University Press.

67. Morrison, D.G.; Brockway, G. A modified beta binomial model with applications to multiple choice and taste tests, Psychometrika 1979, *44*, 427-442.

68. Nadarajah, S. Letter to the editor, to appear in The American Statistician.

69. Neuts, M.F.; Li, J.-M.; Pearce, C.E.M. On an interval-partitioning scheme, Applicationes Mathematicae 1999, *26*, 347-355.

70. Nicholson, W.L. A computing formula for the power of the analysis of variance test, Annals of Mathematical Statistics 1954, *25*, 607-610.

71. Norton, R.M. On properties of the arc-sine law, Sankhyā 1975, *37*, 306-308.

72. Park, J.H. Variations of the non-central t and beta distributions, Annals of Mathematical Statistics **1964**, *35*, 1583-1593.

73. Park, J.H.; Glaser, E.M. The extraction of waveform information by a delay line filter technique, IRE Wescon Convention Record **1957**, *1-2*, 171-184.

74. Patel, I.D.; Khatri, C.G. A Lagrangian beta distribution, South African Statistical Journal **1978**, *12*, 57-64.

75. Patil, G.P.; Boswell, M.T.; Ratnaparkhi, M.V. *Dictionary and Classified Bibliography of Statistical Distributions in Scientific Work: Continuous Univariate Models*; International Cooperative Publishing House: Burtonsville, MD.

76. Patnaik, P.B. The non-central χ^2 and F-distribution and their applications, Biometrika **1949**, *26*, 202-232.

77. Pham-Gia, T. WorkSamp (software section), Mathematical and Computer Modelling **1989**, *12*, I.

78. Pitman, J.; Yor, M. Arcsine laws and interval partitions derived from a stable subordinator, Proceedings of the London Mathematical Society **1992**, *65*, 326-356.

79. Ridout, M.S.; Morgan, B.J.T. Modeling digit preference in fecundability studies, Biometrics **1991**, *47*, 1423-1433.

80. Roy, M.K.; Roy, A.K.; Ali, M.M. Binomial mixtures of some standard distributions, Journal of Information and Optimization Sciences **1993**, *14*, 57-71.

81. Ruben, H. Probability content of regions under spherical normal distributions, III; The bivariate normal integral, Annals of Mathematical Statistics **1961**, *32*, 171-186.

82. Runyan, M.R.; Pearson, C.L.; Wang, C.C.; Hand, R.; Plunkett, S.; Kaplan, C.; Gentry, J.W. The use of the log Beta distribution to characterize atmospheric aerosols, in *Lecture Notes in Physics* **1988**, *309*, 65-68, Springer Verlag: Berlin.

83. Seber, G.A.F. The non-central chi-squared and beta distributions, Biometrika **1963**, *50*, 542-544.

84. Shah, D.A.; Bergstrom, G.C. A rainfall-based model for predicting the regional incidence of wheat seed infection by Stagonospora nodorum in New York, Phytopathology **2002**, *92*, 511-518.

85. Shah, D.A.; Bergstrom, G.C.; Ueng, P.P. Foci of stagonospora nodorum blotch in winter wheat before canopy development, Phytopathology **2001**, *91*, 642-647.

86. Shapiro, S.S.; Wilk, M.B. An analysis of variance test for normality (complete samples), Biometrika **1965**, *52*, 591-611.

87. Shiyomi, M.; Takahashi, S.; Yoshimura, J. A measure for spatial heterogeneity of a grassland vegetation based on the beta-binomial distribution, Journal of Vegetation Science **2000**, *11*, 627-632.

88. Smith, M.C.; Page, W.W.; Holt, J.; Kyetere, D. Spatial dynamics of maize streak virus disease epidemic development in maize fields, International Journal of Pest Management **2000**, *46*, 55-66.

89. Spitzer, F. *Principles of Random Walks*; D. van Nostrand Company: Princeton, 1964.

90. Tang, P.C. The power function of the analysis of variance test with tables and illustrations of their use, Statistical Research Memoirs **1938**, *2*, 126-149.

91. Tiku, M.L. A note on the distribution of the doubly noncentral F distribution, Australian Journal of Statistics **1972**, *14*, 37-40.

92. Tiku, M.L. Doubly noncentral F distribution—tables and applications, in *Selected Tables in Mathematical Statistics* (edited by H.L. Harter and D.B. Owen) **1974**, *2*, 139-176, American Mathematical Society: Providence, RI.

93. Tiku, M.L. Laguerre series forms of the distributions of classical test-statistics and their robustness in non-normal situations, in *Applied Statistics* (edited by R.P. Gupta) **1975**, 333-350, North-Holland: Amsterdam.

94. Tukey, J.W. The future of data analysis, Annals of Mathematical Statistics **1962**, *33*, 1-67.

95. Turechek, W.W.; Ellis, M.A.; Madden, L.V. Sequential sampling for incidence of Phomopsis leaf blight of strawberry, Phytopathology **2001**, *91*, 336-347.

96. Turechek, W.W.; Madden, L.V. Analysis of the association between the incidence of two spatially aggregated foliar diseases of strawberry, Phytopathology **2000**, *90*, 157-170.

97. Van Dorp, J.R.; Kotz, S. A novel extension of the triangular distribution and its parameter estimation, The Statistician **2002**, *51*, 1-17.

98. Van Dorp, J.R.; Kotz, S. The standard two-sided power distribution and its properties: with applications in financial engineering, The American Statistician **2002**, *56*, 90-99.

99. Venette, R.C.; Moon, R.D.; Hutchison, W.D. Strategies and statistics of sampling for rare individuals, Annual Review of Entomology **2002**, *47*, 143-174.

100. Volodin, N. Some generalizations of the beta distribution, Private Communication, 1994.

101. Weinberg, C.R.; Gladen, B.C. The beta-geometric distribution applied to comparative fecundability studies, Biometrics **1986**, *42*, 547-560.

102. Wilfling, B. Lorenz ordering of power-function order statistics, Statistics and Probability Letters **1996**, *30*, 313-319.

103. Williams, T.M. Practical use of distribution in network analysis, Journal of Operations Research Society **1992**, *43*, 265-270.

Beta Distributions in Stochastic Processes

Kosto Mitov[1] and Saralees Nadarajah[2]

[1]Department of Mathematics and Informatics
Airforce academy "G. Benkovski"
Pleven, Bulgaria 5800
[2]Department of Mathematics
University of South Florida
Tampa, Florida 33620

I. Introduction

The pdf of a beta distribution with parameters a and b is:

$$f(x; a, b) = \frac{1}{B(a,b)} x^{a-1}(1-x)^{b-1}, \quad 0 < x < 1, a > 0, b > 0.$$

We denote the corresponding cdf by $B(x; a, b)$ and $Z_{a,b}$ will denote a random variable with this cdf.

The beta distributions with parameters $0 < a < 1$ and $b = 1 - a$ are known as generalized arc sine laws. They have pdf and cdf:

$$f(x; a, 1-a) = \frac{\sin \pi a}{\pi} x^{a-1}(1-x)^{-a},$$

$$B(x; a, 1-a) = \frac{\sin \pi a}{\pi} \int_0^x u^{a-1}(1-u)^{-a} du, \quad 0 < x < 1.$$

The source of the name "arc sine laws" is the well known Lévy's arc sine law, i.e. the beta distribution with parameters $a = 1/2$ and

$b = 1/2$ whose pdf and cdf are

$$f(x; 1/2, 1/2) = \frac{2}{\pi} x^{-1/2}(1-x)^{-1/2},$$

$$B(x; 1/2, 1/2) = \frac{2}{\pi} \arcsin \sqrt{x}, \ 0 < x < 1.$$

The case $a = b = 1$ corresponds to the uniform distribution on the interval $[0, 1]$.

Bingham et al. (1987) mention two main areas where arc sine laws arise in stochastic processes. One is renewal theory. The other is in random walks, dealing with the fraction of time spent in the right half-line. The link between the two is provided by regenerative phenomena of "ladder" type and the classical fluctuation theory for random walks due to E. Sparre Andersen, F. Spitzer and others. We should add Brownian motion (Winner) processes as the third class of stochastic processes, where the beta distributions appear.

Certainly, the beta distributions appear also in the other fields of stochastic processes. The following example is well known.

Example 1. *Let us consider Pólya's urn scheme (Pólya, 1931, page 150). An urn contains W_n white balls and B_n black balls at the instant n. One ball is drawn at random and then replaced, while α balls of the same color are added to the urn. The sequence*

$$\frac{W_n}{W_n + B_n}, \ \ n \to \infty$$

converges with probability 1 to a random variable having beta distribution with parameters $a = W_0/\alpha$, $b = B_0/\alpha$ (see Freedman (1965) for details and generalizations).

The aim of this chapter is to represent some of the main topics in these areas, where the beta distributions appear as limiting distributions or as finite dimensional distributions of certain stochastic processes.

The chapter is organized as follows: Section II deals with the renewal theory; Section III deals with random walks; Section IV deals with Brownian motion and Section V contains some results about occupation times.

II. Renewal Processes

A. Ordinary Renewal Processes

i. Definitions

Suppose that at $t = 0$ a new element is installed in a system. Its lifetime is finite and positive. When the element fails it is immediately replaced by a new one and so on.

To build a corresponding mathematical model let us suppose that the random variables

$$T_1, T_2, \ldots, T_n, \ldots$$

are nonnegative, and they are independently sampled from probability law $F(x)$ on $[0, \infty)$ with mean $\mu = ET_i \in (0, \infty]$.

Define $S_n = \sum_{i=1}^{n} T_i$, $n = 1, 2, \ldots$. Then the points

$$S_0 = 0, S_1, S_2, \ldots, S_n, \ldots,$$

are called *renewal epochs*. The r.v. T_n is considered as the time between the $(n-1)$th and nth renewal epochs.

The process $N(t)$ defined by

$$N(t) = \max\{n : S_n \leq t\}$$

for $t \geq 0$ gives the number of renewal epochs in the closed interval $[0, t]$. Clearly

$$S_{N(t)} \leq t < S_{N(t)+1}, \quad \text{a.s.}$$

The process $(N(t),\ t \geq 0)$ is called *renewal process*. The following three processes are of interest:

- the *spent life time* $Y^-(t) = t - S_{N(t)}$,
- the *residual life time* $Y^+(t) = S_{N(t)+1} - t$,
- the *lifetime* $Y(t) = Y^-(t) + Y^+(t)$.

So, if one observes the system at the instant $t \geq 0$ then $Y^-(t)$ is the time passed from the last replacement of the element; $Y^+(t)$ is the time up to the next replacement and $Y(t)$ is the total time of work of the element which is in the system at that instant. It is clear that this element is $N(t) + 1$ in order.

Denote by

$$U(t) = E\{N(t)\} = \sum_{n=0}^{\infty} F^{*n}(t)$$

the expected value of renewal events during the time interval $[0, t]$. The function $U(t)$ is called *renewal function*. Here, as usual $F^{*0}(t) = 1$ and for $n \geq 1$, $F^{*n}(t)$ is n-fold convolution of the cdf $F(t)$.

To formulate the results we have to discriminate between F *lattice* (concentrated on some set $\{c, 2c, 3c, \ldots\}$ with $c > 0$ maximal, then c is called the span of F) and *non-lattice*. Suppose that F is non-lattice. (The analogs results are valid in lattice case too.)

ii. Case $\mu < \infty$

Assume that $\mu < \infty$. Then the processes $Y^-(t)$, $Y^+(t)$ and $Y(t)$ converge in law (see Feller, 1971, Chapter 11) and

$$\lim_{t \to \infty} \Pr\{Y^-(t) \leq x\} = \lim_{t \to \infty} \Pr\{Y^+(t) \leq x\}$$
$$= \frac{1}{\mu} \int_0^x \{1 - F(y)\}\, dy,$$

$$\lim_{t \to \infty} \Pr\{Y^-(t) > x, Y^+(t) > y\} = \frac{1}{\mu} \int_{x+y}^{\infty} \{1 - F(u)\}\, du$$

and

$$\lim_{t \to \infty} \Pr\{Y(t) \leq x\} = \frac{1}{\mu} \int_0^x y\, dF(y).$$

Moreover, if one supposes that the initial renewal epoch S_0 is not identically equals 0 but S_0 is a r.v. independent of $\{T_n\}$, i.e. a delayed renewal process is considered then if

$$\Pr\{S_0 \leq t\} = \frac{1}{\mu} \int_0^t \{1 - F(u)\}\, du,$$

the renewal function $U(t) \equiv t/\mu$ for all $t \geq 0$.

iii.   Case $\mu = \infty$

The behavior of these processes is quite different in the case when $\mu = E\{T_n\} = \infty$. In this case

$$Y^-(t) \to \infty, \quad Y^+(t) \to \infty, \quad Y(t) \to \infty,$$

as $t \to \infty$. Therefore, non-degenerate limiting distributions without normalization do not exist.

Suppose that $F(t)$ has regularly varying tail with $\beta \in (0,1)$, i.e. it satisfies the condition

$$1 - F(t) \sim \frac{t^{-\beta}L(t)}{\Gamma(1-\beta)}, \quad t \to \infty, \tag{II.1}$$

where $L(\cdot)$ is a slowly varying function at infinity.

Remark 1. *A function $L : (0,\infty) \to (0,\infty)$ is called slowly varying at infinity (svf) if $L(tx)/L(t) \to 1$ as $t \to \infty$ for every $x > 0$.*

In terms of Laplace-Stieltjes transforms this condition can be written in the following equivalent form

$$1 - \widehat{F}(s) \sim s^\beta L(1/s), \quad s \downarrow 0, \tag{II.2}$$

where $\widehat{F}(s) = \int_0^\infty \exp(-st)dF(t)$ is the Laplace-Stieltjes transform of the cdf $F(x)$.

Since $U(t) = \sum_{n=0}^\infty F^{*n}(t)$ the Laplace-Stieltjes transform of $U(t)$ is

$$\sum_{n=0}^\infty \widehat{F}^n(s) = \frac{1}{1 - \widehat{F}(s)}.$$

Karamata's Tauberian theorem shows that this is equivalent to

$$U(t) \sim \frac{t^\beta}{L(t)\Gamma(1+\beta)}, \quad t \to \infty, \tag{II.3}$$

in which case

$$\{1 - F(t)\}\, U(t) \to \frac{1}{\Gamma(1-\beta)\Gamma(1+\beta)} = \frac{\sin \pi\beta}{\pi\beta}, \quad t \to \infty.$$

(See Feller (1971) and Bingham et al. (1987) for the theory of regular variation.)

It turns out that (II.1), (II.2) and (II.3) are equivalent forms of the condition for the processes $Y^-(t)$ and $Y^+(t)$ to have a linear growth to infinity, i.e. the normalized processes $Y^-(t)/t$ and $Y^+(t)/t$ to have non-degenerate limit laws, jointly or separately.

The well known result in this direction is the following theorem, which is usually called Dynkin-Lamperti theorem. It is proved by Dynkin (1955) and Lamperti (1958b).

Theorem 1. *(Dynkin-Lamperti Theorem). The condition (II.1)–(II.3) is necessary and sufficient for existence of a non-degenerate limit law as $t \to \infty$ for each of $Y^-(t)/t$, $Y^+(t)/t$, and $(Y^-(t)/t, Y^+(t)/t)$. The corresponding limit laws are*

$$\lim_{t \to \infty} \Pr\left\{ \frac{Y^-(t)}{t} \leq x \right\} = B(x; 1 - \beta, \beta), \quad 0 < x < 1,$$

$$\lim_{t \to \infty} \Pr\left\{ \frac{Y^+(t)}{t} \leq x \right\} = \frac{\sin \pi \beta}{\pi} \int_0^x u^{-\beta}(1 + u)^{-1} du, \quad x > 0,$$

$$\lim_{t \to \infty} \Pr\left\{ \frac{Y^-(t)}{t} \leq x, \frac{Y^+(t)}{t} \leq y \right\}$$
$$= \frac{\beta \sin \pi \beta}{\pi} \int_0^x \int_0^y (1 - u)^{\beta - 1}(u + v)^{-\beta - 1} du\, dv,$$
$$0 < x < 1, y > 0.$$

Remark 2.   *1. The corresponding result is also true in the lattice case, if only the time parameter t tends to ∞ over the set of numbers nc, $n = 0, 1, 2, \ldots$, where $c > 0$ is the span the cdf $F(x)$.*

2. It is well known that the condition (II.1) characterizes the stable distributions with support $[0, \infty)$ (see e.g. Feller (1971)). It was mentioned in Feller(1971) that the connection between stable distributions with parameter $\beta \in (0, 1)$ and the beta distribution established by the Dynkin-Lamperti theorem gives somewhat explanation of the appearance of the beta distribution in certain stochastic processes with stable one dimensional distributions.

3. It is not evident that the limit laws of $Y^+(t)/t$ and $(Y^-(t)/t,$

$Y^+(t)/t$ are in fact beta distributions, but it is easily seen that the pdf

$$\frac{\sin \pi \beta}{\pi} \frac{1}{x^\beta(1+x)}, \quad x > 0,$$

can be obtained from $f(x; 1 - \beta, \beta)$ by change of variable $x = y/(1 + y)$, and the limit law

$$\frac{\beta \sin \pi \beta}{\pi} \int_0^x \int_0^y (1 - u)^{\beta-1}(u + v)^{-\beta-1} du dv,$$
$$0 < x < 1, y > 0,$$

can be written in the form (see Kovalenko et al. (1983))

$$\lim_{t \to \infty} \Pr \left\{ \frac{Y^-(t)}{t} > x, \frac{Y^+(t)}{t} > y \right\}$$
$$= \frac{\sin \pi \beta}{\pi} \int_{\frac{x+y}{1-x}} u^{-\beta}(1+u)^{-1} du, \quad 0 < x < 1, \ y > 0.$$

To complete the classical results we have to mention the case $\beta = 1$. In this case the linear normalization of $Y^-(t)$ and $Y^+(t)$ is not appropriate. The following theorem proved by Erickson (1970) describes the case.

Theorem 2. *(Theorem 6, Erickson, 1970) Suppose that F is non-lattice,*

$$1 - F(t) \sim t^{-1} L(t), \quad t \to \infty,$$

and

$$m_F(t) = \int_0^t \{1 - F(x)\} \, dx \to \infty, \quad t \to \infty.$$

Then

$$\lim_{t \to \infty} \Pr \left\{ \frac{m_F \left(Y^-(t) \right)}{m_F(t)} \leq x \right\} = x, \quad x \in (0,1).$$

$$\lim_{t \to \infty} \Pr \left\{ \frac{m_F \left(Y^+(t) \right)}{m_F(t)} \leq x \right\} = x, \quad x \in (0,1).$$

$$\lim_{t \to \infty} \Pr \left\{ \frac{m_F \left(Y^-(t) \right)}{m_F(t)} \leq x, \frac{m_F \left(Y^+(t) \right)}{m_F(t)} \leq y \right\} = \max(x,y),$$
$$0 \leq x \leq 1, y > 0.$$

Mohan (1976) has considered the following generalization. Let the distribution function of T_i be either F_1 or F_2. Denote by $\tau(n)$ the number of random variables among $T_1, T_2, \ldots, T_n$ which have F_1 as their distribution function. Suppose that

$$1 - F_1(x) \sim x^{-\beta_1} L_1(x), \ x > 0,$$
$$1 - F_2(x) \sim x^{-\beta_2} L_2(x), \ x > 0,$$
$$0 < \beta_1 < \beta_2 < 1$$

and that there exists two sequences $A(n)$ and $B(n) > 0$ such that

$$\lim_{n \to \infty} \Pr \left\{ (S_n - A(n)) / B(n) \right\} = G_{\beta_1, \beta_2}(x),$$

where $G_{\beta_1, \beta_2}(x)$ is the composition of two stable laws with exponents β_1 and β_2.

Under these conditions the following results are proved (Mohan (1976), Theorem 3.1).

$$\lim_{t \to \infty} \Pr \left\{ \frac{Y^-(t)}{t} \leq x \right\} = b_1 B \left(x; 1 - \beta_1, \beta_1 \right) + b_2 B \left(x; 1 - \beta_2, \beta_2 \right)$$
$$0 < x < 1,$$
$$\lim_{t \to \infty} \Pr \left\{ \frac{Y^+(t)}{t} \leq x \right\} = b_1 \frac{\sin \pi \beta_1}{\pi} \int_0^x u^{-\beta_1} (1+u)^{-1} du$$
$$+ b_2 \frac{\sin \pi \beta_2}{\pi} \int_0^x u^{-\beta_2} (1+u)^{-1} du,$$
$$x > 0,$$

where the positive constants b_1 and b_2 are explicitly given.

B. Alternating Renewal Processes

Suppose that the replacements of the elements in the system take some finite positive time. In other words the working times interchange with independent identically distributed waiting times (or times for installation). This situation can be modeled by the so called alternating renewal processes, which seem more realistic in certain cases (see, e.g., Feller (1971), Kovalenko et al. (1983) and Wolff(1989)). The alternating renewal processes can be defined by a sequence of i.i.d. non-negative random vectors $\{(X_i, T_i)\}_{i=1}^{\infty}$ with independent coordinates, where X_i is interpreted as a *period of installation or repairing (waiting time)* of the ith element and T_i is the *worktime* which follows. Denote $A(x) = \Pr\{X_n \leq x\}$, $\mu_A = EX_n \in (0, \infty]$, $F(x) = \Pr\{T_n \leq x\}$, and $\mu_F = ET_n \in (0, \infty]$.

It is clear now that two types of renewal epochs appear:

$$S_n = \sum_{i=1}^{n} (X_i + T_i), \ n \geq 0,$$

the end of a worktime which coincides with the beginning of the next waiting period, and

$$S'_{n+1} = S_n + X_{n+1}, \ n \geq 0,$$

the end of a waiting period which coincides with the beginning of the next worktime.

The following processes correspond in some sense to the spent lifetime $Y^-(t)$ and the residual lifetime $Y^+(t)$

$$Y_a^-(t) = t - S'_{N(t)+1} = t - S_{N(t)} - X_{N(t)+1}, \ t \geq 0,$$

and

$$Y_a^+(t) = \max \left\{ S_{N(t)+1} - t, T_{N(t)+1} \right\}, \ t \geq 0,$$

where $N(t) = \max\{n : S_n \leq t\}$.

Evidently, if $X_i \equiv 0$ a.s., $i \geq 1$, then $Y_a^-(t) \equiv Y^-(t)$ (which is always non-negative), and $Y_a^+(t) \equiv Y^+(t)$.

If X_i are not identically equal to 0 then the process $Y_a^-(t)$ associated with an alternating renewal process can take both negative and positive values:

$$Y_a^-(t) = Y_{a+}^-(t) - Y_{a-}^-(t),$$

where $Y_{a+}^-(t) = \max\{Y_a^-(t), 0\}$ and $Y_{a-}^-(t) = \max\{-Y_a^-, 0\}$.

The limiting behavior of the processes $Y_a^-(t)$ and $Y_a^+(t)$ is similar to that of the processes $Y^-(t)$ and $Y^+(t)$ if both $\mu_A < \infty$ and $\mu_F < \infty$ (see e.g. Feller (1971), Chapter 11.8 and Sigman and Wolff (1993)). In the case when at least one of these means is infinite, different limiting beta distributions are obtained by Mitov (1999) in the discrete time case and by Mitov and Yanev (2001a) in the non-lattice case. Below we represent the limiting distributions for $Y_a^-(t)$ in the non-lattice case (see Mitov (1999) and Mitov and Yanev (2001a) for more details).

We assume that $A(0) = F(0) = 0$, $A(x)$ and $F(x)$ are non-lattice, and some of the following conditions:

$$m_A = EX_i < \infty; \tag{II.4}$$

$$EX_i = \infty, \ \bar{A}(t) = 1 - A(t) \sim t^{-\alpha}L_A(t), \ t \to \infty, \tag{II.5}$$

where $\alpha \in (\frac{1}{2}, 1]$, $L_A(\cdot)$ is a svf, and for each $h > 0$ fixed $A(t) - A(t - h) = O(1/t)$, $t > 0$;

$$m_F = ET_i < \infty; \tag{II.6}$$

$$ET_i = \infty, \ \bar{F}(t) = 1 - F(t) \sim t^{-\beta}L_F(t), \ t \to \infty, \tag{II.7}$$

where $\beta \in (\frac{1}{2}, 1]$ and $L_F(\cdot)$ is a svf;

$$\lim_{t \to \infty} \frac{\bar{A}(t)}{\bar{F}(t)} = c, \ 0 \leq c \leq \infty. \tag{II.8}$$

Since the assumption is that at least one of μ_A or μ_F is infinite the following hypotheses arise:

(H.1) (μ_F *"is shorter" than μ_A*) *Assume (II.5), $c = \infty$ and either (II.6) or (II.7) is fulfilled.*

or

(H.2) (μ_F *"is longer" than μ_A*) *Assume (II.7), $0 \leq c < \infty$ and either (II.4) or (II.5) is fulfilled.*

As stated by the following theorems, the limiting distributions for $Y_a^-(t)$ are different under the two hypotheses.

Theorem 3. *Assume the hypothesis (H.1). i) If only (II.6) holds then for $x \geq 0$*

$$\lim_{t \to \infty} \Pr \left\{ Y_a^-(t) \leq x \middle| Y_a^-(t) \geq 0 \right\} = \frac{m_F(x)}{m_F},$$

where $m_F(x) = \int_0^x (1 - F(u))du$.
ii) If only (II.7) with $1/2 < \beta < 1$ holds then for $0 < x < 1$

$$\lim_{t \to \infty} \Pr \left\{ \frac{Y_a^-(t)}{t} \leq x \middle| Y_a^-(t) \geq 0 \right\} = B(x; 1 - \beta, \alpha).$$

iii) If only (II.7) with $\beta = 1$ holds then for $0 < x < 1$

$$\lim_{t \to \infty} \Pr \left\{ \frac{m_F\left(Y_a^-(t)\right)}{m_F(t)} \leq x \middle| Y_a^-(t) \geq 0 \right\} = x.$$

Theorem 4. *Assume the hypothesis (H.2). Then*

$$\lim_{t \to \infty} \Pr \left\{ Y_a^-(t) \geq 0 \right\} = \frac{1}{1 + c}.$$

i) If (II.7) with $1/2 < \beta < 1$ holds then for $0 < x < 1$

$$\lim_{t \to \infty} \Pr \left\{ \frac{Y_a^-(t)}{t} \leq x \right\} = \frac{c}{1 + c} + \frac{1}{1 + c} B(x; 1 - \beta, \beta).$$

and

$$\lim_{t \to \infty} \Pr \left\{ \frac{Y_a^-(t)}{t} \leq x \middle| Y_a^-(t) \geq 0 \right\} = B(x; 1 - \beta, \beta).$$

ii) If (II.7) with $\beta = 1$ holds then for $0 < x < 1$

$$\lim_{t \to \infty} \Pr \left\{ \frac{m_F\left(Y_a^-(t)\right)}{m_F(t)} \leq x \right\} = \frac{c}{1 + c} + \frac{x}{1 + c}$$

and

$$\lim_{t \to \infty} \Pr \left\{ \frac{m_F\left(Y_a^-(t)\right)}{m_F(t)} \leq x \middle| Y_a^-(t) \geq 0 \right\} = x.$$

Remark 3. *It is clear that under the hypothesis (H.2) the limiting distributions conditionally on the event $\{Y_a^-(t) \geq 0\}$ coincide with those for $Y^-(t)$ in an ordinary renewal process, and non-conditional* *ones can have an atom at 0, although under the hypothesis (H.1) only the conditional limiting distributions are possible and in the case when both cdf's $A(x)$ and $F(x)$ have regularly varying tails the beta distribution is obtained with two independent parameters $1 - \beta$ and α.*

Let us associate with each T_n a measurable stochastic process $\{z_n(t) : 0 \leq t \leq T_n\}$, called cycle, $n = 1, 2, \ldots$, such that

$$z_n(0) \geq 0, \quad z_n(t) > 0 \text{ for } 0 < t < T_n, \quad z_n(T_n) = 0.$$

The cycles are mutually independent and stochastically equivalent. Also, for each n, $z_n(t)$ may depend on T_n but is independent of $\{T_i : i \neq n\}$.

An alternating regenerative process $\{Z(t)\}$ (Mitov and Yanev (2001b), see also Wolff (1989), Sections 2-11) is defined as follows

$$Z(t) = \begin{cases} z_{N(t)+1}\left(Y_a^-(t)\right) & \text{when } Y_a^-(t) \geq 0, \\ 0 & \text{when } Y_a^-(t) < 0. \end{cases}$$

Assume (II.4)–(II.8) and

$$\lim_{t \to \infty} \Pr\left\{ \frac{z_n(t)}{R(t)} \leq x \,\middle|\, T_n > t \right\} = D(x), \tag{II.9}$$

where $R(t) = L(t)t^\gamma$, $\gamma \geq 0$ for some svf $L(t)$, and $D(x)$ is a proper cdf on $(0, \infty)$, then the limiting behavior of the process $\{Z(t)\}$ is given by the following theorem.

Theorem 5. *(Mitov and Yanev, 2001b) Assume (II.4)–(II.8) and (II.9).*

1. Suppose that (II.7) holds with $\frac{1}{2} < \beta < 1$. Let $x \geq 0$.

a. If $0 \leq c < \infty$, then

$$\lim_{t \to \infty} \Pr\left\{ \frac{Z(t)}{R(t)} \leq x \right\}$$

$$= \frac{c}{c+1} + \frac{1}{(1+c)B(1-\beta, \beta)} \int_0^1 D\left(xu^{-\gamma}\right) u^{-\beta}(1-u)^{\beta-1}du.$$

b. If $c = \infty$, then

$$\lim_{t \to \infty} \Pr \left\{ \frac{Z(t)}{R(t)} \le x \,\middle|\, Z(t) > 0 \right\}$$

$$= \frac{1}{B(1 - \beta, \alpha)} \int_0^1 D\left(x u^{-\gamma}\right) u^{-\beta}(1 - u)^{\alpha-1} du.$$

2. Suppose that (II.7) holds with $\beta = 1$. Assume (II.9) with $D(0) = 0$ and let $0 < x < 1$.

a. If $0 \le c < \infty$ then

$$\lim_{t \to \infty} \Pr \left\{ \frac{m_F\left(R^{-1}(Z(t))\right)}{m_F(t)} \le x \right\} = \frac{c + x}{c + 1},$$

where $R^{-1}(\cdot)$ is the inverse function of $R(\cdot)$.

b. If $c = \infty$ then

$$\lim_{t \to \infty} \Pr \left\{ \frac{m_F\left(R^{-1}(Z(t))\right)}{m_F(t)} \le x \,\middle|\, Z(t) > 0 \right\} = x.$$

Remark 4. The limiting distribution of the process $Z(t)$ is of two types: uniform on the interval $[0, 1]$ or the distribution of the product $Z.Z^\gamma_{a,b}$ of two independent random variables Z with cdfs $D(x)$ and $Z^\gamma_{a,b}$, where a and b take different values depending on the tail parameters α and β.

Pham-Gia and Turkkan (1999) considered alternating renewal processes assuming that $A(x)$ and $F(x)$ are two-parameter Gamma distribution functions, i.e. they have pdfs

$$a(x) = x^{\alpha_X - 1} \exp\left(-x/\beta_X\right) / \left[\beta_X^{\alpha_X} \Gamma(\alpha_X)\right], \quad x > 0$$

and

$$f(x) = x^{\alpha_T - 1} \exp\left(-x/\beta_T\right) / \left[\beta_T^{\alpha_T} \Gamma(\alpha_T)\right], \quad x > 0.$$

They interpreted the times X_i as the off-times for a system and T_i as on-times and solved the question for the proportion of time up to the moment t when the system is on. The authors showed that the proportion of on-time in any one cycle, $T/(X + T)$, has a generalized beta distribution with pdf

$$b(x; \alpha_T, \alpha_X; \beta) = \frac{\beta^{\alpha_T} x^{\alpha_T - 1}(1 - x)^{\alpha_X - 1}}{B(\alpha_T, \alpha_X)[1 - (1 - \beta)x]^{\alpha_T + \alpha_X}}, \quad 0 \le x \le 1,$$

where $\beta = \beta_T / \beta_X$.

To state the next results, the following definition is needed.

Definition 1. *(Bertoin, 1999a) A closed unbounded random set $\mathcal{R}$ $\subseteq [0, \infty)$ is called a regenerative set if it fulfills the regenerative property. That is, if $(\mathcal{F}_t)_{t \ge 0}$ denotes the filtration induced by the characteristic function $1_{\mathcal{R}}$, then for every $\mathcal{F}_t$-stopping time T such that $T \in \mathcal{R}$ a.s., the right-hand portion of $\mathcal{R}$ as viewed from T,*

$$\mathcal{R} \circ \theta_T = \{s \ge 0 : s + T \in \mathcal{R}\}$$

is independent of $\mathcal{F}_T$ and has the same distribution as $\mathcal{R}$.

The set of renewal epochs $\mathcal{R} = \{S_0, S_1, \ldots, S_n, \ldots\}$ of an ordinary renewal process is a regenerative set. This regenerative set is also called the range of the renewal process. In this setting the process $Y^-(t)$ can be defined as follows

$$Y^-(t) = \inf\{s \ge 0 : t - s \in \mathcal{R}\}. \tag{II.10}$$

Bertoin (1999a) considered a finite family of $n \ge 2$ embedded regenerative sets

$$\mathcal{R}^{(1)} \subseteq \mathcal{R}^{(2)} \subseteq \cdots \subseteq \mathcal{R}^{(n)}.$$

For every $i = 1, 2, \ldots, n$ he defines a filtration $(\mathcal{F}_t^{(i)})_{t \ge 0}$ generated by the characteristic functions $1_{\mathcal{R}^{(i)}}, \ldots, 1_{\mathcal{R}^{(n)}}$, after the usual completions.

Definition 2. *(Bertoin, 1999a) The embedding is compatible with the regenerative property if for every $i = 1, \ldots, n$ and every $(\mathcal{F}_t^{(i)})$– stopping time T with $T \in \mathcal{R}^{(i)}$ a.s., the shifted sets*

$$\mathcal{R}^{(j)} \circ \theta_T = \{s \ge 0 : s + T \in \mathcal{R}^{(j)}\}, \quad j = i, \ldots, n$$

are jointly independent of $\mathcal{F}_T^{(i)}$ and have jointly the same law as $\mathcal{R}^{(i)}, \ldots, \mathcal{R}^{(n)}$.

Consider the processes $Y_i^-(t)$, $i = 1, 2, \ldots, n$ defined as in (II.10) for the regenerative sets $\mathcal{R}^{(i)}$, $i = 1, 2, \ldots, n$. The following theorem describes their limiting behavior in the case when their mathematical expectations are infinite.

Theorem 6. *(Bertoin, 1999a) Suppose that*

$$\lim_{t \to \infty} \frac{E\left(Y_i^-(t)\right)}{t} = \alpha_i \in [0, 1],$$

and $0 < \alpha_n < \alpha_{n-1} < \cdots < \alpha_1 < 1$. *Then the vector*

$$\left(\frac{Y_1^-(t) - Y_2^-(t)}{t}, \ldots, \frac{Y_{n-1}^-(t) - Y_n^-(t)}{t}, \frac{Y_n^-(t)}{t} \right)$$

converges in distribution as $t \to \infty$ *towards an n-dimensional Dirichlet distribution with parameters*

$$(\alpha_1 - \alpha_2, \alpha_2 - \alpha_3, \ldots, \alpha_{n-1} - \alpha_n, \alpha_n, 1 - \alpha_1).$$

Remark 5. *Recall that the n dimensional Dirichlet distribution with parameters* $(\beta_1, \beta_2, \ldots, \beta_{n+1}) \in (0, 1)^{n+1}$ *has the pdf*

$$
f(x_1, x_2, \ldots, x_n)
$$
$$
= \frac{\Gamma\left(\beta_1 + \cdots + \beta_{n+1}\right)}{\Gamma\left(\beta_1\right) \cdots \Gamma\left(\beta_{n+1}\right)} (1 - x_1 - x_2 - \cdots - x_n)^{\beta_{n+1}-1} \sum_{i=1}^{n} x_i^{\beta_i - 1}
$$

for $x_i > 0$, $i = 1, 2, \ldots, n$ *and* $x_1 + x_2 + \cdots + x_n < 1$. *It is a multivariate analogue of the one-dimensional beta distribution.*

C. Subordinators

The continuous time analogs of renewal processes are the so-called *subordinators*. A subordinator is a stochastic process $(S = (S(t), t \geq 0)$ on some probability space $(\Omega, \mathcal{A}, \Pr)$ such that

(i) $S(0) = 0$,

(ii) the trajectories are a.s. monotone increasing, right continuous,

(iii) the process has stationary, independent increments.

(For the theory and applications see e.g. Bertoin (1999b).) The distribution of S is characterized by its *Laplace exponent* $\Phi : [0, \infty) \to$

$[0, \infty)$ as follows

$$E \exp\{-\lambda S(t)\} = \exp\{-t\Phi(\lambda)\}, \ t, \lambda \geq 0.$$

The Laplace exponent is given by the Lévy-Khinchine formula

$$\Phi(\lambda) = k + d\lambda + \int_0^\infty \{1 - \exp(-\lambda x)\} \Pi(dx).$$

Here, $k \geq 0$ is the *killing rate*, $d \geq 0$ is the *drift coefficient* and Π is the *Lévy measure* of S. It is a measure on $(0, \infty)$ with

$$\int_0^\infty (1 \bigwedge x)\Pi(dx) < \infty.$$

The closed range $\mathcal{R} = \{S(t), t \geq 0\}^{\text{cl}}$ of a subordinator is a regenerative set in the sense of Definition 1.

The Lévy measure can be thought of as the analog of the distribution $F(x)$ of the times between the renewal epochs in an ordinary renewal process. The renewal measure U of S is given by

$$U(dx) = \int_0^\infty \Pr\{S(t) \in dx\} \, dt,$$

and its Laplace transform is

$$\int_0^\infty \exp(\lambda x) \, U(dx) = 1/\Phi(\lambda).$$

If the Laplace exponent is

$$\Phi(\lambda) = \lambda^\beta = \frac{\beta}{\Gamma(1-\beta)} \int_0^\infty \{1 - \exp(-\lambda x)\} \, x^{-(1+\beta)} dx$$

for a fixed $\beta \in (0, 1)$ the subordinator is called *standard stable subordinator with index* β. It is clear that in this case the Lévy measure is $\Pi(dx) = x^{-(1+\beta)}dx$.

Clearly, this case corresponds to an ordinary renewal process determined by a cdf $F(x)$ such that $1 - F(x) \sim x^{-\beta}, x \to \infty, \beta \in (0, 1)$ and one might expect a result corresponding to Dynkin-Lamperti theorem stated above.

Horowitz (1971) proved a generalization of the Dynkin-Lamperti theorem for continuous time processes.

Suppose that $S = (S(t), t \geq 0)$ is a subordinator with Laplace exponent

$$\Phi(\lambda) = d\lambda + \int_0^\infty \{1 - \exp(-\lambda y)\} \Pi(dy)$$

and let $\mathcal{R}$ be its closed range. Define the random process

$$Y_S^-(t) = t - \sup\{s \leq t : s \in \mathcal{R}\}, \quad t \geq 0.$$

This is analogous to the spent time $Y^-(t)$ in an ordinary renewal process.

Theorem 7. *(Theorem 0.3, Horowitz, 1971) Let*

$$h(x) = \Pi(x, \infty] = x^{-\beta} L(x),$$

where $\beta \in (0, 1)$ and $L(x)$ is a svf at infinity, then

$$\lim_{t \to \infty} \Pr\left\{\frac{Y_S^-(t)}{t} \leq x\right\} = B(x; 1 - \beta, \beta), \quad x \in (0, 1).$$

The converse of this theorem is also proved.

Theorem 8. *(Theorem 3.1, Horowitz, 1971) Let $S = (S(t), t \geq 0)$ be a subordinator with d, Π, h and $Y_S^-(t)$ defined as above. Suppose that $Y_S^-(t)/t$ has a limiting distribution $G(t)$. Then G must be the beta distribution $B(x; 1 - \beta, \beta)$, $0 < x < 1$ for a suitable β and $h(x)$ must satisfy the condition $h(x) \sim t^{-\beta} L(x)$, $x \to \infty$; or $G(x) = H(x)$; or $G(x) = H(x - 1)$, where $H(x - a)$ is a cdf with mass 1 at the point a.*

Bertoin (1999b) gave a similar result. We state it here for comparison with the above theorems.

For a subordinator $S = (S(t), t \geq 0)$ with closed range $\mathcal{R}$ the following processes are defined:

the *last passage time*

$$g(t) = \sup\{s < t : s \in \mathcal{R}\}, \quad t \geq 0,$$

the *first passage time*

$$D(t) = \inf\{s > t : s \in \mathcal{R}\}, \quad t \geq 0.$$

These processes correspond to the renewal epochs $S_{N(t)}$ and $S_{N(t)+1}$ of an ordinary renewal process and they are the closest points as viewed from a fixed point $t \geq 0$. (For more details and different interpretations of these processes see Bertoin (1999b).) The following theorems are valid.

Theorem 9. *(Proposition 3.1, Bertoin, 1999b) Suppose that the Laplace exponent of S is $\Phi(\lambda) = \lambda^\beta$ (i.e. S is a standard stable subordinator). Then $g(1)$ has pdf $f(x; \beta, 1 - \beta)$, $(0 < x < 1)$.*

Theorem 10. *(Theorem 3.2, Bertoin, 1999b) The following assertions are equivalent:*

(i) $g(t)/t$ converges in law as $t \to \infty$,

(ii) $\lim\limits_{t \to \infty} E(g(t))/t = \beta \in [0, 1]$,

(iii) $\lim\limits_{\lambda \to 0+} \dfrac{\lambda \Phi'(\lambda)}{\Phi(\lambda)} = \beta \in [0, 1]$,

(iv) Φ is a regularly varying at $0+$ with index $\beta \in [0, 1]$.

Moreover, when these assertions hold, then the limit distribution of $g(t)/t$ is $H(x)$ for $\beta = 0$; $H(x - 1)$ for $\beta = 1$; and $B(x; \beta, 1 - \beta)$ for $\beta \in (0, 1)$.

D.   Functional Limits

In this last subsection we want to state an earlier result of Lamperti (1962) which gives another limiting behavior of the spent lifetime process $Y^-(t)$ defined for a renewal process generated by the cdf $F(x)$ satisfying (II.1). He considered the time and space scaled process

$$\frac{Y^-(ct)}{c}, t \geq 0, c > 0.$$

The following limit is obtained by Dynkin (1955):

$$\lim_{c \to \infty} \Pr\left\{ \frac{Y^-(ct)}{c} \leq x \right\} = B\left(\min(1, x/t); 1 - \beta, \beta\right). \qquad \text{(II.11)}$$

Lamperti (1962) continued the investigation of the transition probabilities of this process and proved the following theorem.

Theorem 11. *If cdf $F(x)$ satisfies (II.1) and if $y > 0$, then*

$$\lim_{c \to \infty} \Pr \left\{ \frac{Y^-(ct)}{c} \leq x \,\middle|\, \frac{Y^-(0)}{c} = y \right\}$$

$$= \left(\frac{y}{t+y} \right)^{\beta} H(x - y - t)$$

$$+ \frac{\beta t}{y} \int_0^1 B\left(\min\left(1, \frac{x}{t(1-u)} \right); 1 - \beta, \beta \right) \left(\frac{y}{y + ut} \right)^{1+\beta} du$$

$$= p_t^{(\beta)}(y, x),$$

where as above $H(x) = 0$ if $x < 0$ and $H(x) = 1$ if $x \geq 0$. If $y = 0$, $p_t^{(\beta)}(0, x)$ is given by the right hand side of (II.11).

Furthermore he showed that the finite dimensional distributions also converge to ascertainable limits and that – if $Y^{(\beta)}(t)$, $t \geq 0$, $\beta \in (0, 1)$ denotes the Markov process with the transition probabilities obtained above – then the process $Y^{(\beta)}(t)$ can be identified with secondary processes derived from well-known objects.

III.   Random Walks

A.   Definitions

The random walks are stochastic processes defined as a sum of i.i.d. random variables not concentrated on a half line. The case when the random variables are positive or negative is covered by the renewal theory considered in the previous section. There is a large amount of literature concerning random walks. However, the goal of this section is to represent those topics where the beta distributions appear, namely the fluctuations of sums of random variables.

Let on the probability space $(\Omega, \mathcal{A}, \Pr)$ be given a collection

$$X = \{X_n, \ n = 0, 1, 2, \dots\}$$

of i.i.d. random variables with cdf $F(x)$, which is not concentrated on one of the half-lines. The sequence

$$S_0 = 0, \quad S_n = S_{n-1} + X_n, \quad n = 1, 2, \dots,$$

where the index n is interpreted as a time parameter, is called *a one-dimensional random walk*, generated by the distribution $F(x)$, starting at 0.

Another definition in terms of Markov chains is given in the famous Spitzer's book (1964), where only random walks on integers are considered.

The usual interpretation of the sequence $\{S_n\}$ can be given in terms of a particle movement. Let us suppose that a particle moves on the real line starting at the point $S_0 = 0$ at the moment $n = 0$. At the instant $n = 1$ it jumps to the point $S_1 = S_0 + X_1$, at the instant $n = 2$ it jumps to the point $S_2 = S_1 + X_2$ and so on. In this way S_n is the position of the particle on the real line at the moment n.

More generally, one can consider a sum of d-dimensional random vectors with real coordinates. In this case a d-dimensional random walk can be defined.

Define

$$M_n = \max \{0, S_1, \ldots, S_n\}, \quad m_n = \min \{0, S_1, \ldots, S_n\}.$$

One calls n a *strict ascending ladder epoch* if $S_n > M_{n-1}$, a *weak ascending ladder epoch* if $S_n \geq M_{n-1}$, and similarly for *descending ladder epochs*.

Further, define

$$N_n = \sum_{k=0}^{n} 1_{\{S_n > 0\}},$$

i.e. N_n equals the number of $S_0, S_1, \ldots, S_n$ which are positive. Therefore, N_n/n is the fraction of time spent by the random walk in the positive half line $(0, \infty)$.

Define

$$L_n = \min \{k : k = 0, 1, 2, \ldots, n : S_k = M_n\},$$

i.e. L_n equals the index at which S_i, $i = 0, 1, 2, \ldots, n$ attains for the first time the value M_n. So, L_n is the first occurrence up to the time n of a strictly ascending ladder-point. Similarly, define

$$l_n = \max \{k : k = 0, 1, 2, \ldots, n : S_k = m_n\},$$

i.e. l_n equals the index at which S_n, $i = 0, 1, 2, \ldots, n$ attains for the last time the value m_n. So, l_n is the last occurrence up to time n of a weakly descending ladder-point.

B. Main Results

Beta distributions appear in limit theorems for the processes N_n and L_n, as n increases to infinity.

Consider first, the most simple case. Suppose that the distribution of the random variables X_n is

$$\Pr\{X_n = 1\} = p, \quad \Pr\{X_n = -1\} = q = 1 - p, \quad 0 < p < 1.$$

The random walk of this kind is called Bernoulli random walk. If $p = q = 1/2$ we call this random walk the *symmetric simple random walk*. The usual interpretation is in terms of coin tossing (see Feller (1968), Chapters 3 and 14).

Theorem 12. *(Theorem 2, Section 3.6, Theorem, Section 3.8, Feller, 1968). Suppose that the random variables X_n have the distribution $\Pr\{X = 1\} = 1/2$, $\Pr\{X = -1\} = 1/2$. Then for $0 \leq x \leq 1$,*

$$\lim_{n \to \infty} \Pr\left\{\frac{N_n}{n} \leq x\right\} = \frac{2}{\pi} \arcsin \sqrt{x}, \tag{III.1}$$

$$\lim_{n \to \infty} \Pr\left\{\frac{L_n}{n} \leq x\right\} = \frac{2}{\pi} \arcsin \sqrt{x}.$$

Erdös and Kac (1947) proved that (III.1) holds for i.i.d. random variables having mean 0 and variance 1 and satisfying the central limit theorem. Masatomo(1952) proved that the same is true for i.i.d. random variables such that S_n/A_n approaches the normal distribution for some positive constants A_n approaching infinity.

One of the basic tools for the investigation of the fluctuation of sums of random variables is the *equivalence principle* proved by Sparre-Andersen (1953a).

Theorem 13. *(Theorem 8.9.5, Bingham et al., 1987) For each n (separately), the laws of (N_n, S_n), (L_n, S_n) and $(n - l_n, S_n)$ coincide.*

The following equivalent form of the equivalence principle is given in Lemma 1, Section 12.8, Feller (1971).

$$\Pr\{N_n = k\} = \Pr\{L_n = k\} = \Pr\{N_k = k\}\Pr\{N_{n-k} = 0\}$$
$$= \Pr\{S_1 > 0, \ldots, S_k > 0\}\Pr\{S_1 \leq 0, \ldots, S_{n-k} \leq 0\}.$$

The other tool is the regenerative phenomenon which is given by each of the four ladder epochs mentioned above (see e.g. Feller (1971), Chapter 12).

Using the equivalence principle, Sparre-Andersen (1953b, 1954) proved the arc sine law for i.i.d. symmetrically distributed r.v.

Theorem 14. *(Theorem 1, Section 12.8, Feller, 1971). Suppose that the random variables X_n have continuous and symmetric distribution, i.e. $F(-x) = 1 - F(x)$. Then*

$$\lim_{n\to\infty}\Pr\left\{\frac{L_n}{n} \leq x\right\} = \lim_{n\to\infty}\Pr\left\{\frac{N_n}{n} \leq x\right\} = \frac{2}{\pi}\arcsin\sqrt{x}. \text{ (III.2)}$$

Theorem 15. *(Theorem 1^a, Section 12.8, Feller, 1971). Suppose that the cdf $F(x)$ is such that the series*

$$\sum_{n=1}^{\infty}\frac{1}{n}\left[\Pr\{S_n > 0\} - \frac{1}{2}\right] = c \qquad\qquad \text{(III.3)}$$

is convergent, then (III.2) holds.

The condition (III.3) is true if the random variables X_n have zero mean and finite variance (see e.g. Feller (1971), Section 18.5). So the arc sine law is applicable to random walks generated by distributions which satisfy these conditions.

In all these cases the limit distribution is the Lévy's arc sine law. It is noted in Feller (1971) that if the cdf $F(x)$ is such that

$$\Pr\{S_n > 0\} = \delta \in (0, 1)$$

and if it does not depend on n then the limiting distribution in Theorems 14 and 15 has to be replaced by the generalized arc sine distribution with pdf $f(x; \delta, 1 - \delta)$.

Sparre Anderson (Theorem 3, 1954) has proved that if

$$\lim_{n\to\infty}\Pr\{S_n > 0\} = \delta \in (0, 1)$$

then

$$\lim_{n\to\infty} \Pr\left\{\frac{N_n}{n} \leq x\right\} = B(x; \delta, 1-\delta).$$

The following assertion (called Spitzer's arc sine law) is proved by Spitzer(1956) (see also Bingham et al. (1987), Theorem 8.9.9). It generalizes the above result of Sparre Andersen.

Theorem 16. *(Theorem 7.1, Spitzer, 1956) The limit*

$$\lim_{n\to\infty} \Pr\left\{\frac{N_n}{n} \leq x\right\} = B(x; \delta, 1-\delta), \quad 0 < x < 1$$

exists iff

$$\lim_{n\to\infty} \frac{1}{n} \sum_{k=0}^{n} \Pr\{S_n > 0\} = \delta \in [0,1]. \tag{III.4}$$

Hoffmann-Jorgensen (1999) gives a simple proof of the arc sine law for a general class of integer valued sequences of random variables T_n with the distribution given by

$$p_0 = 1, q_0 = 1, \Pr\{T_n = k\} = p_k q_{n-k},$$

where $0 \leq T_n \leq n$ for all $n \geq k \geq 0$. He called these sequences arc sine sequences. It is easy to check that the sequence L_n and N_n are of this kind and the above theorems are consequences of the results proved in Hoffmann-Jorgensen (1999).

C. Symmetrically Dependent Increments

Definition 3. *The sequence of random variables*

$$X_1, X_2, \ldots, X_n, \ldots$$

is called symmetrically dependent if for every n the joint cdf of

$$X_{i_1}, X_{i_2}, \ldots, X_{i_n}$$

is the same for the all $n!$ permutations of the random variables.

Definition 4. *The sequence $X_1, X_2, \ldots, X_n, \ldots$ is S-invariant if every finite dimensional cdf of the sequence is invariant under any* *changes in the signs of* $\{X_n\}$.

Berman(1962) considered a sum of random variables which are symmetrically dependent and S-invariant and proved the following result.

Theorem 17. *(Berman, 1962) Suppose that the sequence*

$$X_1, X_2, \ldots, X_n, \ldots$$

is symmetrically dependent and S-invariant. Then there exists the limit

$$\lim_{n \to \infty} \Pr\{S_n = 0\} = \delta \in [0, 1]$$

and

$$\lim_{n \to \infty} \Pr\left\{\frac{K_n}{n} \leq x\right\} = \begin{cases} 0, & \text{if } x < 0, \\ \delta, & \text{if } x = 0, \\ \delta + (1 - \delta)\dfrac{2}{\pi} \arcsin \sqrt{x}, & \text{if } 0 < x < 1, \\ 1, & \text{if } x \geq 1, \end{cases}$$

where K_n denotes any of the random variables L_n, N_n or $n - l_n$.

D. Left-Continuous Random Walk

Consider a random walk on integers, and suppose that the law F of the step length X_n is concentrated on $\{-1, 0, 1, 2, \ldots\}$, with positive mass on $\{-1\}$. The random walk of this type is called *left-continuous random walk* (see Bingham et al. (1987), Chapter 8.9).

The following result is proved by Bingham and Hawkes (1983); see also Theorem 8.11.7 in Bingham et al. (1987).

Let $S_n, n = 0, 1, 2, \ldots$ be a zero-mean left continuous random walk, such that

$$1 - F(x) \sim x^{-\beta} L(x), \quad x \to \infty,$$

where $1 < \beta \leq 2$ and $\delta = 1 - 1/\beta$. Then Spitzer's condition (III.4) holds or equivalently

$$\lim_{n \to \infty} \Pr\left\{\frac{N_n}{n} \leq x\right\} = B(x; \delta, 1 - \delta).$$

E.   Continuous Time Random Walks

Further generalization of random walks are the so called continuous time random walks (CTRW).

Definition 5. *Suppose that* $\{(T_n, X_n), n = 1, 2, \ldots\}$ *is a sequence of i.i.d. random vectors with state space* $[0, \infty) \times \mathbf{R}^d$ *for* $d \geq 1$. *The random variables* T_n *and* X_n *can be dependent or independent. Now a CTRW process is defined as a sum of random variables* X_n

$$R(0) = 0, \quad R(t) = \sum_{n=1}^{N(t)} X_n$$

governed by the ordinary renewal process

$$S_0 = 0, \quad S_n = \sum_{i=0}^{n} T_i, \quad N(t) = \max\{n : S_n \leq t\}, t \geq 0.$$

The d-dimensional random vector $R(t), t \geq 0$ describes the position of a particle which starts form the origin and moves in the d-dimensional Euclidian space. The jumps occur at the renewal epochs of the process S_n, $n = 0, 1, 2, \ldots$ and the magnitude of the jumps is given by the r.v. X_n. The limiting behavior of $R(t)$ is of interest for many physical processes and it has been investigated widely in the literature. Kotulski (1995) has given a survey of the results in this direction and has proved new and known results using probabilistic methods. He worked out the limiting distributions of $R(t)/t$ as $t \to \infty$ depending on different assumptions about the distribution of (T_n, X_n). In the case when $X_i = T_i$ (in other words the walk is on the real line and the jumps coincide with the waiting times) and $ET_i = \infty$, $\Pr\{T_i > t\} = Ct^{-\beta}$, $0 < \beta < 1$ the following limit is valid (Case 3.2, Kotulski (1995)):

$$\lim_{t \to \infty} \Pr\left\{\frac{R(t)}{t} \leq x\right\} = B(x; \beta, 1 - \beta), 0 < x < 1.$$

Clearly, in this case $R(t) = S_{N(t)}$ and the convergence follows from the Dynkin-Lamperti theorem (Theorem 1).

It is interesting to compare this result with the next one where the r.v.'s X_i are supposed to be independent of T_i but have the same distribution. Then for $x \geq 0$

$$\lim_{t \to \infty} \Pr \left\{ \frac{R(t)}{t} \leq x \right\} = \int_0^\infty G_\beta(xu) dG_\beta(u).$$

So, in this case the limiting distribution is concentrated on $[0, \infty)$ compared to $(0, 1)$ in the previous case. Here $G_\beta(x)$ is a non-negative stable distribution with parameter β.

In the recent paper of Becker-Kern et al. (2002) the functional limit theorems for CTRW $(R(t), t \geq 0)$ were proved. Assuming that $ET_n = \infty$ and certain dependence between T_n and X_n they proved that $B(c)R(ct)$ converges as $c \to \infty$ in M_1 topology (and under weaker conditions) for every t. Here, $B(\cdot)$ is a normalizing function (see Becker-Kern et al. (2002) for details). Let $M(t), t \geq 0$ be the limiting process and let $h(t, x)$ be its one-dimensional density.

Becker-Kern et al. (2002) worked out several examples involving beta distributions. In Example 5.2, T_n has a stable distribution $E \exp(-sT_n) = \exp(s^{-\beta})$, and the conditional distribution of $X_n | T_n = t$ is a normal with mean 0 and variance $2t$, i.e. the CRTW $(R(t), t \geq 0$ is one dimensional. Then

$$h(x, t) = \frac{\sin \pi \beta}{\pi} \int_0^t \frac{1}{\sqrt{4\pi u}} \exp \left(-\frac{x^2}{4u} \right) u^{\beta-1}(t-u)^{-\beta} du.$$

Therefore, $M(t) \stackrel{d}{=} (tZ_{\beta,1-\beta})^{1/2} Z$, where Z is a normally distributed r.v. with mean 0 and variance 2 independent of $Z_{\beta,1-\beta}$.

If T_n has a stable distribution with parameter β, i.e. $E \exp(-sT_n) = \exp(-s^\beta)$ and $X_n = T_n$, then the limiting process $M(t)$ has a one dimensional distribution with density

$$h(x, t) = \frac{\sin \pi \beta}{\pi} x^{\beta-1}(t-x)^{-\beta}.$$

Hence, $M(t) \stackrel{d}{=} tZ_{\beta,1-\beta}$.

In Example 5.6, a d-dimensional CTRW $R(t), t \geq 0$ is constructed with scaling limit $M(t) \stackrel{d}{=} Z(t.Z_{\beta,1-\beta}), t \geq 0$ where $Z(t)$ is an (operator) Lévy motion and $Z_{\beta,1-\beta}$ is a r.v. independent of $Z(t), t \geq 0$.

Hence,

$$h(x,t) = \frac{\sin \pi\beta}{\pi} \int_0^t p(x,u)u^{\beta-1}(t-u)^{-\beta}du,$$

where $p(x,t)$ is the density of $Z(t)$, $t \geq 0$.

F. Continuous Space Markov Chains

Stoyanov and Pirinsky (2000) investigated a class of discrete time Markov chains with continuous state space-the interval $(0,1)$. It is constructed in such way that different beta distributions are obtained as limiting distributions. The construction is as follows:

Suppose the particle D is located at some starting point x_0 in the interval $(0,1)$ and its motion, always along a line segment is determined by the following 2-stage rule:

Stage 1: The particle D chooses to which end of the interval to move from its starting position or from its current location.

Stage 2: The particle D stops at a point on the chosen in Stage 1 segment and either D changes the direction thus going to the opposite end of the interval $(0,1)$, or continues in the same direction.

If the sequence of points visited by the particle D at time n is

$$X_0, X_1, \ldots, X_n, \ldots,$$

it is clear that this sequence is a Markov chain with the state space - the interval $(0,1)$.

Theorem 18. *From the initial position x_0 the particle D moves alternatively in the direction of 0, of 1, of 0, of 1, etc (Stage 1 is deterministic). Every time D changes the direction at a random point chosen uniformly from the corresponding interval (Stage 2 is random). Then the random sequence X_n does not converge. However, it is split into two subsequences $X_{2n+1}, n = 0,1,2,\ldots$ and $X_{2n}, n = 0,1,2,\ldots$ each being convergent in distribution:*

$$\lim_{n\to\infty} \Pr\{X_{2n+1} \leq x\} = B(x;1,2), \quad x \in (0,1),$$

and

$$\lim_{n\to\infty} \Pr\{X_{2n} \leq x\} = B(x; 2, 1), \quad x \in (0, 1).$$

Suppose that D moves from its current position x_0 to 0 with probability p or to 1 with probability $q = 1 - p$ (Stage 1 is random). Then D stops at the point $x_0 + \lambda(x_e - x_0)$, where $\lambda \in (0, 1)$ is a fixed number and x_e denotes the chosen endpoint (Stage 2 is deterministic).

Theorem 19. *Suppose that $p = q = 1/2$ and $\lambda = 1/2$. Then the random sequence X_n is convergent and*

$$\lim_{n\to\infty} \Pr\{X_n \leq x\} = x, \quad x \in (0, 1).$$

Their final result describes the "most" random situation.

Theorem 20. *The particle D starts from point x_0. Stage 1: D moves in the direction of point 0 with probability p and of point 1 with probability $q = 1 - p$. Stage 2: D eventually changes the direction at a point chosen randomly and uniformly from the corresponding interval. Then the random sequence X_n is convergent and*

$$\lim_{n\to\infty} \Pr\{X_n \leq x\} = B(x; q, p), \quad x \in (0, 1).$$

IV. Brownian Motion

There is a large amount of literature concerns Brownian motion and processes related to them (see e.g. Karatzas and Shreve (1991) and Borodin and Salminen (1996)). It is well known that the arc sine law was first proved by Lévy (1939) (see also Lévy (1948), Chapter 6) for the standard Brownian motion.

Definition 6. *A standard (one-dimensional) Brownian motion $B = (B(t), t \geq 0)$ is a stochastic process which satisfies the following conditions:*

1. *$B(0) = 0$;*
2. *$B(t)$ is a continuous function of t a.s.;*
3. *B has independent, normally distributed increments, i.e. for any $0 = t_0 < t_1 < t_2 < \cdots < t_n$:*

   – *the random variables*

$$Y_1 = B(t_1) - B(t_0), \ldots, Y_n = B(t_n) - B(t_{n-1})$$

are independent, normally distributed;

– $EY_i = 0$, $i = 1, 2, \ldots, n$;

– $VarY_i = t_i - t_{i-1}$, $i = 1, 2, \ldots, n$.

Definition 7. *The Brownian bridge* $(B^{a,b}(t), t \in [0, T])$ *from a to b on* $[0, T]$ *is*

$$B^{a,b}(t) \overset{def}{=} a\left(1 - \frac{t}{T}\right) + b\frac{t}{T} + \left(B(t) - \frac{t}{T}B(T)\right), \quad 0 \le t \le T,$$

where $(B(t), t \ge 0)$ *is a standard Brownian motion.*

Definition 8. *The process* $(L(t), t \ge 0)$ *defined by*

$$L(t) = \lim_{\varepsilon \downarrow 0} \frac{1}{2\varepsilon} \int_0^t 1_{\{|B(s)| < \varepsilon\}} ds, \quad t \ge 0,$$

is called Lévy's local time at 0 of the Brownian motion B.

As usual 1_A denotes the indicator of the event A.

Definition 9. *For a positive number* μ *the process* $(|B(t)| - \mu L(t)$, $t \ge 0)$ *is called perturbed reflecting Brownian motion.*

Lévy's arc sine law states that the time spent in $[0, \infty)$ by $(B(t)$, $t \in [0, 1])$ has the distribution $B(x; 1/2, 1/2)$, $0 \le x \le 1$, i.e.

$$\int_0^1 1_{\{B(t) > 0\}} dt \overset{d}{=} Z_{1/2, 1/2}$$

Lévy (1939) has also proved the analog of the above assertion for the standard Brownian bridge $(B^{0,0}(t)$, $0 \le t \le 1)$, i.e. the time spent in $[0, \infty)$ by $(B^{0,0}(t)$, $0 \le t \le 1)$ has a uniform distribution:

$$\int_0^1 1_{\{B^{0,0}(t) \ge 0\}} dt \overset{d}{=} Z_{1,1}.$$

Several generalizations of these assertions have been established.

Petit (1992) proved the corresponding assertions for perturbed reflecting Brownian motion and perturbed reflecting Brownian bridge.

Namely, let $(B(t), \; 0 \le t \le 1)$ be a standard Brownian motion and $L(t), 0 \le t \le 1$ be its local time at 0. Then

$$\int_0^1 1_{\{|B(s)| - \mu L(s) \ge 0\}} ds \overset{d}{=} Z_{1/2\mu, 1/2}$$

and conditionally on the event $\{|B(1)| - \mu L(1) = 0\}$

$$\int_0^1 1_{\{|B(s)| - \mu L(s) \ge 0\}} ds \overset{d}{=} Z_{1/2\nu, 1/2},$$

where $1/\nu = 1 + 1/\mu$. The last identity is equivalent to

$$\int_0^g 1_{\{|B(s)| \le \mu L(s)\}} ds \overset{d}{=} Z_{1/2, 1/2\nu},$$

where $g = \sup\{s < 1 : B(s) = 0\}$ (see also Yor (1992), page 102).

Denote by $l(t)$, $t \in [0, 1]$ the local time process for the Brownian bridge $B^{0,0}(t)$, $t \in [0, 1]$. Petit(1992) also proved that

$$\int_0^1 1_{\{|B^{0,0}(t)| - \nu l(t) \ge 0\}} dt \overset{d}{=} Z_{1/2\nu, 1}.$$

The above results generalize Lévy's results for the standard Brownian motion since in the case $\mu = 1$ the process $(|B(t)| - L(t), \; t \ge 0)$ is a Brownian motion. In this case the distributions of the time spent in $[0, \infty)$ by the process $(|B(t)| - L(t), \; 0 \le t \le 1)$, both non-conditional and conditional on the event $\{|B(1)| - \mu L(1) = 0\}$ are the same as for the standard Brownian motion.

Carmona et al. (1998) obtained beta distributions with all possible values for the parameters $a > 0$ and $b > 0$ as the distributions of occupation measures of some processes which generalize the Brownian motion. They considered the process Y_t which is the solution of the equation

$$Y(t) = B(t) + \alpha M^Y(t) - \beta I^Y(t), \quad t \ge 0, \qquad \text{(IV.1)}$$

for $\alpha < 1$ and $\beta < 1$, where

$$M^Y(t) = \sup_{s \le t} Y(s) \quad \text{and} \quad I^Y(t) = \sup_{s \le t} (-Y(s)).$$

The conditions $\alpha < 1$ and $\beta < 1$ are necessary for the existence of the solution of the above equation. To prove the existence and uniqueness

Carmona et al. (1998) assumed that

$$|\alpha\beta| < (1 - \alpha)(1 - \beta). \tag{IV.2}$$

Under this assumption they proved that

$$\int_0^1 1_{\{Y(t)\geq 0\}} dt \overset{d}{=} Z_{(1-\beta)/2,(1-\alpha)/2} \tag{IV.3}$$

and conditionally on the event $\{Y(1) = 0\}$

$$\int_0^1 1_{\{Y(t)\geq 0\}} dt \overset{d}{=} Z_{(2-\beta)/2,(2-\alpha)/2}. \tag{IV.4}$$

Since the restriction (IV.2) the random variable $Z_{a,b}$ obtained in (IV.3) has the following domain of parameters a and b :

$$D_1 = \left\{ (a,b) : \left| \left(\frac{1}{2} - a \right) \left(\frac{1}{2} - b \right) \right| < ab \right\}$$

or equivalently

1. $a \geq 1/2$ and $b \geq 1/2$;
2. $a \leq 1/2$, $b \leq 1/2$ and $a + b > 1/2$;
3. $a > 1/2$ and $b \geq 1/4$;
4. $b > 1/2$ and $a \geq 1/4$;
5. $1/2 < a < (1 - 2b)/2(1 - 4b)$ and $b < 1/4$;
6. $1/2 < b < (1 - 2a)/2(1 - 4a)$ and $a < 1/4$.

The domain for (IV.4) is $D_2 = \{(a + 1/2, b + 1/2) : (a, b) \in D_1\}$.

Chaumont and Doney (1999) showed that the restriction (IV.2) can be removed (see also Perman and Werner (1997)).

These results generalize the corresponding results for Brownian motion in the following sense: Let $\{\widetilde{B}(t), t \geq 0\}$ be a Brownian motion started from 0, and let $\widetilde{L}(t), t \geq 0$ be its local time at 0. Then $B(t) = |\widetilde{B}(t)| - \widetilde{L}(t)$, $t \geq 0$ is a Brownian motion, and it is easily shown that

$$Y(t) = \left| \widetilde{B}(t) \right| - \mu\widetilde{L}(t), \ t \geq 0$$

is a solution of the equation (IV.1) with the parameters $\alpha = 0$ and $\beta = 1 - 1/\mu$.

Let $(L_Y(t), t \geq 0)$ be the local time at 0 for the process $(Y(t), t \geq 0)$. The inverse process of $L_Y(\cdot)$ is defined as follows

$$\tau(v) = \sup\{u > 0 : L_Y(u) > v\}, \quad v \geq 0.$$

Carmona et al. (1998) considered also the process

$$A^+(t) = \int_0^t 1_{\{Y(u) \geq 0\}} du,$$

which represent the time spent in $[0, \infty)$ by the process $Y(t)$, $t \geq 0$. They showed that

$$\frac{1}{t} A^+(t), \quad \frac{1}{\tau_v} A^+(\tau_v),$$

have the same distribution with cdf $B(x; (1 - \beta)/2, (1 - \alpha)/2)$, $x \in [0, 1]$.

V. Occupation Times

Suppose that one wishes to study the amount of time spent up to time t in a set A by a Markov process $X = (X(t), t \geq 0)$ with stationary transition probabilities $P(x, dy, t)$ and state space I. That is

$$N_t = \int_0^t 1_{\{X^x(u) \in A\}} du,$$

where the upper index x denotes the initial state $x \in I$ of the process. More generally one may take a non negative bounded function V defined on I and consider the "occupation time"

$$N_t^V = \int_0^t V(X^x(u))\, du.$$

Considering the problem for finding the limiting behavior of the occupation time of a set of states of a Markov chain. Lamperti (1958a) obtained a two-parameter distribution on the closed interval $[0, 1]$, which generalizes the beta distributions in certain sense. He considered the process X_n, $n = 0, 1, 2, \ldots$ (not necessarily Markovian), having the property that the states of the process are subdivided into two sets say A and B by a state σ which is *recurrent* (see Feller (1968), Chapter 13 for recurrent events). In other words if $X_{n-1} \in A$ and $X_{n+1} \in B$ or vice versa then $X_n = \sigma$.

Denote by $f(x) = \sum_{n=1}^{\infty} f_n x^n$ the probability generating function of the probabilities f_n that the recurrence time of state σ is n. Denote by N_n the occupation time up to time n of the set A, with the convention that the occupation of the state σ is counted or not according to whether the last other state occupied was in A. Lamperti (1958) proved the following theorem.

Theorem 21. *(Theorem 1, Lamperti, 1958a) Let $X_n, n = 0, 1, \ldots$ be the process described above. Then*

$$\lim_{n \to \infty} \Pr\{N_n/n \le t\} = G(t)$$

exists if and only if

$$\lim_{n \to \infty} E\{N_n/n\} = \alpha$$

exists and

$$\lim_{x \to 1-} \frac{(1-x)f'(x)}{1 - f(x)} = \delta, \quad 0 \le \delta \le 1$$

exists. If both these conditions hold, $G(t) = G_{\alpha,\delta}(t)$ is a distribution on $[0, 1]$ which provided α and $\delta \ne 0$ and 1, has the density

$$G'_{\alpha,\delta}(t) = \frac{A \sin \pi\delta}{\pi} \frac{t^{\delta}(1-t)^{1-\delta} + t^{\delta-1}(1-t)^{\delta}}{A^2 t^{2\delta} + 2At^{\delta}(1-t)^{\delta} \cos \pi\delta + (1-t)^{2\delta}},$$

where $A = (1 - \alpha)/\alpha$, while in other cases

$$G_{0,\delta}(t) = 1, \quad 0 < t < 1,$$

$$G_{1,\delta}(t) = 0, \quad 0 < t < 1,$$

$$G_{\alpha,1}(t) = \begin{cases} 0, t \le \alpha, \\ 1, t \ge \alpha, \end{cases}$$

$$G_{\alpha,0}(t) = 1 - \alpha, \quad 0 \le t < 1.$$

It is clear that if $\alpha = 1/2$ and $\delta = 1/2$ the arc sine distribution is obtained. Continuous time versions of this assertion have been proved by Watanabe (1995) and Khasminskii (1999, 2001).

Let $X^x(t), t \geq 0$ be a Markov process with the generator

$$L(x) \equiv a(x)\frac{d^2}{dx^2}, \quad -\infty < x < \infty, \qquad (V.1)$$

where $a(x)$ is strictly positive and Lipschitz continuous function on any compact set $K \subset (-\infty, \infty)$. Denote by $p(x) = a(x)^{-1}$. Watanabe (1995) proved the necessary and sufficient condition for the convergence of

$$\frac{N_t}{t} = \frac{1}{t}\int_0^t 1_{\{X^x(s)>0\}}ds$$

to a random variable with cdf $G_{\alpha,\delta}(\cdot)$ for $0 < \alpha, \delta < 1$. For nongeneralized diffusion process in canonical scale (the process with generator (V.1)) Watanabe's condition can be written as

$$\left|\int_0^{\pm x} p(y)dy\right| = |x|^{\beta+1}L_{\pm}(|x|),$$

with slowly varying functions $L_+(\cdot)$ and $L_-(\cdot)$ satisfying the condition

$$\lim_{x\to\infty}\frac{L_+(x)}{L_-(x)} = A^{1/\delta},$$

where $\delta(2+\beta) = 1$ and $A = (1-\alpha)/\alpha$.

Khasminskii (2001) found sufficient conditions guaranteeing the convergence of the integral functional

$$\frac{1}{t}\int_0^t V(X^x(u))\,du$$

to the distribution $G_{\alpha,\delta}(\cdot)$ given above.

Theorem 22. *(Theorem 2.1, Khasminskii, 2001) For some constants $p_+ > 0$, $p_- > 0$ and $\beta > -1$, assume there exist the limit*

$$\lim_{x\to\pm\infty}\frac{1}{x}\int_0^x |y|^{-\beta}p(y)dy = p_{\pm},$$

and, moreover, for any $\varepsilon > 0$, assume

$$\sup_{|x|>\varepsilon} p(x)|x|^{-\beta} < \infty.$$

Assume further that $V(x)$ is a piece-wise continuous bounded function and

$$\lim_{x \to \pm \infty} \frac{\int_0^x V(y)|y|^{-\beta}p(y)dy}{\int_0^x |y|^{-\beta}p(y)dy} = V_{\pm},$$

where the constants V_+ and V_- satisfy $V_+ - V_- \neq 0$. Denote $\delta = 1/(2+\beta)$ and $A = (p_+/p_-)^{\delta}$. Then for $0 < x < 1$

$$\lim_{t \to \infty} \Pr\left\{ \frac{\int_0^t (V(X^x(u)) - V_-)\,du}{t(V_+ - V_-)} \leq x \right\} = G_{\alpha,\delta}(x).$$

Acknowledgments

The first author (Kosto Mitov) was partially supported by NFSI-Bulgaria, Grant No. MM-1101/2001.

References

1. Becker-Kern, P.; Meerschaert, M.M.; Scheffler, H.P. Limit theorems for coupled continuous time random walks, Talk on the 17 conference on stable stochastic models, Varna, Bulgaria, June 2002 (Preprint).

2. Berman, S. An extension of the arc sine law, Annals of Mathematical Statistics **1962**, *33*, 681-684.

3. Bertoin, J. Renewal theory for embedded regenerative sets, Annals of Probability **1999a**, *27*, 1523-1535.

4. Bertoin, J. *Subordinators: Examples and Applications, Ecole d'ete de Probabilites de St-Flour XXVII*; Lecture notes in Mathematics, Springer Verlag: Berlin, 1999b.

5. Bingham, N.H.; Goldie, C.M.; Teugels, J.L. *Regular Variation*; Cambridge University Press: Cambridge, 1987.

6. Bingham, N.H.; Hawkes, J. Some limit theorems for occupation times, In: *Probability, Statistics & Analysis* (editors J.F.C. Kingman, G.E.H. Reuter); London Mathematical Society Lecture Notes **1983**, *79*, 46-62.

7. Borodin, A.N.; Salminen, P. *Handbook of Brownian motion - Facts and Formulae*; Birkhaüser: Basel, 1996.

8. Carmona, P.; Petit, F.; Yor, M. Beta variables as times spent in $[0, \infty)$ by certain perturbed Brownian motions, Journal of the London Mathematical Society 1998, *58*, 239-256.

9. Chaumont, L.; Doney, R. Pathwise uniqueness for perturbed versions of Brownian motion and reflected Brownian motion, Probability Theory and Related Fields 1999, *113*, 519-534.

10. Dynkin, E.B. Limit theorems for sums of independent random quantities, Izves. Akad. Nauk U.S.S.R. 1955, *19*, 247-266.

11. Erdös, P.; Kac, M. On the number of positive sums of independent random variables, Bulletin of the American Mathematical Society 1947, *53*, 1011-1020.

12. Erickson, K.B. Strong renewal theorems with infinite mean, Transactions of the American Mathematical Society 1970, *151*, 263-291.

13. Feller, W. *An Introduction to Probability Theory and Its Applications* (volume 1, third edition); Wiley: New York, 1968.

14. Feller, W. *An Introduction to Probability Theory and Its Applications* (volume 2, second edition); Wiley: New York, 1971.

15. Freedman, D.A. Bernard Friedman's urn, Annals of Mathematical Statistics 1965, *36*, 956-970.

16. Hoffmann-Jorgensen, J. The arc sine law, Journal of Theoretical Probability 1999, *12*, 131-145.

17. Horowitz, J. A note on the arc-sine law and Markov random sets, Annals of Mathematical Statistics 1971, *42*, 1068-1074.

18. Karatzas, I.; Shreve, S.E. *Brownian Motion and Stochastic Calculus* (second edition); Springer: New York, 1991.

19. Khasminskii, R. Arc sine law and one generalization, Acta Appl. Math. 1999, *58*, 151-157.

20. Khasminskii, R. Limit distributions of some integral functionals for null recurrent diffusions, Stochastic Processes and Their Applications 2001, *92*, 1-9.

21. Kotulski, M. Asymptotic distributions of continuous time random walks: A probabilistic approach, Journal of Statistical Physics *1995*, *81*, 777-792.

22. Kovalenko, I.N.; Kuznetzov, N.U.; Shurenkov, V.M. *Stochastic Processes: Handbook*; Naukova Dumka: Kiev, 1983 (In Russian).
23. Lamperti, J. An occupation time theorem for a class of stochastic processes, Transactions of the American Mathematical Society **1958a**, *88*, 380-387.
24. Lamperti, J. Some limit theorems for stochastic processes, J. Math. and Mech. **1958b**, *7*, 433-450.
25. Lamperti, J. An invariance principle in renewal theory, Annals of Mathematical Statistics **1962**, *33*, 685-696.
26. Lévy, P. Sur certains processus stochastiques homogenes, Compositio Math. **1939**, *7*, 91-111.
27. Lévy, P. *Processus Stochastiques et Mouvement Brownien*; Gauthier-Villars: Paris, 1948.
28. Masatomo, U. On number of positive sums of independent random variables, Kodai Math. Sem. Rep. **1952**, *1952*, 42-50.
29. Mitov, K.V. Limit theorems for regenerative excursion processes, Serdica Mathematical Journal **1999**, *25*, 19-40.
30. Mitov, K.V.; Yanev, N.M. Limit theorems for alternating renewal processes in the infinite mean case, Advances in Applied Probability **2001a**, *33*, 896-911.
31. Mitov, K.V.; Yanev, N.M. Regenerative processes in the infinite mean cycle case, Journal of Applied Probability **2001b**, *38*, 165-179.
32. Mohan, H.R. Limit distributions of the number of renewals and waiting times, Journal of Applied Probability **1976**, *13*, 301-312.
33. Perman, M.; Werner, W. Perturbed Brownian motions, Probability Theory and Related Fields **1997**, *108*, 357-383.
34. Petit, F. Quelques extensions de la loi de l'arcsinus, C. R. Acad. Sci. Paris Ser. I Math. **1992**, *315*, 855-858.
35. Pham-Gia, T.; Turkkan, N. System availability in a Gamma Alternating renewal process, Naval Research Logistics **1999**, *46*, 822-844.
36. Pólya, G. Sur quelque points de la théorie des probabilités, Ann. Inst. H. Poincaré **1931**, *1*, 117-161.
37. Sigman, K.; Wolff, R.W. A review of regenerative processes,

SIAM Review **1993**, *35*, 269-288.

38. Sparre Andersen, E. On sums of symmetrically dependent random variables, Scandinavisk Aktuarietidskrift **1953a**, *36*, 123-138.

39. Sparre Andersen, E. On the fluctuations of sums of random variables, Math. Scand. **1953b**, *1*, 263-285.

40. Sparre Andersen, E. On the fluctuations of sums of random variables II, Math. Scand. **1954**, *2*, 195-223.

41. Spitzer, F. A combinatorial lemma and its application to probability theory, Transactions of the American Mathematical Society **1956**, *82*, 323-339.

42. Spitzer, F. *Principles of Random Walk*; Princeton: New Jersey, 1964.

43. Stoyanov, J.; Pirinsky, C. Random motions, classes of ergodic Markov chains, and beta distributions, Statistics and Probability letters **2000**, *50*, 293-304.

44. Watanabe, S. Generalized arc-sine laws for one dimensional diffusion processes and random walks, Proc. Symp. Pure Math. **1995**, *57*, 157-174.

45. Wolff, R.W. *Stochastic Modelling and the Theory of Queues*; Prentice Hall: Englewood Cliffs, 1989.

46. Yor, M. *Some Aspects of Brownian Motion, Part I: Some Special Functionals*; Lecture Notes in Mathematics, Birkhaüser: ETH Zurich, 1992.

Approximations and Tables of Beta Distributions

Narayan Giri

Département de Mathématiques et de Statistique
Université de Montréal
C.P. 6128, succursale Centre-ville
Montréal, Canada H3C 3J7

I. Introduction

In univariate and multivariate testing of statistical hypotheses linear combination of independent (or dependent) central beta distributions plays an important role in deriving appropriate test statistics and their power functions.

In this chapter we deal with the distribution of linear combination of central beta and non-central beta distributions and their approximations along with tables of comparison of different methods of approximations.

Monti and Sen (1976) give a test based on a linear combination of the type $W = \sum_{i=1}^{k} C_i X_i$ where $X_i's$ are independently distributed central beta random variables with parameters (b, a_i), $i = 1, \ldots, k$. Giri (1968), Sinha, Clément and Giri (1985), Giri (1988), among others, have encountered linear combination of the type $Z = \sum_{i=1}^{k} d_i Y_i$

where Y_i's are independently distributed central beta random variables with parameters (b_i, a_i), $i = 1, \ldots, k$. Monti and Sen (1976) gave a procedure for finding critical values of W with $k = 2$ and supplied tables of critical values for $b = 1$, 2 and various values of a_1, a_2. Their procedure is somewhat cumbersome in the sense that iterative solution is necessary in most cases and expressions become very complicated for values of b other than 1 or 2 when $a_i \geq 3$. Using simulation, Jóhannesson and Giri (1983) have given tables of critical values for different values of b, ranging from $b = 0.5$ to $b = 30$. Though these tables are adequate for statistical testing purposes, there are many instances like power calculations etc, where they will be of little use. Therefore it is of great help to have close approximations involving well known distribution. We consider here two methods whereby the linear combination of central beta random variables is approximated by a single beta random variable. They involve the first two moments and three moments of the central beta random variable, respectively. For the simplicity of calculation we give tables involving two independent random variables only. This can, of course, be used for linear combinations of more than two independent beta random variables as well.

Linear combination of non-central beta random variables also occur in statistical testing problems. As examples we refer to Sinha, Clément and Giri (1985), Giri (1968, 1988). We also compare the above two methods of approximation and two other approximations for the non-central beta distribution.

To investigate the theoretical properties of these approximations we also include an integral representation for the moments of the non-central beta distribution.

Extensive numerical computations are done to show that in most cases the approximation is sufficiently good for practical purposes. Tables of comparison of percentile points of linear combination of two independent beta random variables X_1, X_2 where X_i is $beta(a_i, b_i)$, $i = 1, 2$ for method 1 and 2 along with simulated values and values calculated by Monti and Sen (1976) for different values of b_1, b_2 for both central and non-central beta random variables will be reproduced here with the kind permission of the publisher, Marcel Dekker, Inc., from the paper: On approximations involving the beta distribution by Jóhannesson and Giri, 1994, *Communications in Statistics–Simulation and Computation*, **9**, pages 489-503.

II. A Multivariate Statistic and Its Distribution with Applications

For any p-vector $Y = (Y_1, \ldots, Y_p)'$ and any $p \times p$ matrix $A = (a_{ij})$ we write for $i = 1, \ldots, k$, $k \leq p$,

$$Y = \left(Y'_{(1)}, \ldots, Y'_{(k)} \right)',$$

$$Y_{[i]} = \left(Y'_{(1)}, \ldots, Y'_{(i)} \right)',$$

$$A = \begin{pmatrix} A_{(11)}, \ldots, A_{(1k)} \\ \cdots \\ A_{(k1)}, \ldots, A_{(kk)} \end{pmatrix},$$

$$A_{[ii]} = \begin{pmatrix} A_{(11)}, \ldots, A_{(1i)} \\ \cdots \\ A_{(i1)}, \ldots, A_{(ii)} \end{pmatrix},$$

$$A_{[ij]} = \left(A_{(i1)}, \ldots, A_{(ij)} \right),$$

$$A_{[ji]} = \left(A_{(1i)}, \ldots, A_{(ji)} \right),$$

where $Y_{(i)}$ are sub-vectors of Y of dimension $p_i \times 1$ and $A_{(ii)}$ are sub-matrices of A of dimension $p_i \times p_i$ where p_i are arbitrary integers including zero such that $\sum_{i=1}^{k} p_i = p$.

Let $X^\alpha = (X_{\alpha 1}, \ldots, X_{\alpha p})'$, $\alpha = 1, \ldots, N$ ($N > p$) be a sample of size N from $N_p(\mu, \Sigma)$ and let

$$N\bar{X} = \sum_{\alpha=1}^{N} X^\alpha, \quad S = \sum_{\alpha=1}^{N} (X^\alpha - \bar{X})(X^\alpha - \bar{X})'.$$

Define $R_1, \ldots, R_k$ by

$$\sum_{j=1}^{i} R_j = N\bar{X}'_{[i]} \left(S_{[ii]} + N\bar{X}_{[i]}\bar{X}'_{[i]}\right)^{-1} \bar{X}_{[i]}$$

$$= \frac{N\bar{X}'_{[i]}S_{[ii]}^{-1}\bar{X}_{[i]}}{1 + N\bar{X}'_{[i]}S_{[ii]}^{-1}\bar{X}_{[i]}}, \quad i = 1, \ldots, k. \tag{II.1}$$

From Giri (1996a), the joint probability density function of $R_1, \ldots, R_k$ is given by

$$f(r_1, \ldots, r_k) = \Gamma\left(\frac{1}{2}N\right) \left[\left(\prod_{i=1}^{k} \Gamma\left(\frac{1}{2}p_i\right)\right) \Gamma\left(\frac{1}{2}(N-p)\right)\right]^{-1}$$

$$\times \left[\prod_{i=1}^{k} (r_i)^{\frac{1}{2}p_i - 1}\right] \left(1 - \sum_{i=1}^{k} r_i\right)^{\frac{1}{2}(N-p)-1}$$

$$\times \exp\left\{-\frac{1}{2}\sum_{i=1}^{k} \delta_i^2 + \frac{1}{2}\sum_{j=1}^{k} r_j \sum_{i>j}^{k} \delta_i^2\right\}$$

$$\times \prod_{i=1}^{k} \phi\left(\frac{1}{2}(N-\sigma_{i-1}), \frac{1}{2}p_i, \frac{1}{2}r_i\delta_i^2\right),$$

where

$$\sigma_i = \sum_{j=1}^{i} p_j, \quad \sigma_0 = 0,$$

$$\sum_{1}^{i} \delta_j^2 = N\mu'_{[i]}\Sigma_{[ii]}^{-1}\mu_{[i]}, \quad i = 1, \ldots, k$$

and $\phi(a, b; x)$ is the confluent hyper-geometric function given by

$$\phi(a, b; x) = 1 + \frac{a}{b}x + \frac{a(a+1)}{b(b+1)}\frac{x^2}{2!} + \cdots.$$

Note that this is a non-central multivariate beta distribution.

III. Applications

$k = 1$: For testing $H_0 : \mu = 0$ against $H_1 : \mu \neq 0$, the optimum-test rejects H_0 when $R_1 = N\bar{X}'S^{-1}\bar{X}(1 + N\bar{X}'S^{-1}\bar{X})^{-1}$ is large. The probability density function of R_1 $(0 \leq r_1 \leq 1)$

$$f(r_1) = \Gamma\left(\frac{1}{2}N\right)\left\{\Gamma\left[\frac{1}{2}(N-p)\right]\Gamma\left(\frac{1}{2}p\right)\right\}^{-1}$$
$$\times r_1^{\frac{1}{2}p-1}(1-r_1)^{\frac{1}{2}(N-p)-1}$$
$$\times \exp\left(-\frac{1}{2}\delta_1^2\right)\phi\left(\frac{1}{2}N, \frac{1}{2}p; \frac{1}{2}r_1\delta_1^2\right).$$

$k = 2$: For testing $H_0 : \mu = 0$ against $H_1 : \mu_{(1)} = 0$, Giri (1968) has shown that the locally minimax test as $\sigma_2^2 \to 0$ in the sense of Giri and Kiefer (1964) rejects H_0 for large values of $T = R_1 + \frac{N-p_1}{p_2}R_2$. This problem is known as the problem of mean with covariates. The joint probability density function of R_1, R_2 is

$$f(r_1, r_2) = \Gamma\left(\frac{1}{2}N\right)\left\{\Gamma\left[\frac{1}{2}(N-p)\right]\Gamma\left(\frac{1}{2}p_1\right)\Gamma\left(\frac{1}{2}p_2\right)\right\}^{-1}$$
$$\times r_1^{\frac{1}{2}(p_1-2)}r_2^{\frac{1}{2}(p_2-2)}(1-r_1-r_2)^{\frac{1}{2}(N-p-2)}$$
$$\times \exp\left[\frac{1}{2}\delta_2^2(-1+r_1)\right]$$
$$\times \phi\left(\frac{1}{2}(N-p_1), \frac{1}{2}p_2; \frac{1}{2}r_2\delta_2^2\right).$$

Lemma III.1. *Let* $c = \frac{N-p_1}{p_2}$. *Then*

$$F_T(t) = \Pr(T \leq t)$$
$$= \int_0^{\frac{t-1}{c-1}} dr_2 \int_0^{1-r_2} f(r_1, r_2)\, dr_1$$
$$+ \int_{\frac{t-1}{c-1}}^{\frac{t}{c}} dr_2 \int_0^{t-cr_2} f(r_1, r_2)\, dr_1$$

Proof From (II.1) $0 \leq r_i \leq 1$, $i = 1, 2$, $\quad 0 \leq r_1 + r_2 \leq 1$. Since

$r_1 + cr_2 \leq t$ we get

$$0 \leq r_1 \leq \min(1 - r_2, t - cr_2).$$

If $1 - r_2 < t - cr_2$, $\min((1 - r_2), (t - cr_2)) = 1 - r_2$. But $1 - r_2 \leq t - cr_2$ implies that $r_2(c - 1) \leq t - 1$ and hence $0 \leq r_2 \leq \frac{t-1}{c-1}$. Hence, $1 - r_2 > t - cr_2$ implies $r_2(c - 1) > t - 1$. Hence, $r_2 \geq \frac{t-1}{c-1}$, thus proving the lemma. $\triangle$

Cléroux and Giri (2002) have compared numerically the powers of the locally minimax test, the likelihood ratio test which rejects $H_0 : \mu = 0$ when $Z = \frac{1 - R_1 - R_2}{1 - R_1}$ is small and the Hotelling's T^2 test which rejects H_0 when $T^2 = R_1 + R_2$ is large. Note that under H_0, Z is distributed as central beta with parameter $\left(\frac{N-p}{2}, \frac{p_2}{2}\right)$.

They calculated that the locally minimax test and the likelihood ratio test are superior to Hotelling's T^2 test for testing $H_0 : \mu = 0$ against $H_1 : \mu_{(1)} = 0$. The locally minimax test is superior to the likelihood ratio test for small values of δ_2^2.

The locally minimax test shows almost no difference in power when the power is ≥ 0.9 and a choice cannot be made between these two tests in the middle range of the power function.

For the application of this distribution in tests of hypotheses concerning the discriminant coefficients and multiple-correlation coefficients with partial information we refer to Giri (1966a, 1966b). We refer to Jóhannesson and Giri (1983) for more relevant results in this context.

IV. Approximation of a Linear Combination of Beta Random Variables

A.   Method 1

We propose here approximating $Z = \sum_{i=1}^{k} d_i Y_i$, where $Y_1, \ldots, Y_k$ are independently distributed central beta random variables with parameters (b_i, a_i), $i = 1 \ldots, k$, by $\rho Z'$ where Z' is a central beta random variable with parameter (e, f) (see Sinha, Clément and Giri

(1985)).

Monti and Sen (1976) gave a test based on the linear combination $W = \sum_{i=1}^{k} c_i X_i$ where $X_1, \ldots, X_k$ are independently distributed central beta random variables with parameters (b, a_i), $i = 1, \ldots, k$. They gave a procedure for finding the critical values of W with $k = 2$ and gave tables of values of W for $b = 1, 2$ and different values of a_1 and a_2. Their procedure is somewhat cumbersome in the sense that iterative solution is necessary in most cases and the expression became very complicated for values of $b > 2$ when $a_i \geq 3$. Jóhannesson and Giri (1982) gave tables of critical values for different values of b ranging from $b = 0.5$ to $b = 30$. These tables are computed by simulation.

The parameters (e, f) are obtained by equating the mean and the variance of Z with those of $\rho Z'$ where $\rho = \sum_{1}^{k} d_i$.

Since Z' is a central beta random variable with parameter (e, f) we get $E(Z') = \frac{e}{e+f}$. Equating it to the mean of $\frac{Z}{\rho}$ we get

$$
E\left(\frac{Z}{\rho}\right) = \rho^{-1} \sum_{i=1}^{k} d_i E(Y_i) = E \text{ (say)},
$$

$$
\text{Var}\left(Z'\right) = ef\left[(e+f)^2(e+f+1)\right]^{-1}
$$

$$
= \text{Var}\left(\frac{Z}{\rho}\right)
$$

$$
= \rho^{-2} \sum_{i=1}^{k} d_i \text{Var}(Y_i)
$$

$$
= S \text{ (say)}.
$$

Let $F = E(1 - E)^{-1}$. Then

$$
e = Ff,
$$
$$
f = F\left[S(1 + F)^3\right]^{-1} - (1 + F)^{-1}.
$$

Henceforth we refer it as Method 1 approximation.

B.   Method 2

Here we approximate Z by the random variable $\frac{\rho Z^*}{\gamma}$ with Z^* a central beta random variable with parameter (g, h). The constants g, h, γ are obtained by equating the first 3 moments of Z with those of $\frac{\rho Z^*}{\gamma}$. Let

$$E(Z) = M_1, \ E\left(Z^2\right) = M_2 \text{ and } E\left(Z^3\right) = M_3.$$

Define

$$K_1 = M_1, \ K_2 = \frac{M_2}{M_1}, \ K_3 = \frac{M_3}{M_1}.$$

Equating the first 3 moments and solving we obtain

$$A\gamma^2 + B\gamma + c = 0$$

where

$$A = K_3^2 K_2 + K_3 K_2 K_1 - 2K_3^3 K_1,$$
$$B = K_3^2 - 3K_3 K_2 + 3K_3 K_1 - K_2 K_1,$$
$$C = 2K_2 - K_1 - K_3.$$

By substitution we obtain

$$\gamma = \left[-B - \left(B^2 - 4AC\right)^{\frac{1}{2}}\right] / 2A,$$
$$g = (\gamma K_3 - 1)(\gamma K_2 - 1) / \gamma (K_3 - K_2),$$
$$h = (\gamma K_3 g + 2\gamma K_3 - 2) / (1 - \gamma K_3).$$

Henceforth we refer this as Method 2 approximation.

V.   Approximation for Non-Central Beta

We compare here Method 1 and Method 2 approximations of Section III to two other approximations for the non-central beta random variables. We want to find a simple approximation for the non-central beta distribution which can be used for theoretical purposes, for example, power calculations, when one faces linear combination of non-central beta random variables (see Sinha, Clément and Giri (1985),

Giri (1968, 1988)). If one can approximate the non-central beta distribution by a central beta distribution, the methods of Section III can be successfully applied for a good approximation.

The probability density function of a non-central beta random variable with parameters (a, b, λ) where λ is the non-centrality parameter is given by

$$f(x, a, b, \lambda) = \exp(-\lambda) \sum_{i=0}^{\infty} \frac{\lambda^i}{i!} B(a + i, b),$$

where $B(a, b)$ is the density of a central beta distribution.

Das Gupta (1968) suggested two approximations for the non-central beta distribution. The first is based on the fact that, if X_1 is a non-central chi-square with $2a$ degrees of freedom and non-centrality parameter 2λ and X_2 is a central chi-square with $2b$ degrees of freedom, then $Z = X_1/(X_1 + X_2)$ is non-central beta with parameters (a, b, λ).

He then uses Patnaik's approximation for the χ^2-distribution (Patnaik, 1949) to get the following approximation for Z,

$$Z \approx \rho B(\nu, b)$$

with $\nu = (2a + \lambda)^2/4(a + \lambda)$ and $\rho = (2a + 2\lambda)/(2a + \lambda)$.

Though Das Gupta calls this Patnaik's approximation, we refer it as Das Gupta's approximation which we will give below and which also uses Patnaik's approximation for the chi-square distribution.

The second approximation (Table 1), suggested by Das Gupta, is based on equating the first two moments of Z to those of a central beta distribution. It is equivalent to method 1 of Section III.

The next approximation is based on Patnaik's approximation to the chi-squared distribution and rewriting the probability density in terms of the density function of an F-distribution.

Write

$$Z = X_1/(X_1 + X_2) = (X_1/X_2)/(1 + X_1/X_2)^{-1}.$$

Patnaik (1949) suggests approximating X_1 by cX_1' where $c = (a + 2\lambda)/(a + \lambda)^{-1}$ and X_1' has a central chi-squared distribution with $\nu = 2(a + \lambda)^2/(a + 2\lambda)$ degrees of freedom. Since the F-random variable is a ratio of the two independent chi-square random variables multiplied by the ratio of their degrees of freedom we conclude that Z is approximately as $\frac{\gamma F}{1+\gamma F}$, where $\gamma = (a + \lambda)/b$ and F is a central F with $2(a + \lambda)^2/(a + 2\lambda)$ and $2b$ degrees of freedom. Hence,

$$\Pr(Z \leq x) \approx \Pr(F \leq y)$$

with $y = x/(1 - x)\gamma$. We refer this as F-approximation in table 4.

It may be remarked that we did try other approaches to using Patnaik's approximation, but found them less satisfactory.

The last approximation (Table 4) parallels our method 2 in the Section III, which uses an approximation by equating the first three moments of Z and Z'/γ, where Z' has a central beta distribution and the parameters (a, b, γ) are found as before.

The first three moments about the origin of a non-central beta random variable Z are given by

$$E(Z) = M_1 = \exp(-\lambda) \sum_{i=0}^{\infty} \frac{\lambda^i}{i!} \frac{a+i}{a+b+i}$$

$$E(Z^2) = M_2 = \exp(-\lambda) \sum_{i=0}^{\infty} \frac{\lambda^i}{i!} \frac{(a+i)(a+1+i)}{(a+b+i)(a+b+1+i)}$$

$$E(Z^3) = M_3 = \exp(-\lambda) \sum_{i=0}^{\infty} \frac{\lambda^i}{i!} \frac{(a+i)(a+i+1)}{(a+b+i)(a+b+1+i)} \cdot$$
$$\times \frac{a+i+2}{a+b+2+i}.$$

The parameters are found exactly as described in Section III.

VI.   Integral Representation for the Moments of Non-Central Beta

The moments of non-central beta random variable sometimes play a role in approximating a non-central beta of a linear combination of independent non-central beta random variables by a central beta random variable. We give here the integral representation for the moments of a non-central beta which may be of use in investigating the theoretical properties of such procedures. We show here that the moments of a non-central beta random variable may be, in general, represented by a simple integral. Explicit formulas are given for the first two moments. The non-central beta probability density function is given by an infinite sum as (Section IV)

$$f(x, a, b, \lambda) = \exp(-\lambda) \sum_{i=0}^{\infty} \frac{\lambda^i}{i!} B(a+i, b).$$

The ith moment of a central beta random variable Y with parameter (α, β) is given by

$$E(Y^i) = \frac{\alpha(\alpha+1)\cdots(\alpha+i-1)}{(\alpha+\beta)(\alpha+\beta+1)\cdots(\alpha+\beta+i-1)}.$$

If X follows a non-central beta with parameter (a, b, λ), then

$$E(X) = \int_0^1 x \left[\exp(-\lambda) \sum_{i=0}^{\infty} \frac{\lambda^i}{i!} B(a+i, b) \right] dx$$

$$= \exp(-\lambda) \sum_{i=0}^{\infty} \frac{\lambda^i}{i!} \int_0^1 x B(a+i, b) dx$$

$$= \exp(-\lambda) \sum_{i=0}^{\infty} \frac{\lambda^i}{i!} \frac{a+i}{a+b+i}$$

$$= \exp(-\lambda) \sum_{i=0}^{\infty} \frac{\lambda^i}{i!} \left(1 - \frac{b}{a+b+i} \right)$$

$$= 1 - b \exp(-\lambda) \sum_{i=0}^{\infty} \frac{\lambda^i}{i!} \frac{1}{a+i+1}$$

$$= 1 - b \exp(-\lambda) \sum_{i=0}^{\infty} \frac{1}{i!} \lambda^{-(a+b)} \int_0^\lambda x^{a+b+i-1} dx$$

$$= 1 - b \exp(-\lambda) \lambda^{-(a+b)} \int_0^\lambda x^{a+b-1} \left(\sum_{i=0}^{\infty} \frac{x^i}{i!} \right) dx$$

$$= 1 - b \exp(-\lambda) \lambda^{-(a+b)} \int_0^\lambda x^{a+b-1} \exp(x) dx.$$

Alternatively one may get

$$E(X) = \exp(-\lambda) \sum_{i=0}^{\infty} \frac{\lambda^i}{i!} \left(\frac{a+i}{a+b+i} \right)$$

$$= \exp(-\lambda) \sum_{i=0}^{\infty} \frac{\lambda^i}{i!} \left(\frac{a}{a+b+i} \right) + \sum_{i=0}^{\infty} \frac{\lambda^i}{i!} \left(\frac{i}{a+b+i} \right)$$

$$= \exp(-\lambda) \lambda^{-(a+b)} \int_0^{\lambda} x^{a+b-1}(x+a)\exp(x)dx.$$

Note that if a is an integer, then

$$\int x^a \exp(x)dx = \sum_{i=0}^{a} x^{a-i} \exp(x)dx + \text{ constant.}$$

Hence, if $a + b$ is an integer $E(X)$ can be expressed as a finite sum. For the second moment we get

$$E(X^2) = \exp(-\lambda) \sum_{i=0}^{\infty} \frac{\lambda^i}{i!} \frac{(a+i)(a+i+1)}{(a+b+i)(a+b+i+1)}$$

$$= \exp(-\lambda) \sum_{i=0}^{\infty} \frac{\lambda^i}{i!} \left(1 - \frac{b}{a+b+i} \right)$$

$$\times \left(1 - \frac{b}{a+b+i+1} \right)$$

$$= E(X) - \exp(-\lambda) b \sum_{i=0}^{\infty} \frac{\lambda^i}{i!} \frac{1}{a+b+i+1}$$

$$- \sum_{i=0}^{\infty} \frac{\lambda^i}{i!} \frac{b}{(a+b+i)(a+b+i+1)}.$$

The second sum can be written as

$$\sum_{i=0}^{\infty} b\lambda^{-(a+b+1)} \frac{1}{i} \int_0^\lambda \int_0^y x^{a+b+i-1} dxdy$$

$$= b\lambda^{-(a+b+1)} \int_0^\lambda \int_0^y x^{a+b-1} \left(\sum_{i=0}^{\infty} \frac{x^i}{i!} \right) dxdy$$

$$= b\lambda^{-(a+b+1)} \int_0^\lambda \int_0^y x^{a+b-1} \exp(x) dxdy$$

$$= b\lambda^{-(a+b+1)} \int_0^\lambda (\lambda - x) x^{a+b-1} \exp(x) dx.$$

Thus,

$$E\left(X^2\right) = E(X) - \exp(-\lambda)b\exp\left\{-(a+b+1)\right\}$$
$$\times \int_0^\lambda \left[x - b(\lambda - x) \right] x^{a+b-1} \exp(x) dx.$$

The higher moments can be expressed as simple integral of multiple power of x and an exponential by using similar techniques.

VII.   Numerical Comparison

A.   Approximation of a Linear Combination of Central Beta

Extensive numerical comparison are performed for the two approximations in Section III. As usual, it is difficult to obtain the exact values for the percentiles. We have side-stepped this problem by using simulated values for most of these comparisons. The simulation involved 10,000 observations from a randomly generated distribution. In table 1 we compare results for Method 1 and 2 to simulated values and the values obtained by Monti and Sen (1976). In most cases the simulation is accurate to three or even four significant digits. Hence, one seems to be justified in using the simulated values for reference. We also refer to Jóhannesson and Giri (1983) for further details in this context.

As pointed out earlier, all tables involve linear combination of two independent beta random variables.

This was done only for the simplicity of calculations and this can be done for linear combinations of more than two independent beta random variables.

Table 1 gives comparison for the 95th percentile for the case $b_1 = b_2 = b$. These two methods are compared to values calculated by Monti and Sen (1976). As expected, Method 2 gives generally more accurate results. Table 2 compares various percentiles for the two methods to simulated values and the same conclusion is evident here also. The approximations seem to do worst when the difference between a_1 and a_2 is large. However, Method 2 appears to give a very close approximation and would in most cases seem to be sufficiently accurate for all practical purposes. Table 3 gives comparison when $b_1 \neq b_2$. This approximations seem a bit less accurate in this case and Method 2 does not appear to be as consistently dominant as before.

Method 2 often overestimates the 99th percentile even by values outside the range of $VX_1 + X_2$. This occurs when $\delta < 1$, since in this case Method 2 assigns a non zero probability to the interval $(V + 1, \frac{V+1}{\delta})$. However, Method 2 still seems to be more accurate in most cases. But it is difficult to spot patterns in this table. Generally the approximations appear worse when $V = 5$. The question, whether they are good enough, can, of course, be answered in terms of what they should be used for. It seems that for approximate power computations, for example, they would in most cases suffice. In some cases, Method 2 may not be applicable, since equating the first three moments might lead to negative values of g and h. This cases have been marked NA in the tables. In such situations one would naturally use Method 1.

B. Approximation of a Linear Combination of Non-Central Beta

Table 4 gives the numerical results for the four approximations of Section IV. It is evident that Method 1, Method 2 and the F-

approximation give consistently good results. The table gives probabilities $\Pr(Z \leq K)$ for four different K-values. Das Gupta's approximation, on the other hand, was found to be rather unreliable. In some cases it is excellent, but in some others the approximations are rather poor. It seems to work best when $a > \lambda$, but even then, it was not as consistent as the other methods.

On the whole, Method 2 and the F-approximation appear to give the best results, with Method 1 a close competitor. Any of these methods would suffice for most applications. However, if one were, for example, faced with the problem of approximating the distribution for a linear combination of non-central beta variables, only Method 1 or 2 would do, since with these approximations the methods of Section III can be applied.

Note. Tables 1-4 are reproduced here with the kind permission of the publisher - Marcel Dekker, Inc. from the paper: On approximation involving the beta distribution by Jòhannesson, B. and Giri, N., *Communications in Statistics–Simulation and Computation*, 1994, pages 489-503.

Table I: Comparison of 95th percentile of $VX_1 + X_2$ $(b_1 = b_2 = b)$

b	(a_1, a_2)	V	Method 1	Method 2	Simulation	Values from Monti and Sen
1	(1,1)	1.0	1.669	1.669	1.685	1.684
		2.5	3.018	3.018	3.012	3.000
	(1,5)	1.0	1.245	1.205	1.186	1.182
		2,5	2.708	2.598	2.563	2.547
	(1,20)	1.0	1.088	1.014	0.999	0.998
		2.5	2.598	2.437	2.428	2.423
	(5,1)	1.0	1.245	1.205	1.186	1.182
		2.5	1.753	1.749	1.752	1.742
	(5,5)	1.0	0.712	0.712	0.710	0.709
		2.5	1.314	1.314	1.320	1.326
	(5,20)	1.0	0.502	0.502	0.511	0.507
		2.5	1.166	1.166	1.186	1.178
	(20,1)	1.0	1.088	1.014	0.999	0.998
		2.5	1.196	1.135	1.103	1.101
	(20,5)	1.0	0.502	0.502	0.516	0.507
		2.5	0.633	0.633	0.643	0.632
	(20,20)	1.0	0.219	0.219	0.221	0.219
		2.5	0.406	NA*	0.411	0.407
2	(1,1)	1.0	1.821	1.828	1.829	1.832
		2.5	3.245	3.242	3.231	3.233
	(1,5)	1.0	1.427	1.397	1.391	1.396
		2.5	2.923	2.804	2.781	2.785
	(1,20)	1.0	1.177	1.095	1.084	1.084
		2.5	2.732	2.547	2.524	2.528
	(5,1)	1.0	1.427	1.397	1.395	1.396
		2.5	2.179	2.190	2.207	2.199
	(5,5)	1.0	0.975	0.974	0.975	0.972
		2.5	1.774	1.776	1.789	1.782
	(5,20)	1.0	0.690	0.687	0.688	0.688
		2.5	1.551	1.546	1.549	1.551

NA* = not applicable. See Section VI.

Table II: Comparison of various (x) percentiles of $VX_1 + X_2 (b_1 = b_2 = b)$

b	(a_1, a_2)	V	x	Method 1	Method 2	Simulation
0.5	(1,1)	1.0	0.1	0.152	0.144	0.128
			0.4	0.489	0.490	0.523
			0.7	0.878	0.889	0.897
			0.9	1.278	1.280	1.247
			0.95	1.449	1.438	1.440
			0.99	1.703	1.654	1.723
		3.0	0.1	0.205	0.205	0.223
			0.4	0.902	0.902	0.898
			0.7	1.809	1.809	1.843
			0.9	2.735	2.735	2.784
			0.95	3.106	3.106	3.085
			0.99	3.600	3.600	3.558
0.5	(1,30)	1.0	0.1	0.041	0.017	0.023
			0.4	0.201	0.185	0.180
			0.7	0.451	0.504	0.509
			0.9	0.782	0.820	0.830
			0.95	0.954	0.922	0.918
			0.99	1.267	1.022	0.998
		3.0	0.1	0.071	0.037	0.046
			0.4	0.540	0.501	0.505
			0.7	1.376	1.483	1.494
			0.9	2.385	2.439	2.454
			0.95	2.826	2.728	2.725
			0.99	3.449	2.984	2.959
0.5	(5,5)	1.0	0.1	0.024	0.022	0.022
			0.4	0.104	0.103	0.104
			0.7	0.227	0.229	0.236
			0.9	0.402	0.406	0.410
			0.95	0.503	0.504	0.509
			0.99	0.711	0.699	0.699
		3.0	0.1	0.031	0.034	0.038
			0.4	0.182	0.185	0.186
			0.7	0.452	0.447	0.446
			0.9	0.858	0.848	0.858
			0.95	1.094	1.088	1.102
			0.99	1.581	1.603	1.638
1.0	(10,5)	1.0	0.1	0.076	0.074	0.073

Table II: (continued)

b	(a_1, a_2)	V	x	Method 1	Method 2	Simulation
			0.4	0.188	0.188	0.182
			0.7	0.319	0.321	0.317
			0.9	0.483	0.485	0.481
			0.95	0.572	0.571	0.565
			0.99	0.751	0.741	0.742
1.0	(30,1)	2.0	0.1	0.218	0.159	0.161
			0.4	0.449	0.468	0.465
			0.7	0.691	0.756	0.769
			0.9	0.971	0.973	0.966
			0.95	1.116	1.040	1.018
			0.99	1.397	1.110	1.116
2.0	(10,30)	1.0	0.1	0.100	0.099	0.099
			0.4	0.186	0.186	0.185
			0.7	0.275	0.276	0.275
			0.9	0.381	0.381	0.381
			0.95	0.437	0.437	0.438
			0.99	0.552	0.549	0.552
2.0	(30,5)	1.0	0.1	0.153	0.147	0.145
			0.4	0.286	0.287	0.280
			0.7	0.420	0.425	0.421
			0.9	0.573	0.576	0.579
			0.95	0.654	0.650	0.657
			0.99	0.811	0.786	0.789
		2.0	0.1	0.197	0.190	0.187
			0.4	0.344	0.346	0.341
			0.7	0.488	0.495	0.491
			0.9	0.655	0.657	0.656
			0.95	0.743	0.738	0.745
			0.99	0.919	0.889	0.889
10.0	(1,10)	1.0	0.90	1.581	1.580	1.579
			0.99	1.696	1.687	1.696
		2.0	0.90	2.564	2.546	2.546
			0.99	2.712	2.630	2.675
10.0	(1,30)	1.0	0.90	1.296	1.288	1.283
			0.99	1.401	1.351	1.377
10.0	(30,10)	1.0	0.9	0.917	0.917	0.918
			0.99	1.058	1.049	1.042
		2.0	0.9	1.227	1.227	1.225
			0.99	1.423	1.422	1.412

Table II: (continued)

b	(a_1, a_2)	V	x	Method 1	Method 2	Simulation
	(30,30)	1.0	0.9	0.626	0.625	0.625
			0.99	0.734	0.738	0.734
		2.0	0.9	0.949	0.949	0.945
			0.99	1.129	1.131	1.126
30.0	(30,10)	1.0	0.9	1.369	1.368	1.368
			0.99	1.459	1.456	1.457
		2.0	0.9	1.935	1.936	1.933
			0.99	2.078	2.081	2.082
	(30,30)	1.0	0.9	1.116	1.116	1.117
			0.99	1.210	1.210	1.211
		2.0	0.9	1.684	1.684	1.684
			0.99	1.831	1.831	1.839

Table III: Comparison for various (x) percentiles of $VX_1 + X_2$ when $b_1 \neq b_2$

(b_1, a_1)	(b_2, a_2)	V	x	Method 1	Method 2	Simulation
(1,1)	(5,1)	1.0	0.1	0.888	0.901	0.923
			0.4	1.274	1.253	1.238
			0.7	1.535	1.515	1.547
			0.9	1.734	1.750	1.773
			0.95	1.806	1.852	1.849
			0.99	1.903	2.019	1.938
		5.0	0.1	1.286	1.365	1.338
			0.4	2.954	2.903	2.855
			0.7	4.295	4.213	4.356
			0.9	5.253	5.301	5.348
			0.95	5.547	5.702	5.608
			0.99	5.855	6.223	5.844
(1,5)	(5,20)	1.0	0.1	0.174	0.180	0.183
			0.4	0.308	0.306	0.302
			0.7	0.438	0.432	0.428
			0.9	0.586	0.583	0.591
			0.95	0.662	0.665	0.678
			0.99	0.813	0.838	0.841
		5.0	0.1	0.246	0.267	0.301
			0.4	0.720	0.722	0.679
			0.7	1.305	1.282	1.265
			0.9	2.029	2.007	2.063
			0.95	2.409	2.410	2.490
			0.99	3.138	3.242	3.220
(1,20)	(5,1)	1.0	0.1	0.691	0.671	0.670
			0.4	0.841	0.872	0.877
			0.7	0.958	0.981	0.975
			0.9	1.073	1.045	1.034
			0.95	1.128	1.063	1.065
			0.99	1.228	1.081	1.134
		5.0	0.1	0.741	NA	0.782
			0.4	0.987	NA	0.995
			0.7	1.199	NA	1.163
			0.9	1.425	NA	1.417
			0.95	1.540	NA	1.554
			0.99	1.764	NA	1.914
(1,1)	(20,1)	1.0	0.1	1.045	1.073	1.054
			0.4	1.409	1.376	1.357

Table III: (continued)

(b_1, a_1)	(b_2, a_2)	V	x	Method 1	Method 2	Simulation
			0.7	1.640	1.610	1.658
			0.9	1.806	1.832	1.858
			0.95	1.863	1.934	1.913
			0.99	1.936	2.113	1.965
		5.0	0.1	1.394	1.500	1.464
			0.4	3.101	3.030	2.963
			0.7	4.425	4.317	4.469
			0.9	5.337	5.405	5.464
			0.95	5.608	5.818	5.711
			0.99	5.881	6.374	5.901
(1,20)	(20,20)	1.0	0.1	0.434	0.434	0.435
			0.4	0.522	0.522	0.524
			0.7	0.593	0.590	0.592
			0.9	0.666	0.666	0.663
			0.95	0.702	0.702	0.699
			0.99	0.770	0.771	0.768
		5.0	0.1	0.447	NA	0.499
			0.4	0.656	NA	0.635
			0.7	0.847	NA	0.803
			0.9	1.058	NA	1.060
			0.95	1.169	NA	1.209
			0.99	1.389	NA	1.530
(5,1)	(1,20)	1.0	0.1	0.691	0.671	0.679
			0.4	0.841	0.872	0.876
			0.7	0.958	0.981	0.974
			0.9	1.073	1.045	1.033
			0.95	1.128	1.063	1.064
			0.99	1.228	1.081	1.128
		5.0	0.1	3.257	3.202	3.221
			0.4	4.083	4.028	4.210
			0.7	4.639	4.704	4.707
			0.9	5.090	4.948	4.944
			0.95	5.270	5.003	4.995
			0.99	5.542	5.046	5.067
(5,5)	(1,1)	1.0	0.1	0.568	0.568	0.570
			0.4	0.910	0.910	0.905
			0.7	1.184	1.184	1.203
			0.9	1.432	1.432	1.439
			0.95	1.537	1.537	1.534

Table III: (continued)

(b_1, a_1)	(b_2, a_2)	V	x	Method 1	Method 2	Simulation
			0.99	1.702	1.702	1.660
		5.0	0.1	1.941	1.941	1.955
			0.4	2.784	2.784	2.798
			0.7	3.446	3.446	3.455
			0.9	4.060	4.060	4.081
			0.95	4.332	4.332	4.353
			0.99	4.787	4.787	4.800
(20,1)	(1,1)	1.0	0.1	1.045	1.073	1.055
			0.4	1.409	1.376	1.362
			0.7	1.640	1.610	1.659
			0.9	1.806	1.832	1.854
			0.95	1.863	1.934	1.908
			0.99	1.936	2.113	1.966
		5.0	0.1	4.763	4.774	4.804
			0.4	5.221	5.193	5.175
			0.7	5.491	5.474	5.498
			0.9	5.688	5.715	5.746
			0.95	5.761	5.819	5.827
			0.99	5.864	5.991	5.923
(20,1)	(1,5)	1.0	0.1	0.927	NA	0.966
			0.4	1.083	NA	1.054
			0.7	1.198	NA	1.172
			0.9	1.309	NA	1.331
			0.95	1.360	NA	1.411
			0.99	1.451	NA	1.560
		5.0	0.1	4.575	4.566	4.583
			0.4	4.877	4.901	4.913
			0.7	5.082	5.095	5.060
			0.9	5.259	5.235	5.227
			0.95	5.336	5.286	5.310
			0.99	5.467	5.358	5.482
(5,5)	(20,5)	1.0	0.1	1.076	1.080	1.079
			0.4	1.262	1.258	1.258
			0.7	1.396	1.392	1.395
			0.9	1.515	1.518	1.524
			0.95	1.568	1.577	1.541
			0.99	1.659	1.683	1.677
		5.0	0.1	2.299	2.312	2.311
			0.4	3.112	3.100	3.107

Table III: (continued)

(b_1, a_1)	(b_2, a_2)	V	x	Method 1	Method 2	Simulation
(5,20)	(20,5)	1.0	0.7	3.726	3.713	3.738
			0.9	4.282	4.288	4.326
			0.95	4.525	4.550	4.569
			0.99	4.930	5.002	4.968
			0.1	0.857	0.857	0.860
			0.4	0.972	0.972	0.972
			0.7	1.059	1.059	1.058
			0.9	1.143	1.143	1.140
			0.95	1.183	1.183	1.183
			0.99	1.256	1.256	1.255
		5.0	0.1	1.296	NA	1.317
			0.4	1.681	NA	1.663
			0.7	2.000	NA	1.980
			0.9	2.327	NA	2.327
			0.95	2.487	NA	2.506
			0.99	2.793	NA	2.873

Table IV: Comparison of various approximations to the noncentral beta distribution

K	a	b	λ	Exact	DasGupta's approx.	F-approx.	Method 1	Method 2
.2800	1.0000	1.0000	1.0000	.1363	.1728	.1216	.1347	.1396
.2800	1.0000	7.0000	1.0000	.6996	.7741	.7063	.7117	.7009
.2800	7.0000	1.0000	1.0000	.0001	.0001	.0001	.0001	.0001
.2800	7.0000	7.0000	1.0000	.0259	.0307	.0257	.0250	.0257
.2800	1.0000	1.0000	7.0000	.0018	.0093	.0005	.0004	.0009
.2800	1.0000	7.0000	7.0000	.0698	.2388	.0589	.0618	.0719
.2800	7.0000	1.0000	7.0000	.0000	.0000	.0000	.0000	.0000
.2800	7.0000	7.0000	7.0000	.0009	.0038	.0007	.0005	.0007
.5200	1.0000	1.0000	1.0000	.3218	.3467	.3139	.3345	.3278
.5200	1.0000	7.0000	1.0000	.9611	.9611	.9613	.9572	.9594
.5200	7.0000	1.0000	1.0000	.0064	.0064	.0063	.0061	.0063
.5200	7.0000	7.0000	1.0000	.4600	.4599	.4601	.4620	.4603
.5200	1.0000	1.0000	7.0000	.0181	.0445	.0138	.0157	.0195
.5200	1.0000	7.0000	7.0000	.4809	.6162	.4862	.5006	.4801
.5200	7.0000	1.0000	7.0000	.0004	.0006	.0003	.0002	.0003
.5200	7.0000	7.0000	7.0000	.1091	.1403	.1075	.1089	.1108
.6000	1.0000	1.0000	1.0000	.4022	.4073	.3969	.4154	.4042
.6000	1.0000	7.0000	1.0000	.9861	.9808	.9858	.9824	.9867
.6000	7.0000	1.0000	1.0000	.0188	.0175	.0187	.0185	.0188
.6000	7.0000	7.0000	1.0000	.6911	.6748	.6913	.6923	.6906
.6000	1.0000	1.0000	7.0000	.0365	.0640	.0314	.0367	.0408
.6000	1.0000	7.0000	7.0000	.6816	.7196	.6869	.6900	.6706
.6000	7.0000	1.0000	7.0000	.0017	.0019	.0016	.0013	.0015
.6000	7.0000	7.0000	7.0000	.2765	.2697	.2764	.2823	.2787
.7200	1.0000	1.0000	1.0000	.5442	.5000	.5419	.5516	.5360
.7200	1.0000	7.0000	1.0000	.9984	.9944	.9982	.9970	.9994
.7200	7.0000	1.0000	1.0000	.0758	.0631	.0757	.0762	.0763
.7200	7.0000	7.0000	1.0000	.9298	.9069	.9298	.9287	.9295
.7200	1.0000	1.0000	7.0000	.1014	.1015	.0967	.1102	.1082
.7200	1.0000	7.0000	7.0000	.9135	.8391	.9141	.9022	.9103
.7200	7.0000	1.0000	7.0000	.0141	.0078	.0137	.0135	.0144
.7200	7.0000	7.0000	7.0000	.6789	.5280	.6799	.6817	.6754

References

1. Cléroux, R.; Giri, N. Power comparison of some optimum invariant tests for means with covariates (to be published).

2. Das Gupta, P. Two approximations for the distribution of double non-central beta, Sankhyā, B **1968**, *30*, 83-88.

3. Giri, N. Locally and asymptotically minimax tests of a multivariate problem, Annals of Mathematical Statistics **1968**, *29*, 171-178.

4. Giri, N. Robust test of independence, Canadian Journal of Statistics **1988**, *16*, 419-428.

5. Giri, N. *Multivariate Statistical Analysis*; Marcel Dekker: New York, 1996a.

6. Giri, N. *Group Invariance in Statistical Inference*; World Scientific: Singapore, 1996b.

7. Giri, N.; Kiefer, J. Local and asymptotic minimax properties of multivariate tests, *Annals of Mathematical Statistics* **1964**, *39*, 21-35.

8. Jóhannesson, B.; Giri, N. Critical values of the locally optimum combination of two independent test statistics, Journal of Statistical Computation and Simulation **1983a**, *16*, 251-286.

9. Jóhannesson, B.; Giri, N. Comparison of power functions of some step-down tests for means with additional data, Journal of Statistical Computation and Simulation **1983b**, *16*, 1-30.

10. Jóhannesson, B.; Giri, N. On approximations involving the beta distribution, Communications in Statistics–Simulation and Computation **1994**, *9*, 489-503.

11. Monti, K.L.; Sen, P.K. Locally optimum combination of independent test statistics, Journal of the American Statistical Association **1976**, *71*, 903-911.

12. Patnaik, P.B. The non-central χ^2 and F-distribution and their applications, Biometrika **1949**, *26*, 202-232.

13. Sinha, B.K.; Clément, B.; Giri, N. Tests for means with additional information, *Communications in Statistics–Theory and Methods* **1985**, *14*, 1427-1451.

Maximum Likelihood Estimators of the Parameters in a Beta Distribution

Truc T. Nguyen

Department of Mathematics and Statistics
Bowling Green State University
Bowling Green, Ohio 43403

I. Introduction

The Maximum Likelihood Estimators (MLE's) for the parameters of Beta distributions were studied by a number of authors. The close algebraic forms of MLE's for the parameters of Beta distributions do not exist. For the case when the end points a and b are known, Johnson and Kotz (1970) gave the maximum likelihood equations for the two shape parameters p and q. Beckman and Tietjen (1978) developed a method that is fast and free from convergence problems and does not require starting values for solving these system of maximum likelihood equations.

Gnanadesikan, Pinkham, and Hughes (1967) discussed the estimation problem for ordered samples. For the case when the end points are not known, Johnson and Kotz (1970) suggested an iterative scheme involving guessing the end points, solving the maximum likelihood equation, and searching for a maximum for the likelihood function. This technique was used by Koshal (1933, 1935) using the estimates derived by the method of moments as initial values. The other results in estimation for the parameters of Beta distributions were given by Fielitz and Myers (1975, 1976), Romesburg (1976), Beckman and Tietjen (1978), Kottes and Lau (1978), Dishon and Weiss (1980), Farnum and Stanton (1987), Carnahan (1989), Lau and Lau (1991). In all the studies of these MLE's, the common purpose is how to find the maximum points of the likelihood function in a fast and accurate numerical way.

II. Likelihood Equations

Assume that a random sample $x_1, \ldots, x_n$ has been collected for a random variable X which follows the distribution defined by the density function

$$f_X(x; a, b, p, q) = \begin{cases} \dfrac{\Gamma(p+q)(x-a)^{p-1}(b-x)^{q-1}}{\Gamma(p)\Gamma(q)(b-a)^{p+q-1}}, & \text{if } a \leq x \leq b, \\ 0, & \text{otherwise,} \end{cases}$$

(II.1)

where $p > 0$; $q > 0$; and, $a < b$ are two real numbers. This density function is of a Beta distribution with four parameters a, b, p, q, denoted Beta(a, b, p, q).

If a and b are known, the transformation

$$Y = \frac{X - a}{b - a}$$

is used and Y has the density function

$$f_Y(y) = \begin{cases} \dfrac{\Gamma(p+q)}{\Gamma(p)\Gamma(q)} y^{p-1}(1-y)^{q-1}, & \text{if } 0 \leq y \leq 1, \\ 0, & \text{otherwise,} \end{cases}$$

where $p > 0$ and $q > 0$. This is the standard form of the Beta distribution with parameters p and q, denoted Beta(p, q).

(i) In the case a and b are known, the study of MLE's of unknown parameters is based on the density function (II.1). In this case the loglikelihood function is

$$\log L(p, q; x) = n(\log \Gamma(p + q) - \log \Gamma(p) - \log \Gamma(q))$$
$$+ (p - 1) \sum_{i=1}^{n} \log y_i$$
$$+ (q - 1) \sum_{i=1}^{n} \log (1 - y_i),$$

and the system of maximum likelihood equations for finding the respective MLE's $\hat{p}$ and $\hat{q}$ of p and q is

$$\psi(\hat{p}) - \psi(\hat{p} + \hat{q}) = n^{-1} \sum_{i=1}^{n} \log y_i, \qquad (II.2)$$

$$\psi(\hat{q}) - \psi(\hat{p} + \hat{q}) = n^{-1} \sum_{i=1}^{n} \log (1 - y_i), \qquad (II.3)$$

where $\psi(\cdot)$ is the digamma function, defined by

$$\psi(x) = \frac{d}{dx} \log \Gamma(x) = \frac{\Gamma'(x)}{\Gamma(x)} \text{ for } x > 0,$$

and $y_i = \frac{x_i - a}{b - a}$ for $i = 1, \ldots, n$.

Newton method is used to find a solution of this system of maximum likelihood equation.

Beckman and Tiegjen (1978) showed that to solve the system of equations (II.2) and (II.3) is equivalent to solve the equation

$$\psi(\hat{q}) - \frac{1}{n} \sum_{i=1}^{n} \log (1 - y_i)$$
$$- \psi \left[\psi^{-1} \left(n^{-1} \left\{ \sum_{i=1}^{n} \log y_i - \sum_{i=1}^{n} \log (1 - y_i) \right\} + \psi(\hat{q}) \right) + \hat{q} \right]$$

and developed a method which is rapid, free from the convergence problem and does not require starting values for solving the system of equations (II.2) and (II.3). Gnanadesikan, Pinkham and Hughes (1967) studied the MLE's of p and q under the censoring of type 2, that is, only the M smallest observations are obtained in a random

sample size $n \geq M$. In this case the system of maximum likelihood
equations becomes

$$\psi(\hat{p}) - \psi(\hat{p} + \hat{q}) - \left(1 - \frac{M}{n}\right) \frac{I_1\left(y_{(M)}; \hat{p}, \hat{q}\right)}{I\left(y_{(M)}; \hat{p}, \hat{q}\right)}$$

$$= n^{-1} \sum_{i=1}^{M} \log y_{(i)},$$

$$\psi(\hat{q}) - \psi(\hat{p} + \hat{q}) - \left(1 - \frac{M}{n}\right) \frac{I_2\left(y_{(M)}; \hat{p}, \hat{q}\right)}{I\left(y_{(M)}; \hat{p}, \hat{q}\right)}$$

$$= n^{-1} \sum_{i=1}^{M} \log\left(1 - y_{(i)}\right),$$

where

$$I(x; p, q) = \int_x^1 t^{p-1}(1 - t)^{q-1} dt$$

(for $0 \leq x \leq 1$),

$$I_1(x; p, q) = \frac{\partial}{\partial p} I(x; p, q) = \int_x^1 t^{p-1}(1 - t)^{q-1} \log t \, dx$$

(for $0 \leq x \leq 1$),

$$I_2(x; p, q) = \frac{\partial}{\partial q} I(x; p, q) = \int_x^1 t^{p-1}(1 - t)^{q-1} \log(1 - t) dt$$

(for $0 \leq x \leq 1$), $y_{(1)}, \ldots, y_{(M)}$ are the first M order statistics of y_1,
$\ldots, y_n$.

The comparisons between the MLE's and the method-of-moment
estimators (MME's) for the parameters of Beta(p, q) distributions
were studied by Fielitz and Myers (1975, 1976), Romesburg (1976),
then clarified and summarized by Kottas and Lau (1978). Lau and
Lau (1991) compared several methods for computing the MLE's in
this case and identified which is the most effective method, then
proposed a practical way to establish confidence intervals for the
parameters.

(ii) In the case a and b also are unknown the study of MLE's of
parameters is based on the density function (II.1). In this case the

loglikelihood function is

$$\log L(a,b,p,q;x) = n\Big[\log\gamma(p+q) - \log\Gamma(p) - \log\Gamma(q)$$
$$+(1-p-q)\log(b-a)\Big]$$
$$+(p-1)\sum_{i=1}^{n}\log(x_i+a)$$
$$+(q-1)\sum_{i=1}^{n}\log(b-x_i).$$

Differentiating the likelihood function with respect to a, b, p, q, respectively, the following system of maximum likelihood equations for finding the respective MLE's $\hat{a}$, $\hat{b}$, $\hat{p}$ and $\hat{q}$ of a, b, p, q is obtained.

$$\psi(\hat{p}) - \psi(\hat{p}+\hat{q}) = n^{-1}\sum_{i=1}^{n}\log(x_i-\hat{a}) - n\log(\hat{b}-\hat{a})$$

$$\psi(\hat{q}) - \psi(\hat{p}+\hat{q}) = n^{-1}\sum_{i=1}^{n}\log\left(\hat{b}-x_i\right) - n\log(\hat{b}-\hat{a}),$$

$$\frac{\hat{p}+\hat{q}-1}{\hat{p}-1} + \hat{b} - \hat{a} = n^{-1}\sum_{i=1}^{n}(x_i-\hat{a})^{-1},$$

$$\frac{\hat{p}+\hat{q}-1}{\hat{q}-1} + \hat{b} - \hat{a} = n^{-1}\sum_{i=1}^{n}\left(\hat{b}-x_i\right)^{-1}.$$

To find appropriate numerical methods solving this system of maximum likelihood equations, that is, to find the MLE's for a, b, p, q is the subject of the papers of Koshal (1933), and then Carnahan (1989). Carnahan gave the regularity condition for the Beta distribution and asymptotic properties of MLE's. He also numerically studied the bias and variance of parameter estimators. He pointed out that there is a problem of global maximum which plagues the numerical scheme to maximize the likelihood function especially when there is a non well defined local maximum. According to his numerical study, he concluded that the MLE's appear to be unbiased and have variance approximately equal to the Cramer-Rao lower bound only at sample size approaching 1000, and there are some cases the likelihood function does not have a well defined local maximum at smaller sample. He suggested that we modify the search algorithm for these cases, such as using some of the approaches that have been effective in MLE for other distributions with a threshold or shift parameter. He suggested to employ the first and nth order statistics to improve the estimates of the end points. Carnahan also studied the asymptotic properties of these MLE's, and gave the conclusion it might be stressed here that, although such asymptotic properties are desirable, in practice, sample sizes are discouragingly small.

References

1. Beckman, R.; Tietjen, G. (1978). Maximum likelihood estimation for the Beta distribution, Journal of Statistical Computation and Simulation **1978**, *7*, 253-258.

2. Carnahan, J.V. Maximum likelihood estimation for the four parameter beta distribution, Communications in Statistics–Simulation and Computation **1989**, *18*, 513-536.

3. Dishon, M.; Weiss, G.H. Small sample comparison of estimation methods for the Beta distribution, Journal of Statistical Computation and Simulation **1980**, *11*, 1-11.

4. Farnum, N.R.; Stanton, L.W. Some results concerning the estimation of Beta distribution parameters in PERT, Journal of the Operations Research Society **1987**, *38*, 287-290.

5. Fielitz, B.D.; Myers, B.L. Estimation of parameters in the Beta distribution, Decision Sciences **1975**, *6*, 1-13.

6. Fielitz, B.D.; Myers, B.L. Estimation of parameters in the Beta distribution: reply, Decision Sciences **1976**, *7*, 163-164.

7. Gnanadesikan, R.; Pinkham, R.S.; Hughes, L.P. Maximum likelihood estimation of the parameters of the Beta distribution from smallest order statistics, Technometrics **1967**, *9*, 607-620.

8. Johnson, N.L.; Kotz, S. *Continuous Univariate Distributions* (volume 2); Houghton Mifflin Company: Boston, 1970.

9. Johnson, N.L.; Kotz, S.; Balakrishnan, N. *Continuous Univariate Distributions* (volume 2); John Wiley and Sons: New York, 1995.

10. Koshal, R.S. Application of the method of maximum likelihood to the derivation of efficient statistics for fitting frequency curves, Journal of the Royal Statistical Society, A **1935**, *98*, 265-272.

11. Koshal, R.S. Application of the method of maximum likelihood to the improvement of curves fitted by the method of moments, Journal of the Royal Statistical Society, A **1993**, *96*, 303-313.

12. Kottes, J.F.; Lau, H.S. On estimating parameters for Beta distributions, Decision Sciences **1978**, *9*, 526-531.

13. Lau, H.S.; Lau, A.H.L. Effective procedures for estimating Beta distributions's parameters and their confidence intervals, Journal of Statistical Computation and Simulation **1991**, *38*, 139-150.

14. Romesburg, H.C. Estimation of parameters in the Beta distribution, Decision Sciences **1976**, *7*, 162.

Goodness-of-Fit Testing of the Beta-Binomial Model

Steven T. Garren

Department of Mathematics and Statistics
James Madison University
Harrisonburg, Virginia 22807

I. Introduction

The beta-binomial distribution is commonly used to model overdispersed data. Testing goodness-of-fit to this distribution is problematic due to varying sample sizes, and several testing procedures from the literature are reviewed herein. The simple procedure proposed by Neerchal and Morel (1998) is based on a Pearson statistic. A more involved procedure, which is also based on Pearson statistics but in addition is based on bootstrapping, is by Garren, Smith, and Piegorsch (2001), and is discussed in much detail. These latter two procedures test goodness-of-fit of the beta-binomial model against all alternative models, and are both quite reasonable testing procedures.

We now discuss extra-binomial variability, and how it relates to the beta-binomial model. In many experiments, binary data are collected in the form of sample proportions, Y/n, where $Y = 0, 1, \ldots, n$ and n is a positive integer. For a given n, the number, Y, of responses might be modeled as binomial(n, p). If the binary data are correlated, then the binomial model is not suitable. Haseman and Piegorsch (1994) discuss such correlated data, where fetuses of laboratory rodents exposed to some toxic stimulus have binary responses. Since the probability of response may vary across the pregnant rodents, then the binary data are correlated within each litter, so the binomial model should not be used. Hence, a *litter effect* results, and the data may be modeled hierarchically by the following: For a given pregnant rodent, let p be the probability of response for any particular offspring. The parameter p is allowed to vary according to a beta distribution, and values of p are independent across litters. For a given litter size n and response proportion p, the number, Y, of responses among the n pups in the litter is modeled as binomial(n, p). Hence, unconditional on p, the number, Y, of responses is beta-binomial (Williams, 1975; Haseman and Kupper, 1979). The mathematical details of the beta-binomial model are provided in Section II. For simplicity, throughout this chapter, we tend to restrict our discussions to these types of experiments involving litters of rodents. However, the mathematical details are, of course, valid for other types of experiments. For example, Madden and Hughes (1994) discuss the beta-binomial model with regards to Phomopsis leaf blight of strawberries, where multiple strawberry leaves at each of many locations are sampled to determine the disease status.

When data show evidence of extra-binomial variability, then the beta-binomial model might be preferred over the binomial model. Various methods for testing for departure from the binomial model have been reviewed by Risko and Margolin (1996). Testing for the departure from beta-binomial models is more complicated, since the litter sizes, n, may vary across litters.

II.   Beta-Binomial Model

Suppose J litters are sampled, and n_i is the littersize of the ith litter, for $i = 1, \ldots, J$. Let y_i be the number of litter-mates which are malformed in the ith litter, where the term "malformed" must be defined prior to the experiment. The data pairs (y_1, n_1), (y_2, n_2), $\ldots$, (y_J, n_J) are assumed to be independent.

Suppose that Y_i conditional on n_i and p_i is binomial(n_i, p_i). Now, allow p_i conditional on α and β to have the beta density

$$f(p_i|\alpha, \beta) = [B(\alpha, \beta)]^{-1} p_i^{\alpha-1} (1 - p_i)^{\beta-1},$$

where $0 < p_i < 1$, and α and β are both positive. Unconditional on p_i, the variable Y_i given n_i, α and β is said to be beta-binomial and has probabilities

$$\Pr(Y_i = y \mid n_i, \alpha, \beta) = \binom{n_i}{y} \frac{B(\alpha + y, \beta + n_i - y)}{B(\alpha, \beta)}, \qquad \text{(II.1)}$$

for $y = 0, \ldots, n$, where $B(\cdot, \cdot)$ is the beta function.

Note that the J data pairs are based on the common values of α and β. These two unknown parameters may be reparameterized as the *mean* parameter

$$\mu = \frac{\alpha}{\alpha + \beta} \qquad \text{(II.2)}$$

and the *dispersion* parameter

$$\theta = \frac{1}{\alpha + \beta}. \qquad \text{(II.3)}$$

The beta-binomial probabilities in (II.1) may be conveniently written as

$$\Pr(Y_i = y \mid n_i, \mu, \theta) = \binom{n_i}{y} \prod_{k=0}^{y-1} (\mu + k\theta)$$

$$\times \prod_{k=0}^{n_i-y-1} (1 - \mu + k\theta) \left/ \prod_{k=0}^{n_i-1} (1 + k\theta),\right.$$

for $y = 0, \ldots, n_i$. The mean and variance of the beta-binomial random variables are

$$E(Y_i|n_i, \mu, \theta) = n_i \mu$$

and

$$\mathrm{Var}(Y_i|n_i, \mu, \theta) = n_i \mu (1 - \mu)(1 + \theta)^{-1} (1 + n_i \theta).$$

Therefore, the variance of a beta-binomial distribution is monotonically increasing in θ for $n_i \geq 2$. In the limiting case as $\theta \downarrow 0$, the beta-binomial distribution converges to a binomial distribution. By allowing the limiting case when $\theta = 0$, the binomial distribution is a special case of the beta-binomial distribution. When θ is strictly positive, the data are said to be *overdispersed*, since the variance is larger than than from a binomial distribution. *Underdispersion* (Prentice, 1986; Engel and te Brake, 1993) occurs when the variance of Y_i is smaller than that for the binomial distribution, but does not commonly occur in developmental toxicology and is not examined further in this chapter.

In most any practical example, not only is the distribution unknown, but the parameters of the distribution are also unknown. Under the beta-binomial model, the two unknown parameters may be estimated by maximum likelihood, since analytical solutions are not possible. Ennis and Bi (1998) discuss numerically estimating the maximum likelihood estimators (MLE) of the beta-binomial distribution. Furthermore, computer programs for determining the MLE were written by Erdfelder (1993) using Basic code and by Madden and Hughes (1994) using Fortran code. The MLE can be shown to be consistent (Lehmann, 1983, pages 409-413) as J goes to infinity.

III. Approaches to Goodness-of-Fit

As in Section II, assume that our independent pairs of observations are of the form (y_i, n_i) for $i = 1, \ldots, J$, where y_i is the number of responses out of n_i trials. We also consider μ and θ to be the unknown parameters of the beta-binomial distribution, as defined by (II.2) and (II.3). The reason why standard goodness-of-fit tests fail in this scenario is that n_i is not constant.

A. Mantel and Paul (1987)

To resolve the problem of varying values of n_i, Mantel and Paul (1987) assume that the n_i are from some known distribution. They consider the distribution of Y_i unconditional on n_i, estimated the MLE of (μ, θ), and used the Pearson chi-squared statistic to determine goodness-of-fit. The drawback is that information about the n_i is lost under their approach. For example, the two (y_i, n_i) data pairs $\{(0, 10), (3, 3)\}$ have greater overdispersion than the two data pairs $\{(0, 3), (3, 10)\}$, but these two data sets are mathematically equivalent according to the approach of Mantel and Paul (1987). Hence, this approach is recommended only when the values of n_i are constant or at least close to constant.

B. Erdfelder (1993)

The computer code written by Erdfelder (1993) not only estimates the MLE of (μ, θ), but also determines goodness-of-fit to the beta-binomial distribution and goodness-of-fit to a binomial mixture distribution. Their goodness-of-fit testing is based on the Pearson chi-squared statistics and the likelihood ratio statistics. The drawback is that this goodness-of-fit testing works only when n_i is constant. However, the computing algorithm for estimating (μ, θ) is valid even when n_i is allowed to vary.

C. Liang and McCullagh (1993)

Instead of directly testing goodness-of-fit of the beta-binomial model, Liang and McCullagh (1993) test if the ratio of the mean to the variance is consistent with that of the beta-binomial model. The model parameters are estimated by quasi-likelihood, since the beta-binomial distribution is not the null model. Since some distributions may have a mean-variance ratio structure similar to that of the beta-binomial, then their test may have difficulty distinguishing certain alternative distributions from beta-binomial distributions.

D.   Neerchal and Morel (1998)

The approach by Neerchal and Morel (1998) is both simplistic and useful, in that the Pearson chi-squared statistic is applied to the sample proportions, Y_i/n_i. Nonoverlapping and exhaustive intervals between zero and one constitute the categories for these sample proportions. The expected number of observations in a particular category depends on the MLE of (μ, θ). After the Pearson chi-squared statistic is calculated, the P-value is determined based on the chi-squared distribution, where the number of degrees of freedom is three fewer than the number of categories.

As with any Pearson statistic, the selection of the categories may affect the P-value somewhat, since information is lost whenever data are categorized. However, the particular selection of categories becomes less relevant, as the number of data pairs and the number of categories increase appropriately.

Careless selection of the categories can lead to an unreasonable conclusion regarding goodness-of-fit. Consider the following hypothetical example, which is unlikely to occur as a real data set. Suppose $n_i = 5$ for all observations, and the four selected categories are $[0, 0.3)$, $[0.3, 0.5)$, $[0.5, 0.7)$, and $[0.7, 1]$. Suppose that J is a multiple of 16, and that the first 16 observations of Y_i are $\{1, 1, 1, 2, 2, 2, 2, 2, 3, 3, 3, 3, 3, 4, 4, 4\}$. Also, assume that the next 16 observations are the same as the first sixteen observations, and so on. Therefore, the proportion of ones is 3/16, the proportion of twos is 5/16, the proportion of threes is 5/16, and the proportion of fours is 3/16. Since Y_i never takes on the values 0 and 5, the distribution which generates Y_i cannot be beta-binomial. Nevertheless, the MLE of (μ, θ) is $(0.5, 0)$ for all J (which is a multiple of 16). Therefore, the binomial$(5, 0.5)$ distribution is the best fit to the beta-binomial distribution. Under the binomial$(5, 0.5)$ distribution, the *expected* number of observations in the category $[0, 0.3)$ is $J \times [\Pr(Y = 0) + \Pr(Y = 1)] = 3J/16$, where $Y \sim$ binomial$(5, 0.5)$. However, the *observed* number of observations in the category $[0, 0.3)$ is also $3J/16$. Similarly, the *expected* number of observations is the same as the *observed* number of observations in each of the other three selected categories. Therefore, the Pearson chi-squared statistic is zero for this example, so this Pearson statistic, based on one degree of freedom, would fail to reject the null hypothesis that the data are beta-binomial, even as J goes to infinity.

Although the above example illustrates a situation where the approach by Neerchal and Morel (1998) might fail, that example is based on a poor selection of categories and should not be interpreted as a counterexample to their approach. In the above example, as J gets large, then the number of categories should increase, eventually to six categories corresponding to the six possible values of Y_i/n_i. After all six categories are used, then with a moderate value of J, their test statistic will show with high probability that the data come from a distribution which is not beta-binomial. Therefore, whenever the categories are selected reasonably, this goodness-of-fit approach is quite satisfactory.

E. Brooks, Morgan, Ridout, and Pack (2000)

Brooks et al. (1997) originally proposed a statistic for testing goodness-of-fit to the beta-binomial distribution using a likelihood test statistic, rather than a likelihood ratio test statistic. Garren et al. (2000) showed that this likelihood test statistic will have rather low power under a particular model which is not beta-binomial. In the response to Garren et al. (2000), Brooks et al. (2000) discussed using the deviance instead, as suggested by Pack and Morgan (1990). Brooks et al. (2000) asserted that using the deviance is, indeed, appropriate when testing for goodness-of-fit, when a reasonable alternative distribution is specified. Lockhart et al. (1992) also used a deviance statistic, but their alternative model allowed each litter to be independent binomial(n_i, p_i); however, since the number of unknown parameters p_i gets large as J gets large, then technical difficulties arise when attempting to interpret their results.

IV. A Bootstrap Test Statistic by Garren, Smith, and Piegorsch (2001)

A bootstrap test procedure, as discussed by Garren et al. (2001), is based on bootstrapping Pearson chi-squared statistics. The data sets and results published in this section as Tables 1 through 9 are also available in Garren et al. (2001), and these tables are reproduced with permission from the publisher (http://www.tandf.co.uk). Our statistic, τ, for testing goodness-of-fit to the beta-binomial model is a P-value, and will be constructed below. Thus, small values of τ will be strong evidence that the data are from a distribution which is not beta-binomial.

A. Construction of the Test Statistic τ

For simplicity, we again consider the example where n_i is the litter size of the ith litter, Y_i is the number of malformed pups in the ith litter, and J is the total number of litters. Let K be the number of litter sizes, and denote the K litter sizes by $n_1, n_2, \ldots, n_K$.

Let n be in $\{n_1, \ldots, n_K\}$ and let y be a nonnegative integer no larger than n. The observed number of litters of size n with y malformations is denoted by $O_{y,n}$. Determine the expected number of litters of size n with y malformations, based on the MLE of (μ, θ) in the beta-binomial model (II.1); denote this expectation by $E_{y,n}$. An individual Pearson statistic is defined by

$$q_n = \sum_{y=0}^{n} (O_{y,n} - E_{y,n})^2 \Big/ E_{y,n}. \qquad (\text{IV.1})$$

In many practical data sets, quite a few values of $E_{y,n}$ are close to zero, so approximating the distribution of q_n using a chi-squared distribution is not at all reasonable.

Now, we introduce a parametric bootstrap aspect of the algorithm, noting that bootstrapping test statistics is suggested by Romano (1988). Using the MLE of (μ, θ) and the J litter sizes, generate beta-binomial pseudo-random variates. Compute a bootstrapped

Pearson statistic in (IV.1) for each n, based on these pseudo-random variates. Perform this bootstrapping procedure M times, and label the bootstrapped Pearson statistics as $Q_{n,1}^*$, $Q_{n,2}^*$, ..., $Q_{n,M}^*$. Notice that the MLE of (μ, θ) needs to be re-estimated for each of the M bootstrapped Pearson statistics. The distribution of the non-bootstrapped q_n is estimated by

$$\rho_n = M^{-1} \sum_{m=1}^{M} I\left(Q_{n,m}^* < q_n\right),$$

where $I(\cdot)$ is the indicator function. The distribution of ρ_n can be regarded as approximately uniform$(0, 1)$, but not exactly uniform, not even in the limit as $M \to \infty$ since q_n is discrete.

Since each ρ_n is approximately uniform$(0, 1)$ under the beta-binomial model, then our overall test statistic,

$$\tau = 1 - \left(\max_{k=1,\dots,K} \rho_{n_k}\right)^K,$$

is also approximately uniform$(0, 1)$ under the beta-binomial model. The null hypothesis that the data come from a beta-binomial distribution is rejected in favor of the alternative hypothesis that the data come from a distribution which is not beta-binomial, whenever τ is suitably small. Hence, the statistic τ is the P-value for this testing procedure.

B. A Monte Carlo Study

To examine the size and power of our test statistic, a simulation study was performed using $J = 50$ litters. The 50 litter sizes used are from the actual data set in Table 1, and they vary in size from 6 to 18. To study the size and power of our test statistic, beta-binomial pseudo-random variates were generated using all nine pairings of the parameters $\mu \in \{0.05, 0.1, 0.15\}$ and $\theta \in \{0, 0.05, 0.1\}$. To study the power of our test statistic, the alternative distribution defined is the beta-binomial distribution mixed with a binomial distribution with parameter $p \in \{0.7, 0.8, 0.9\}$, such that the beta-binomial distribution is selected with mixing probability $\{0.85, 0.9, 0.95\}$. When the mixing probability is large, the alternative model might be viewed as a beta-binomial distribution contaminated with a few outliers. The number of bootstrap samples is $M = 1000$, and the number of independent replications is 2000.

The estimated size or power is the proportion, γ, of independent replications whose value of τ is smaller than the nominal values $\tau_0 = \{0.1, 0.05, 0.025, 0.01\}$. The standard error is γ is approximated by $\sqrt{\gamma(1-\gamma)/2000}$. Duplicating the litter sizes in Table 1 results in $J = 100$ litters. The estimated sizes for $J = 50$ and $J = 100$ litters are shown in Tables 4 and 5, respectively. The estimated powers for $J = 50$ and $J = 100$ litters are shown in Tables 6 and 7, respectively.

The estimated sizes, τ, are shown to be close to their nominal values, τ_0, in Tables 4 and 5. Since Tables 4 and 5 are similar, then some of the error in estimating the sizes may be from using only 2000 bootstrap samples and only $M = 1000$ independent replications, rather than from the limited number of litters. Power increases somewhat by increasing the number of litters from 50 to 100, as noted when comparing Tables 6 and 7. The test statistic shows much power when the mixing probability is large and when the binomial probability is also large, since at least one value of y may be extreme and its corresponding value of $E_{y,n}$ is approximately zero. Thus, one outlier can inflate a Pearson statistic. Therefore, our test statistic is quite powerful against the alternative model of a beta-binomial contaminated with at least one outlier. The Fortran program which produced Tables 4 through 7 may be found at http://www.stat.unc.edu/postscript/rs/betabin.

C. Examples

We applied our goodness-of-fit test statistic to data sets involving pregnant mice, originally studied by Lockhart et al. (1992). Those experiments involved matings between parent mice to examine damage in the resulting embryos based on dominant lethal mutations. Such damage was assessed by sacrificing the pregnant females approximately two weeks after mating to examine the uterine contents. For each litter the number of viable implants and the number of non-viable implants were determined (Lockhart et al., 1991). These parent mice were not exposed to any toxic chemicals before or during the experiment.

Our goodness-of-fit test statistic was applied to the data sets listed in Tables 1 through 3, to test whether the data were beta-binomial against all alternative models. The results are summarized in Table 8. The number of bootstrap samples used was $M = 100,000$ for determining the P-value, τ. At the 0.05 level, the data sets from Tables 2 and 3 are statistically significant, whereas the data set from Table 1 is not significant, as observed from Table 8. The data set from Table 3 has the P-value $\tau = 0.000$, which may be explained by the three outlying observations of the (y, n) pairs $(7, 7)$, $(9, 9)$, and $(5, 8)$. Under the beta-binomial model, those three observations are quite unlikely, considering that the MLE of (μ, θ) is $(0.068, 0.064)$. When those three observations are removed from the data set, the new P-value becomes $\tau = 0.054$, which is still rather small. Hence, these three observations account for perhaps part of the reason why the data set shows departure from the beta-binomial model.

Our goodness-of-fit test statistic was also applied to six data sets which were examined and published by Brooks et al. (1997). The results are summarized in Table 9, again using $M = 100,000$ bootstrap samples. The tables in Brooks et al. (1997) contain some minor topographical errors: In their Table 1 the entry at position $(8, 14)$ should be moved to $(9, 14)$; in Table 2 the entry of "one" should appear at position $(11, 16)$; in Table 5 the entry at position $(9, 10)$ should be moved to $(10, 10)$. The P-values, τ, of the six data sets numbered 1 through 6 are 0.144, 0.231, 0.000, 0.000, 0.009, and 0.375, respectively. Thus, data sets #3, #4, and #5 are highly statistically significant, showing strong departure from the beta-binomial model, whereas the other three data sets do not show such departure. The Fortran program which produced Tables 8 through 9 is available at the same website listed in Section IV.B.

Brooks et al. (1997) also had tested goodness-of-fit to the beta-binomial model on the six data sets analyzed in Table 9. However, their use of the likelihood statistic rather than a deviance statistic proved to be unsatisfactory, as discussed in Section III.E.

Table I. Frequency of litter sizes with number of non-viable implants in Swiss CD-1 mice, from Lockhart et al. (1992)

		\multicolumn{5}{c}{y, number of non-viable implants}
		0	1	2	3	4
	6	-	1	-	-	-
	7	-	-	-	-	-
	8	-	-	-	-	-
	9	1	-	-	-	-
	10	-	-	-	-	-
n,	11	1	4	-	-	1
litter	12	4	2	1	-	-
size	13	-	2	1	1	-
	14	4	2	2	-	1
	15	4	6	2	1	-
	16	3	2	-	-	1
	17	-	1	-	-	-
	18	1	1	-	-	-

Note: This data set consists of $J = 50$ litters.

Table II. Frequency of litter sizes with number of non-viable implants in Swiss CD-1 mice after sham intraperitoneal injection, from Lockhart et al. (1992)

		\multicolumn{5}{c}{y, number of non-viable implants}
		0	1	2	3	4
	2	1	-	-		
	3	2	1	-	-	
	4	-	-	-	1	-
	5	2	-	-	-	-
	6	-	1	-	-	-
	7	3	2	-	-	-
n,	8	3	-	-	1	-
litter	9	9	3	-	-	-
size	10	20	9	1	1	-
	11	42	15	4	-	-
	12	26	15	4	-	-
	13	10	5	4	-	2
	14	5	2	2	-	1
	15	-	1	-	-	2
	16	1	-	-	-	-

Note: This data set consists of $J = 201$ litters.

Table III. Frequency of litter sizes with number of non-viable implants in $(SEC \times C57L)F_1 \times [(SEC \times C57L)F_1 \times C3H \times 101]$ mice, from Lockhart et al. (1992)

		y, number of non-viable implants									
		0	1	2	3	4	5	6	7	8	9
	4	1	-	-	-	-					
	5	-	1	-	-	-	-				
	6	1	2	-	-	-	-	-			
	7	3	2	-	-	1	-	-	1		
	8	3	-	-	-	-	1	-	-	-	
	9	4	2	1	-	-	-	-	-	-	1
n,	10	12	4	1	1	-	1	-	-	-	-
litter	11	12	9	4	-	1	-	-	-	-	-
size	12	20	11	8	-	-	-	-	-	-	-
	13	38	20	12	2	-	-	-	-	-	-
	14	20	17	5	1	1	-	-	-	-	-
	15	12	10	2	-	-	-	-	-	-	-
	16	4	3	3	1	-	-	-	-	-	-
	17	1	-	1	-	-	1	-	-	-	-
	18	-	-	1	-	-	-	-	-	-	-

Note: This data set consists of $J = 263$ litters.

Table IV. Estimated sizes of proposed test, based on simulations from beta-binomial distributions using 50 litters

		Estimated size at nominal level τ_0			
μ	θ	$\tau_0 = 0.1$	$\tau_0 = 0.05$	$\tau_0 = 0.025$	$\tau_0 = 0.01$
0.05	0.00	0.0765	0.0420	0.0180	0.0060
0.05	0.05	0.1155	0.0675	0.0305	0.0210
0.05	0.10	0.1225	0.0710	0.0395	0.0250
0.10	0.00	0.0855	0.0450	0.0265	0.0160
0.10	0.05	0.1100	0.0655	0.0330	0.0195
0.10	0.10	0.1180	0.0635	0.0260	0.0190
0.15	0.00	0.0835	0.0465	0.0230	0.0155
0.15	0.05	0.1165	0.0625	0.0285	0.0160
0.15	0.10	0.1025	0.0585	0.0300	0.0210

Table V. Estimated sizes of proposed test, based on
simulations from beta-binomial distributions using 100
litters

		Estimated size at nominal level τ_0			
μ	θ	$\tau_0 = 0.1$	$\tau_0 = 0.05$	$\tau_0 = 0.025$	$\tau_0 = 0.01$
0.05	0.00	0.0775	0.0365	0.0185	0.0125
0.05	0.05	0.1270	0.0770	0.0395	0.0250
0.05	0.10	0.1155	0.0615	0.0295	0.0195
0.10	0.00	0.0925	0.0505	0.0285	0.0180
0.10	0.05	0.0980	0.0570	0.0295	0.0220
0.10	0.10	0.1170	0.0600	0.0280	0.0190
0.15	0.00	0.0860	0.0500	0.0255	0.0160
0.15	0.05	0.1110	0.0610	0.0335	0.0275
0.15	0.10	0.0985	0.0605	0.0310	0.0220

Table VI. Estimated power of proposed test, based on simulations
from mixture models using 50 litters

beta-binomial		binomial	mixing	Estimated power at nominal level τ_0			
μ	θ	p	probability	$\tau_0 = 0.1$	$\tau_0 = 0.05$	$\tau_0 = 0.025$	$\tau_0 = 0.01$
0.05	0.00	0.7	0.85	0.2530	0.1495	0.0785	0.0545
0.05	0.05	0.7	0.90	0.2510	0.1445	0.0790	0.0595
0.05	0.10	0.7	0.95	0.3475	0.2420	0.1470	0.1070
0.10	0.00	0.8	0.85	0.4200	0.2550	0.1465	0.1115
0.10	0.05	0.8	0.90	0.4635	0.3185	0.1990	0.1360
0.10	0.10	0.8	0.95	0.5215	0.3845	0.2500	0.1875
0.15	0.00	0.9	0.85	0.5465	0.3915	0.2555	0.1970
0.15	0.05	0.9	0.90	0.5720	0.4355	0.2855	0.2225
0.15	0.10	0.9	0.95	0.6600	0.5295	0.3675	0.2910

Table VII. Estimated power of proposed test, based on simulations
from mixture models using 100 litters

beta-binomial		binomial	mixing	Estimated power at nominal level τ_0			
μ	θ	p	probability	$\tau_0 = 0.1$	$\tau_0 = 0.05$	$\tau_0 = 0.025$	$\tau_0 = 0.01$
0.05	0.00	0.7	0.85	0.2850	0.1410	0.0705	0.0445
0.05	0.05	0.7	0.90	0.2370	0.1300	0.0605	0.0430
0.05	0.10	0.7	0.95	0.3195	0.1960	0.1145	0.0725
0.10	0.00	0.8	0.85	0.5735	0.3850	0.2005	0.1340
0.10	0.05	0.8	0.90	0.4880	0.3070	0.1685	0.1065
0.10	0.10	0.8	0.95	0.5685	0.4025	0.2475	0.1775
0.15	0.00	0.9	0.85	0.8050	0.6615	0.4825	0.3800
0.15	0.05	0.9	0.90	0.7045	0.5700	0.3975	0.3040
0.15	0.10	0.9	0.95	0.7850	0.6350	0.4500	0.3520

Table VIII. Analysis of dominant lethal
data

Table number	Number of litters	MLE $\hat{\mu}$	$\hat{\theta}$	P-value τ
1	50	0.075	0.021	0.333
2	201	0.051	0.041	0.010
3	263	0.068	0.064	0.000

Table IX. Analysis of data sets published
by Brooks et al. (1997)

Data set number	Number of litters	MLE $\hat{\mu}$	$\hat{\theta}$	P-value τ
1	205	0.090	0.074	0.144
2	211	0.112	0.111	0.231
3	524	0.090	0.073	0.000
4	1328	0.109	0.045	0.000
5	554	0.074	0.081	0.009
6	127	0.069	0.063	0.375

Acknowledgment

The author is very grateful to Walter W. Piegorsch, Richard L. Smith, Elizabeth R. Hume, and Glen R. Leppert for useful conversations on this subject.

References

1. Brooks, S.P.; Morgan, B.J.T.; Ridout, M.S.; Pack, S.E. Finite mixture models for proportions, Biometrics **1997**, *53*, 1097-1115.

2. Brooks, S.P.; Morgan, B.J.T.; Ridout, M.S.; Pack, S.E. Response to reader reaction by Garren, Smith, and Piegorsch, Biometrics **2000**, *56*, 950.

3. Engel, B.; te Brake, J. Analysis of embryonic development with a model for underdispersion or overdispersion relative to binomial variation, Biometrics **1993**, *49*, 269-279.

4. Ennis, D.M.; Bi, J. The beta-binomial model: accounting for inter-trial variation in replicated difference and preference tests, Journal of Sensory Studies **1998**, *13*, 389-412.

5. Erdfelder, E. Binomix: A Basic program for maximum likelihood analyses of finite and beta-binomial mixture distributions, Behavior Research Methods, Instruments, and Computers **1993**, *25*, 416-418.

6. Garren, S.T.; Smith, R.L.; Piegorsch, W.W. Reader reaction: On a likelihood-based goodness-of-fit test of the beta-binomial model, Biometrics **2000**, *56*, 947-949.

7. Garren, S.T.; Smith, R.L.; Piegorsch, W.W. Bootstrap goodness-of-fit test for the beta-binomial model, Journal of Applied Statistics **2001**, *28*, 561-571.

8. Haseman, J.K.; Kupper, L.L. Analysis of dichotomous response data from certain toxicological experiments, Biometrics **1979**, *35*, 281-293.

9. Haseman, J.K.; Piegorsch, W.W. Statistical analysis of developmental toxicity data. In: *Developmental Toxicology* (editors C. Kimmel and J. Buelke-Sam) (second edition) **1994**, pp. 349-361, Raven Press: New York.

10. Lehmann, E.L. *Theory of Point Estimation*; Wiley: New York, 1983.

11. Liang, K.-Y.; McCullagh, P. Case studies in binary dispersion, Biometrics **1993**, *49*, 623-630.

12. Lockhart, A.-M.; Bishop, J.B.; Piegorsch, W.W. Issues regarding data acquisition and analysis in the dominant lethal assay, Proceedings of the American Statistical Association, Biopharmaceutical Section **1991**, 234-237.

13. Lockhart, A.-M.; Piegorsch, W.W.; Bishop, J.B. Assessing overdispersion and dose response in the male dominant lethal assay, Mutation Research **1992**, *272*, 35-58.

14. Madden, L.V.; Hughes, G. BBD–Computer software for fitting the beta-binomial distribution to disease incidence data, Plant Disease **1994**, *78*, 536-540.

15. Mantel, N.; Paul, S.R. Goodness-of-fit issues in toxicological experiments involving litters of varying size. In: *Biostatistics* (editors I.B. MacNeill and G.J. Umphrey) **1987**, pp. 169-176, Reidel Publishing Company: New York.

16. Neerchal, N.K.; Morel, J.G. Large cluster results for two parametric multinomial extra variation models, Journal of the American Statistical Association **1998**, *93*, 1078-1087.

17. Pack, S.E.; Morgan, B.J.T. A mixture model for interval-censored time-to-response quantal assay data, Biometrics **1990**, *46*, 749-758.

18. Prentice, R.L. Binary regression using an extended beta-binomial distribution, with discussion of correlation induced by covariate measurement errors, Journal of the American Statistical Association **1986**, *81*, 321-327.

19. Risko, K.J.; Margolin, B.H. Some observations on detecting extra-binomial variability within the beta-binomial model. In: *Statistics in Toxicology* (editor B. J. T. Morgan) **1996**, pp. 57-65, Clarendon Press: Oxford.

20. Romano, J.P. A bootstrap revival of some nonparametric distance tests, Journal of the American Statistical Association **1988**, *83*, 698-708.

21. Williams, D.A. The analysis of binary responses from toxicological experiments involving reproduction and teratogenicity, Biometrics **1975**, *31*, 949-952.

Distributions with Beta Conditionals

Barry C. Arnold[1], Enrique Castillo[2] and José M. Sarabia[3]

[1]Department of Statistics
University of California
Riverside, California 92521, USA
[2]Departamento de Matemática Aplicada y
Ciencias de la Computación
University of Cantabria
39005 Santander, Spain
[3]Departamento de Economía
University of Cantabria
39005 Santander, Spain

I. Motivation

Multivariate analogs of the Beta distribution will clearly be useful tools for data analysts and modelers. However it is not immediately obvious how such distributions should be related to the basic family of Beta distributions. One common approach is to consider multivariate distributions whose marginal densities are of the Beta form. The best known such distribution is the Dirichlet distribution. It, together with other multivariate distributions with Beta marginals, will be discussed in detail in Chapter 19. A different approach will be considered in the present chapter. Our problem is to visualize suitable models for k-dimensional random vectors, each of whose coordinates has support in the unit interval. Multivariate densities are most easily described in terms of cross sections (think of how the topography of the continents is frequently illustrated in atlases by cross sectional drawings rather than 3 dimensional diagrams). Indeed the only serious competitors to cross sectional depiction is one involving contour maps. The contour approach to modeling has been fruitfully implemented in cases where the support of the marginal variables is the entire real line. This has led, for example, to the study of elliptically contoured variables. Such a contour focussed modeling approach seems difficult to implement effectively in the present setting where the marginal variables have unit interval support. What shape might the contours take here? However the cross-sectional approach admits a straightward and potentially fruitful implementation. For cross sections of the joint density are merely un-normalized conditional densities. Consequently, our modeling approach will involve postulating appropriate structure for various conditional distributions and considering the spectrum of multivariate models with the given conditional specifications. In particular, we will focus on families of multivariate distributions with support $(0,1)^k$ which have univariate conditional densities of the Beta form. We will begin discussion in the bivariate case which is discussed in some detail. Subsequently higher dimensional extensions will be considered. A convenient source for a survey of conditionally specified models of this genre is Arnold, Castillo and Sarabia (1999).

II.   The Bivariate Beta Conditionals Density

Let (X,Y) denote a two dimensional random variable with support $(0,1)^2$. We wish to consider all possible joint distributions for (X,Y) with the following properties:

(i)   For each $y \in (0,1)$, the conditional distribution of X given $Y = y$ is a Beta distribution with parameters $\alpha(y)$ and $\beta(y)$ which may depend on y.

(ii)   For each $x \in (0,1)$, the conditional distribution of Y given $X = x$ is a Beta distribution with parameters $\gamma(x)$ and $\delta(x)$

which may depend on x.

Such distributions will be called Beta conditionals distributions.

Since the family of Beta distributions is a two parameter exponential family we can make use of a theorem due to Arnold and Strauss (1991) dealing with bivariate distributions with conditionals in prescribed exponential families.

Consider two possibly different exponential families of densities $\{f_1(x;\underline{\theta}) : \underline{\theta} \in \Theta \subset \mathbf{R}^{\ell_1}\}$ and $\{f_2(y;\underline{\tau}) : \underline{\tau} \in T \subset \mathbf{R}^{\ell_2}\}$ where

$$f_1(x;\underline{\theta}) = r_1(x)\beta_1(\underline{\theta}) \exp\left[\sum_{i=1}^{\ell_1} \theta_i q_{1i}(x)\right] \qquad (II.1)$$

and

$$f_2(y;\underline{\tau}) = r_2(y)\beta_2(\underline{\tau}) \exp\left[\sum_{j=1}^{\ell_2} \tau_j q_{2j}(y)\right]. \qquad (II.2)$$

We can identify all bivariate densities $f(x,y)$ with conditionals in these prescribed exponential families as follows.

Theorem 1. *Let $f(x,y)$ be a bivariate density whose conditional densities satisfy*

$$f(x|y) = f_1(x;\underline{\theta}(y))$$

and

$$f(y|x) = f_2(y;\underline{\tau}(x))$$

for every x and y for some functions $\underline{\theta}(y)$ and $\underline{\tau}(x)$ where f_1 and f_2 are as defined in (II.1) and (II.2). It follows that $f(x,y)$ is of the form

$$f(x,y) = r_1(x)r_2(y) \exp[\underline{q}^{(1)}(x)M\underline{q}^{(2)}(y)'] \qquad (II.3)$$

in which

$$\underline{q}^{(1)}(x) = (1, q_{11}(x), \ldots, q_{1\ell_1}(x))$$

and

$$\underline{q}^{(2)}(y) = (1, q_{21}(y), \ldots, q_{2\ell_2}(y))$$

and M is a matrix of parameters of dimension $(\ell_1 + 1) \times (\ell_2 + 1)$ subject to the requirement that

$$\int\int_{\mathbf{R}^2} f(x,y)dxdy = 1. \qquad (II.4)$$

The factor $\exp(m_{00})$ in (II.3) is a normalizing constant that is a function of the other m_{ij}'s determined by the constraint (II.4).

It may thus be observed that (II.3) (the class of densities with conditionals in the prescribed families) is itself an exponential family with $(\ell_1 + 1) \times (\ell_2 + 1) - 1$ parameters. Upon partitioning the matrix M in (II.3) in the following manner

$$M = \begin{pmatrix} m_{00} & m_{01} & \ldots & m_{0\ell_2} \\ m_{10} & & & \\ \vdots & & \widetilde{M} & \\ m_{\ell_1 0} & & & \end{pmatrix},$$

it can be verified that independent marginals will be encountered iff $\widetilde{M} \equiv 0$. The elements $\widetilde{M}$ thus determine the dependence structure in the joint density.

We may apply this theorem directly to the case of Beta conditionals, i.e. the case in which (II.1) and (II.2) correspond to Beta families. In this situation, $\ell_1 = \ell_2 = 2$ and the functions r_1, r_2, q_{11}, q_{22}, q_{21} and q_{22} in (II.1) and (II.2) are as follows.

$$r_1(x) = [x(1-x)]^{-1}I(0 < x < 1),$$
$$r_2(y) = [y(1-y)]^{-1}I(0 < y < 1),$$
$$q_{11}(x) = \log x,$$
$$q_{12}(x) = \log(1-x),$$
$$q_{21}(y) = \log y,$$
$$q_{22}(y) = \log(1-y).$$

Substituting these functions in (II.3), we may write the general class of bivariate densities with Beta conditionals as

$$f(x,y) = [x(1-x)y(1-y)]^{-1} \exp \Big[m_{00} + m_{10} \log x + m_{01} \log y$$
$$+ m_{20} \log(1-x) + m_{02} \log(1-y)$$
$$+ m_{11} \log x \log y + m_{12} \log x \log(1-y)$$
$$+ m_{21} \log(1-x) \log y + m_{22} \log(1-x) \log(1-y) \Big]$$
$$\times I(0 < x, y < 1). \tag{II.5}$$

In order for this to be a proper density (integrate to 1) we need to impose the parameter restrictions:

$$m_{10}, m_{20}, m_{01}, m_{02} > 0$$

and

$$m_{11}, m_{12}, m_{21}, m_{22} \le 0.$$

The family (II.5) includes all densities with Beta conditionals.

Let us denote:

$$a_1(y) = m_{10} + m_{11} \log y + m_{12} \log(1-y)$$
$$b_1(y) = m_{20} + m_{21} \log y + m_{22} \log(1-y)$$
$$a_2(x) = m_{01} + m_{11} \log x + m_{21} \log(1-x)$$
$$b_2(x) = m_{02} + m_{12} \log x + m_{22} \log(1-x)$$

Then, we have:

$$X|Y = y \sim Beta\,(a_1(y), b_1(y)) \tag{II.6}$$
$$Y|X = x \sim Beta\,(a_2(x), b_2(x)) \tag{II.7}$$

and the marginal densities of X and Y are given by:

$$f_X(x) = \exp\,(m_{00})\, x^{m_{10}-1}(1-x)^{m_{20}-1} B\,(a_2(x), b_2(x))$$
$$f_Y(y) = \exp\,(m_{00})\, y^{m_{01}-1}(1-y)^{m_{02}-1} B\,(a_1(y), b_1(y))$$

The necessity of requiring that $m_{ij} \le 0$ ($i \ge 1$, $j \ge 1$) is evident upon perusing these conditional specifications. As remarked at the end of Theorem 1, it is these m_{ij}'s ($i \ge 1$, $j \ge 1$) which control the dependence between X and Y. Independence will be encountered only if all these m_{ij}'s $(i \geq 1, j \geq 1)$ are zero. Expressions for conditional moments of the Beta conditionals density (II.5), follow readily from (II.6) and (II.7) (since moments of Beta variables are simple functions of the corresponding parameters). Thus, as an example,

$$E(X|Y=y) = \frac{a_1(y)}{a_1(y) + b_1(y)}$$

$$= \frac{m_{10} + m_{11}\log y + m_{12}\log(1-y)}{m_{10}+m_{20}+(m_{11}+m_{21})\log y+(m_{12}+m_{22})\log(1-y)}$$

If we introduce two auxiliary random variables

$$U \sim Beta(m_{10}, m_{20})$$

and

$$V \sim Beta(m_{01}, m_{02})$$

then we can write two alternative formulas for the normalizing constant $\exp(m_0)$:

$$\exp(m_{00}) = [B(m_{10}, m_{20}) E(B(a_2(U), b_2(U)))]^{-1}$$
$$= [B(m_{01}, m_{02}) E(B(a_1(V), b_1(V)))]^{-1}$$

Moments of X, Y can be expressed in terms of expectations of the random variables U, V. Thus, for $(n = 1, 2, \ldots)$:

$$E(X^n) = \frac{E\{U^n B(a_2(U), b_2(U))\}}{E\{B(a_2(U), b_2(U))\}},$$

$$E(Y^n) = \frac{E\{V^n B(a_1(V), b_2(V))\}}{E\{B(a_1(V), b_1(V))\}}.$$

These formulas can be used to obtain values of the moments by simulation (note that only simulation of univariate Beta random variables is required). An expression for the mixed moment $E(XY)$ is available, by conditioning on Y;

$$E(XY) = E(Y E(X|Y)) = E\left[\frac{Y a_1(Y)}{a_1(Y) + b_1(Y)}\right]$$

From this we can obtain an expression involving the random variable V (an analogous expression involving U can also be obtained).

$$E(XY) = \frac{E\left[\frac{Va_1(V)B(a_1(V),b_1(V))}{a_1(V)+b_1(V)}\right]}{E\left[B\left(a_1(V),b_1(V)\right)\right]}$$

What do such beta-conditionals densities look like? They constitute a flexible family of densities of a broad variety of shapes. They can be unimodal or multimodal, bounded or unbounded. A small catalog of these densities is displayed in Figures 1 to 4.

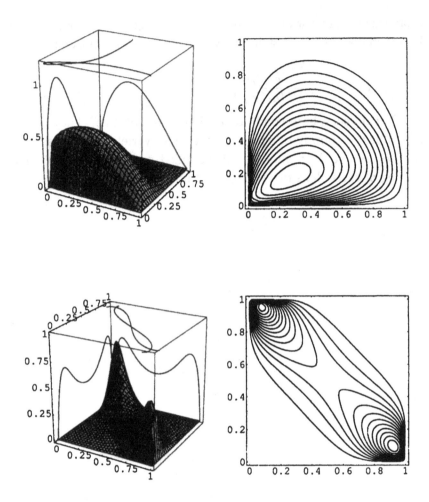

Figure 1. Joint and marginal densities, and conditional modes for the beta conditionals models with $m_{11} = 0$, $m_{12} = -1.4$, $m_{21} = -0.5$, $m_{22} = -2.3$, $m_{10} = 1.1$, $m_{20} = 0.7$, $m_{01} = 1.3$, $m_{02} = 0.6$ and $m_{11} = -3.7$, $m_{12} = -0.4$, $m_{21} = -0.5$, $m_{22} = -4$, $m_{10} = 0.6$, $m_{20} = 0.1$, $m_{01} = 0.4$, $m_{02} = 0.1$.

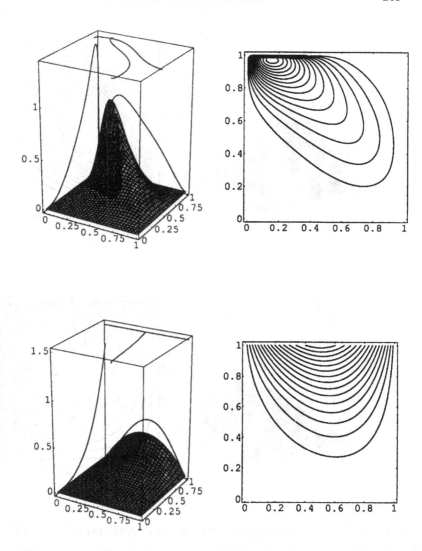

Figure 2. Joint and marginal densities, and conditional modes for the beta conditionals models with $m_{11} = -3.7$, $m_{12} = -0.4$, $m_{21} = -1.5$, $m_{22} = -2$, $m_{10} = 0.6$, $m_{20} = 0.1$, $m_{01} = 0.4$, $m_{02} = 0.1$ and $m_{11} = -2$, $m_{12} = 0$, $m_{21} = -1$, $m_{22} = 0$, $m_{10} = 2$, $m_{20} = 2$, $m_{01} = 1.5$, $m_{02} = 1$.

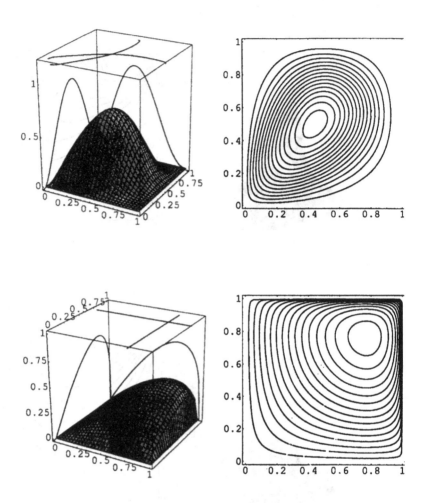

Figure 3. Joint and marginal densities, and conditional modes for the beta conditionals models with $m_{11} = -0.2$, $m_{12} = -1.7$, $m_{21} = -2$, $m_{22} = -2$, $m_{10} = 1.5$, $m_{20} = 0.5$, $m_{01} = 1.5$, $m_{02} = 1$ and $m_{11} = 0$, $m_{12} = 0$, $m_{21} = 0$, $m_{22} = 0$, $m_{10} = 2$, $m_{20} = 1.3$, $m_{01} = 2$, $m_{02} = 1.3$.

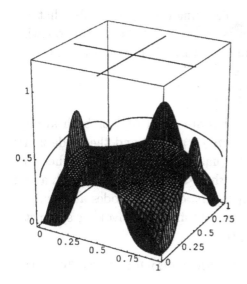

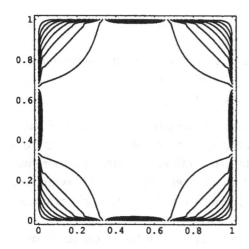

Figure 4. Joint and marginal densities, and conditional modes for the beta conditionals models with $m_{11} = m_{12} = m_{21} = m_{22} = -0.4$, $m_{10} = m_{20} = m_{01} = m_{02} = 0.4$.

In order to visualize the possible configurations of modes that can arise for these joint densities it is convenient to recall the following expression for the mode of a random variable $U \sim Beta(\alpha, \beta)$.

$$mode(U) = \frac{(\alpha - 1)^+}{(\alpha - 1)^+ + (\beta - 1)^+}$$

provided that not both of α and β are equal to 1 or both are less than 1. If $\alpha = \beta = 1$, then $U \sim uniform(0, 1)$ and the mode is not unique, any value between $(0, 1)$ qualifies as a mode. The other non-standard case is that in which both $\alpha < 1$ and $\beta < 1$. In such a case the corresponding Beta density is bimodal with modes at 0 and 1.

Some simplified submodels are obtained by invoking symmetry and/or exchangeability assumptions.

(i) Symmetry in x: Here we set: $m_{10} = m_{20}, m_{11} = m_{21}$ and $m_{12} = m_{22}$.

(ii) Symmetry in y: Here we set: $m_{01} = m_{02}, m_{11} = m_{12}$ and $m_{21} = m_{22}$.

(iii) Exchangeability: For this we set: $m_{10} = m_{01}, m_{20} = m_{02}$ and $m_{12} = m_{21}$.

(iv) Homogeneity: ((i), (ii) and (iii)): in this setting we have $m_{01} = m_{02} = m_{10} = m_{20}$ and $m_{11} = m_{12} = m_{21} = m_{22}$.

(v) Complete homogeneity: $m_{ij} = k, \forall i, j$.

The corresponding dimensions of the parameter spaces of these submodels are 5, 5, 5, 2 and 1 representing a notable reduction from the 8 dimensions in the full beta-conditionals model. Another submodel which may be included in this hierarchy is double-symmetry in which we invoke (i) and (ii), yielding a 3 dimensional parameter space.

There is still considerable flexibility even in these restricted models. See Figure 4 for a sample density.

III. Estimation and Inference

The beta-conditionals density (II.5) is specified only up to a complicated normalizing coefficient $\exp(m_0)$ that is a function of the other m_{ij}'s. This clearly will make it impossible to obtain maximum likelihood estimates via consideration of likelihood equations. However (II.5) does represent a relatively simple exponential family of densities so that numerical implementation of maximum likelihood is definitely feasible. For most such implementations a fairly good set of starting values or preliminary estimates of the m_{ij}'s will be desirable. Note also that we are dealing with an 8 dimensional likelihood surface so that anomalous surface behavior can be expected to arise for many data configurations. There is definitely interest in developing alternative estimation strategies for our Beta conditionals density (even if only to get us some reliable initial values for implementation of maximum likelihood if that is desired).

Several alternative strategies for estimation in general conditionally specified models are outlined in Arnold, Castillo and Sarabia (1999, Chapter 9). Specific implementation of these approaches in the Beta conditionals case will not be detailed here. Instead we will focus on an approach based on a multivariate version of Stein's identity introduced in Arnold, Castillo and Sarabia (2001).

Our data is presumed to be of the form: (X_1, Y_1), (X_2, Y_2), ..., (X_n, Y_n) i.i.d. with common density (II.5) and we wish on the basis of this data to estimate the m_{ij}'s (elements of the parameter matrix M, in the general expression (II.3)). We will develop a system of equations relating the parameters and expectations of certain functions of (X_1, Y_1). Using a method of moments approach, sample versions of these expectations are substituted and the system of equations is solved for the unknown parameters. We begin by briefly reviewing the multivariate Stein identity.

Consider a continuous k-dimensional random variable $\underline{X} = (X_1, \ldots, X_k$, with absolutely continuous density function with respect to Lebesgue measure given by the s-parameter exponential family

$$f(\underline{x}; \underline{\eta}) = h(\underline{x}) \exp\left\{ \sum_{i=1}^{s} \eta_i T_i(\underline{x}) - A(\underline{\eta}) \right\}, \qquad \text{(III.1)}$$

where $\underline{x} = (x_1, \ldots, x_k)$ and $\underline{\eta} = (\eta_1, \ldots, \eta_s)$. The support of (III.1) is given by $(a_1, b_1) \times \ldots \times (a_k, b_k)$, where $a_i < b_i$ and $a_i \in \mathbf{R} \cup \{-\infty\}$, $b_i \in \mathbf{R} \cup \{+\infty\}$, $i = 1, \ldots, k$. The functions $T_i(\underline{x})$, $i = 1, \ldots, s$ are assumed to be linearly independent.

Theorem 2. *Let* $\underline{X} = (X_1, \ldots, X_k)$ *be a* k-*dimensional random vector with density (III.1) and* $g : \mathbf{R}^k \to \mathbf{R}$ *a differentiable function such that* $E_{\underline{\eta}}|\partial g(\underline{X})/\partial X_i| < \infty$, *for* $i \in \{1, 2, \ldots, k\}$, $\forall \underline{\eta}$. *Then for each* j *we have*

$$E_{\underline{\eta}}\left(g_j(\underline{X})\right) = -E_{\underline{\eta}}\left\{g(\underline{X})\left[\sum_{i=1}^{s} \eta_i T_{ij}(\underline{X}) + \frac{h_j(\underline{X})}{h(\underline{X})}\right]\right\}, \; \forall \underline{\eta}$$

where $g_j(\underline{x}) = \frac{\partial}{\partial x_j}g(\underline{x})$, $T_{ij}(\underline{x}) = \frac{\partial}{\partial x_j}T_i(\underline{x})$ *and* $h_j(\underline{x}) = \frac{\partial}{\partial x_j}h(\underline{x})$ *if*

$$\lim_{x_i \to b_i} \left[g(\underline{x})f(\underline{x}; \underline{\eta})\right] = \lim_{x_i \to a_i} \left[g(\underline{x})f(\underline{x}; \underline{\eta})\right] = 0, \; \forall \underline{\eta}. \qquad (III.2)$$

The identity (III.1) is especially appealing, since the integration constant, $A(\underline{\eta})$, does not appear in it. This makes it particularly useful in the case of multivariate distributions where the normalization constant is difficult to obtain. The identity is useful for two purposes: (a) obtaining recurrence formulas for moments, and (b) obtaining consistent estimators based on moments.

Now if a random vector (X, Y) has all its conditional densities in specified exponential families then its density will be of the form (II.3). We will assume that all the functions $r_i(\cdot), q_{ij}(\cdot)$ appearing in (II.3) are differentiable. The family (II.3) is clearly an exponential family and is a particular case of (III.1) with $k = 2$ (using the notation (X, Y) instead of (X_1, X_2)).

Theorem 2 may thus be applied to the conditionally specified density (II.3). Consequently, if g and $\tilde{g}$ are differentiable functions defined on $\mathbf{R}^2$ satisfying (III.2), we have

$$E\left[\frac{\partial g(X, Y)}{\partial X}\right] = -E\left\{g(X, Y)\left[\sum_{i=1}^{\ell_1} m_{i0}q'_{1i}(X)\right.\right.$$
$$\left.\left. + \sum_{i,j} m_{ij}q'_{1i}(X)q_{2j}(Y) + s_1(X)\right]\right\} \qquad (III.3)$$

and

$$E\left[\frac{\partial \widetilde{g}(X,Y)}{\partial Y}\right] = -E\left\{\widetilde{g}(X,Y)\left[\sum_{j=1}^{\ell_2} m_{0j}q'_{2j}(Y)\right.\right.$$

$$\left.\left.+ \sum_{i,j} m_{ij}q_{1i}(X)q'_{2j}(Y) + s_2(Y)\right]\right\}, \quad \text{(III.4)}$$

where $s_i(u) = r'_i(u)/r_i(u)$, $i = 1, 2$.

The beta conditionals model corresponds to the expression (II.3) with:

$$q_{11}(u) = q_{21}(u) = \log u$$

$$q_{12}(u) = q_{22}(u) = \log(1-u)$$

and

$$r_i(u) = [u(1-u)]^{-1}, \quad i = 1, 2.$$

Now, using formulas (III.3) and (III.4) we have:

$$E\left[\frac{\partial g}{\partial X}\right] = -E\left\{g\left[\frac{m_{10}}{X} - \frac{m_{20}}{1-X} + \frac{m_{11}\log Y}{X}\right.\right.$$

$$\left.\left. - \frac{m_{12}\log(1-Y)}{X} - \frac{m_{21}\log Y}{1-X} - \frac{m_{22}\log(1-Y)}{1-X} + s_1(X)\right]\right\}$$

$$\text{(III.5)}$$

and

$$E\left[\frac{\partial \widetilde{g}}{\partial X}\right] = -E\left\{\widetilde{g}\left[\frac{m_{01}}{Y} - \frac{m_{02}}{1-Y} + \frac{m_{11}\log X}{Y}\right.\right.$$

$$\left.\left. + \frac{m_{21}\log(1-X)}{Y} - \frac{m_{12}\log X}{1-Y} - \frac{m_{22}\log(1-X)}{1-Y} + s_2(Y)\right]\right\},$$

$$\text{(III.6)}$$

where now

$$s_i(u) = \frac{r'(u)}{r_i(u)} = -\frac{1-2u}{u(1-u)}, \quad i = 1, 2.$$

We may then use the following 4 choices for $g(x, y)$,

$$\Big\{ x(1 - x)\exp(x), x^2(1 - x)\exp(x), x^2(1 - x)^2\exp(x),$$

$$xy(1 - x)(1 - y)\exp(x) \Big\}, \qquad (\text{III.7})$$

and the following 4 choices for $\tilde{g}(x, y)$

$$\Big\{ y(1 - y)\exp(y), y^2(1 - y)\exp(y), y^2(1 - y)^2\exp(y),$$

$$xy(1 - x)(1 - y)\exp(y) \Big\}, \qquad (\text{III.8})$$

in equations (III.5) and (III.6) to obtain a system of 8 equations in the 8 unknown m_{ij}'s. Substituting sample moments for population moments and solving yields consistent asymptotically normal estimates of the m_{ij}'s. Experience, based on simulation studies, suggests that quite large sample sizes are however necessary in order to obtain good estimates.

Choices of the g's and $\tilde{g}$'s used (i.e. (III.7) and (III.8)) undoubtedly affect the performance of this estimation approach. Further investigation of alternatives to (III.7) and (III.8) may lead to estimates with better small sample properties.

IV. In a Bayesian Setting

In this section, we present an application of the beta conditionals model to Bayesian analysis. Beta conditionals models turn out to be quite natural conjugate prior distributions in settings in which the likelihood function will admit a beta conjugate prior for each of its parameters, assuming that the remaining parameters are known. The general method is described in Arnold, Castillo and Sarabia (1999, Chapter 13).

A remarkably large number of inferential problems involve Bernoulli trials with more than one success probability. Such scenarios call for priors supported by $[0, 1]^k$ where k is the number of potentially distinct success probabilities involved in the problem. Most of the key ideas are apparent in the setting in which $k = 2$ and it is this situation which we focus on in this section. Discussion of higher dimensional parameter spaces is deferred until the end of Section V.

Let us consider a fairly typical problem in which two treatments are compared by applying them to experimental units. A "success" criterion is specified and our data consists of information regarding how many successes and how many failures have been observed for each treatment (the treatments could be brands of medicine, the criterion for success might be disappearance of symptoms in the patients, the experimental units). Thus our model for the data is two independent random variables X_1, X_2 where $X_1 \sim binomial(n_1, p_1)$ and $X_2 \sim binomial(n_2, p_2)$. A joint prior needs to be placed on (p_1, p_2).

For the prior specification for (p_1, p_2), note that a natural conjugate prior for p_1, assuming that p_2 is known, is a beta prior. The same is of course true for p_2, assuming p_1 is known. It is then natural to look for the most general density for (p_1, p_2) whose conditional densities satisfy:

$$p_1|p_2 \sim Beta\left(a_1(p_2), b_1(p_2)\right)$$
$$p_2|p_1 \sim Beta\left(a_2(p_1), b_2(p_1)\right)$$

As we have earlier observed, this means that necessarily the joint density of (p_1, p_2) will be of the form (II.5). It is readily verified that (II.5) constitutes a conjugate family of prior densities for our problem in which likelihood is of the form

$$\ell(p_1, p_2) \propto p_1^{x_i} (1 - p_1)^{n_1 - x_1} p_2^{x_2} (1 - p_2)^{n_2 - x_2} \qquad (IV.1)$$

It is a conditionally conjugate prior in the terminology of Arnold, Castillo and Sarabia (1999). We will use $m_{ij}^{(0)}$ to denote the parameters in the beta conditionals prior for (p_1, p_2) (refer to equation (II.5)) and analogously $m_{ij}^{(1)}$ will denote the parameters in the posterior density, which is again of the beta conditionals form.

When combining the prior (II.5) with the likelihood (IV.1), it is evident that only 4 of the parameters are affected by the data. The posterior values of the parameters are related to the prior values as follows:

$$m_{10}^{(1)} = m_{10}^{(0)} + x_1,$$
$$m_{20}^{(1)} = m_{20}^{(0)} + n_1 - x_1,$$
$$m_{01}^{(1)} = m_{01}^{(0)} + x_2,$$
$$m_{02}^{(1)} = m_{02}^{(0)} + n_2 - x_2,$$
$$m_{ij}^{(1)} = m_{ij}^{(0)}, \; i,j = 1,2.$$

Note that the natural conjugate joint prior would have independent marginals. It may be argued that, when we compare drugs in an experiment such as this, our prior beliefs about the efficacies of the treatments might well be dependent. The prior (II.5) allows us to accommodate dependent as well as independent prior beliefs. If we use the prior (II.5), the resulting posterior density will also have beta conditionals and simulation of realizations from the posterior density is readily implemented using the Gibbs sampler. Decisions regarding the relationship between p_1 and p_2, given the data, will then be based on this posterior density.

Interest frequently is directed to the ratio of the corresponding odds ratios, i.e. the cross-product ratio:

$$\psi(\underline{p}) = \frac{p_1(1-p_2)}{p_2(1-p_1)} \tag{IV.2}$$

If (p_1, p_2) are distributed as in (II.5) the mean value of (IV.2) is given by ($m_{01}, m_{20} > 1$):

$$E\{\psi(\underline{p})\} = \frac{m_{10}}{m_{20}-1} \cdot \frac{E[B\left(a_2(\widetilde{U}), b_2(\widetilde{U})\right)]}{E[B(a_2(U), b_2(U))]}$$
$$= \frac{m_{02}}{m_{01}-1} \cdot \frac{E\left[B\left(a_1(\widetilde{V}), b_1(\widetilde{V})\right)\right]}{E[B(a_1(V), b_1(V))]} \tag{IV.3}$$

where $\widetilde{U} \sim Beta(m_{10}+1, m_{20}-1)$ and $\widetilde{V} \sim Beta(m_{01}-1, m_{02}+1)$. From (IV.3), and considering a quadratic loss function, the Bayes estimator of (IV.2) is given by (using the first expression in (IV.3)):

$$E\left\{\psi(\underline{p})|\underline{x}\right\} = \frac{m_{10} + x_1}{m_{20} + n_1 - x_1 - 1} \cdot \frac{E\left[B\left(a_2(\widetilde{W}), b_2(\widetilde{W})\right)\right]}{E\left[B\left(a_2(W), b_2(W)\right)\right]}$$

in which:

$$W \sim Beta\left(m_{10} + x_1, m_{20} + n_1 - x_1\right)$$
$$\widetilde{W} \sim Beta\left(m_{10} + x_1 + 1, m_{20} + n_1 - x_1 - 1\right) \qquad \text{(IV.4)}$$

We can obtain an approximate posterior distribution for $\psi(\underline{p})$ by simulating realizations from the joint posterior density of (p_1, p_2). As remarked earlier, such simulations are readily implemented using a Gibbs sampler routine.

V. Multivariate Extensions

A straightforward k-dimensional extension of Theorem 1 is available. It may be used to generate k-dimensional joint densities with beta conditionals. We consider a k-dimensional random vector $\underline{X} = (X_1, \ldots, X_k)$ and introduce the notation $\underline{X}_{(i)}$ to denote the vector $\underline{X}$ with the ith coordinate, X_i, deleted. An analogous convention is used to define $\underline{x}_{(i)}$ as the real vector $\underline{x}$ with its ith coordinate deleted. We are then led to consider joint densities for $\underline{X}$ for which for each i and each $\underline{x}_{(i)} \in \mathbb{R}^{k-1}$

$$X_i | \underline{X}_{(i)} = \underline{x}_{(i)} \sim Beta\left(\alpha_i\left(\underline{x}_{(i)}\right), \beta_i\left(\underline{x}_{(i)}\right)\right)$$

for some functions $\alpha(\cdot)$, $\beta_i(\cdot)$; $i = 1, 2, \ldots, k$. The resulting class of k-dimensional beta-conditional densities is of the form:

$$f_{\underline{X}}(\underline{x}) = \left[\prod_{i=1}^{k} x_i(1 - x_i)\right]^{-1}$$

$$\times \exp\left\{\sum_{i_1=0}^{2} \sum_{i_2=0}^{2} \cdots \sum_{i_k=0}^{2} m_{\underline{i}} \left[\sum_{j=1}^{k} q_{i_j}(x_j)\right]\right\}$$

$$\times I\left\{\underline{0} < \underline{x} < \underline{1}\right\},$$

where

$$q_0(t) = 1$$

$$q_{i1}(t) = \log t, \ i = 1, 2, \ldots, k$$

$$q_{i2}(t) = \log(1 - t), \ i = 1, 2, \ldots, k$$

and where $m_{\underline{0}}$ is a function of the other $m_{\underline{i}}$'s, chosen so that the density integrates to 1. There are constraints on the $m_{\underline{i}}$, needed to ensure that the conditional densities are proper beta densities. Thus if $\underline{i}$ has one non zero coordinate, then the corresponding $m_{\underline{i}}$ must be positive. More generally if $\underline{i}$ has an odd number of non-zero coordinates, the corresponding $m_{\underline{i}}$ must be positive. In other cases the $m_{\underline{i}}$'s must be negative. Densities such as (IV.4) clearly can be used as joint priors for k Bernoulli parameters. They will be conjugate priors and, as in Section IV, their corresponding posterior densities permit ready simulation using a Gibbs sampler.

VI. Scaled Beta Conditionals

We revert once more to the two dimensional setting recognizing that k-dimensional extensions can be readily developed, albeit with complicated notation. In this section we consider two dimensional random vectors (X, Y) with scaled beta conditionals. Thus for each y we assume that X given $Y = y$ has a scaled beta distribution with parameters $a_1(y)$ and $b_1(y)$ and support on the interval $(0, c_1(y))$. Analogously for each x, we assume that Y given $X = x$ has a scaled beta distribution with parameters $a_2(x)$ and $b_2(x)$ and support on the interval $(0, c_2(x))$. For compatibility the functions $c_1(y)$ and $c_2(x)$ must be such that the sets $\{(x, y) : 0 < x < c_1(y)\}$ and $\{(x, y) : 0 < y < c_2(x)\}$ coincide.

To achieve this, consider a decreasing function $c_2(x)$ with $c_2(0) = c_2 > 0$ and $c_2(_1) = 0$ for some $0 < 1 < \infty$. Then define $c_1(y) = c_2^{-1}(y)$. A classic example of such a function $c_2(x)$ is $1 - x$, for which $c_2 = 1$ and $c_1 = 1$.

If we are to have scaled beta conditionals (using $c_1(y) = c_2^{-1}(y)$

as described above), we may write the joint density of (X, Y) in the following two equal forms:

$$\left[\frac{y}{c_2(x)}\right]^{a_2(x)-1} \left[1 - \frac{y}{c_2(x)}\right]^{b_2(x)-1} B^{-1}(a_2(x), b_2(x)) \frac{f_X(x)}{c_2(x)}$$
$$\times I\left(0 < y < c_2(x)\right) \tag{VI.1}$$

By construction $\{(x, y) : 0 < x < c_1(y)\} = \{(x, y) : 0 < y < c_2(x)\}$ and for (x, y) in this support set we may rewrite (VI.1) in a more convenient form:

$$x^{\alpha_1(y)} \left(c_1(y) - x\right)^{\beta_1(y)} \psi_1(y) = y^{\alpha_2(x)} \left(c_2(x) - y\right)^{\beta_2(x)} \psi_2(x) \tag{VI.2}$$

where $\alpha_1(y) = a_1(y) - 1, \beta(y) = b_1(y) - 1, \alpha_2(x) = a_2(x) - 1, \beta_2(y) = b_2(y) - 1$ and $\psi_1(y)$ and $\psi_2(x)$ are appropriately defined functions of a single variable.

We then need to solve (VI.2) for the unknown functions α_1, β_1, c_1, ψ_1, α_2, β_2, c_2 and ψ_2 in order to determine the most general class of densities with scaled beta conditionals.

A general solution to (VI.2) appears to be elusive. However certain special cases are amenable to solution. For example, if $c_1(y) \equiv c_1$ and $c_2(x) \equiv c_2$, then the resulting densities will just be rescaled versions of the beta conditionals densities given in (II.5) (the support set will now be $0 < x < c_1$, $0 < y < c_2$). Another easily resolved case is that in which $\beta_1(y) \equiv \beta_2(x) = \beta > 0$, $c_2(x) = 1 - x$ and $c_1(y) = 1 - y$. In such a situation we may cancel $(1 - x - y)^\beta$ from both sides of (VI.2) to obtain the simpler equation:

$$x^{\alpha_1(y)} \psi_1(y) = y^{\alpha_2(x)} \psi_2(x) \tag{VI.3}$$

to be solved for α_1, α_2, ψ_1 and ψ_2. This, after taking logarithms, is a standard Stephanos-Levi-Civita-Suto functional equation (see e.g. Arnold, Castillo and Sarabia (1999), page 13). The general solution of (VI.3) yields

$$\alpha_1(y) = c_{01} + c_{11} \log y$$

and

$$\alpha_1(y) = c_{01} + c_{11} \log y$$

from which we get the following family of joint densities with support $\{(x, y) : x > 0, y > 0, x + y < 1\}$:

$$f(x, y) \propto \left[\frac{1}{xy}\right] \exp\{d_{00} + d_{10} \log x + d_{01} \log y + d_{11} \log x \log y\}$$

$$\times [1 - x - y]^{b-1} I(x > 0, y > 0, x + y < 1) \qquad \text{(VI.4)}$$

It will be observed that (VI.4) includes as a special case the classical two dimensional Dirichlet density with parameters (d_{10}, d_{01}, b) when d_{11} is set equal to 0. It may be observed that the marginal densities corresponding to (VI.4) will be beta densities if and only if $d_{11} = 0$. This observation was the basis of a Dirichlet characterization result presented in James(1975, page 683). A small amount of additional flexibility over the the classical Dirichlet model is provided by (VI.4) since d_{11}, of course, need not equal 0.

A. Multinomial Data

Suppose now that our data consists of the results of n independent trials each with $k + 1$ possible outcomes $1, 2, \ldots, k + 1$. For $i = 1, 2, \ldots, k$ let X_i denote the number of outcomes of type i observed in the n trials. Then $\underline{X}$ has a multinomial distribution with parameters n and $\underline{p} = (p_1, p_2, \ldots, p_k)$. Based on $\underline{X}$, we wish to make inferences about $\underline{p}$ (note that $\sum_{i=1}^{k} p_i < 1$ and for convenience we define $p_{k+1} = 1 - \sum_{i=1}^{k} p_i$). If $\underline{p}_{(1)}$ (i.e. $(p_2, \ldots, p_k)$) were known, the natural conjugate prior for p_1 would be a scaled Beta density.

Considerations such as this will lead us to a joint prior for $\underline{p}$ which has scaled Beta conditionals, i.e. such that for $i = 1, 2, \ldots, k$

$$p_i | \underline{p}_{(i)} \sim \left(1 - \sum_{j \neq i} p_j\right) \times Beta\left(a_i(p_i), b_i(p_{(i)})\right), \ i = 1, 2, \ldots, k,$$

where $p_i > 0$, $i = 1, 2, \ldots, k$ and $p_1 + \cdots + p_k < 1$.

Such distributions were discussed by James (1975) and were described in two dimensions in the previous section.

A general form for such densities may be written as

$$f(p_1, \ldots, p_k) \propto \prod_{i=1}^{k} p_i^{\alpha_i - 1} [1 - (p_1 + \cdots + p_k)]^{\alpha_{k+1} - 1}$$
$$\times \exp\{\phi(p_1, \ldots, p_k)\} \tag{VI.5}$$

where

$$\phi(p_1, \ldots, p_k) = \sum_{i<j} a_{ij} \log p_i \log p_j + \sum_{i<j<k} a_{ijk} \log p_i \log p_j \log p_k$$
$$+ \cdots + a_{12\cdots k} \log p_1 \cdots \log p_k.$$

If p has a scaled Beta conditionals distribution, i.e. has (VI.5) as its joint density, we write

$$p \sim SBC(\underline{\alpha}, A). \tag{VI.6}$$

(here $\underline{\alpha}$ is of dimension $k + 1$). Note that if $A \equiv 0$ this reduces to the standard Dirichlet density, often used as a prior in multinomial settings. Since our likelihood is of the form

$$\ell(\underline{p}) \propto \prod_{i=1}^{k} p_i^{x_i} \left(1 - \sum_{i=1}^{k} p_i\right)^{n - \sum_{i=1}^{k} x_i} I\left(p_i > 0, \forall i, \sum_{i=1}^{k} p_i < 1\right),$$

it follows immediate that the family (VI.6) is a conjugate family and that the posterior distribution of p given $\underline{X} = \underline{x}$ will be in the same family. Specifically we will have, introducing the notation $x_{k+1} = n - \sum_{i=1}^{k} x_i$ and $\widetilde{\underline{x}} = (\underline{x}, x_{k+1})$,

$$\underline{p}|\underline{X} = \underline{x} \sim SBC(\underline{\alpha} + \widetilde{\underline{x}}, A). \tag{VI.7}$$

Gibbs sampler simulations using the posterior density will be readily accomplished since simulation of univariate scaled Beta variables is a straightforward.

VII. Improper Models

It is possible to encounter improper (non-integrable) measures with scaled uniform conditionals. For example, if a bivariate "density" has the following structure

$$f(x,y) \propto I(x > 0, y > 0, xy < 1) \qquad \text{(VII.1)}$$

then the corresponding conditional distributions would be of the form

$$X|Y = y \sim Uniform\,(0, 1/y)\,, \forall y > 0, \qquad \text{(VII.2)}$$

and

$$Y|X = x \sim Uniform(0, 1/x), \forall x > 0, \qquad \text{(VII.3)}$$

These are perfectly legitimate conditional densities, even though the joint "density" integrates to infinity. It corresponds to a uniform distribution over a region of infinite area. Similar models can be constructed using other support sets. Suppose that $A \subset \mathbb{R}^2$ is such that each x-cross-section and each y-cross-section is a finite interval. Then consider

$$f(x,y) \propto I((x,y) \in A). \qquad \text{(VII.4)}$$

If A has infinite area, then this model will also be improper despite having proper uniform conditionals (of X given Y and of Y given X).

Examples such as these are perhaps of only pedagogical interest. They do warn us that improper joint densities (i.e. non-integrable ones) can have proper scaled beta conditionals.

A more general version of (VII.1) with scaled beta conditionals in the same support set is of the form

$$f(x,y) \propto x^{a-1} y^{b-1} (1 - xy)^{c-1} \exp(-d \log x \log y)$$
$$\times I(x > 0, y > 0, xy < 1) \qquad \text{(VII.5)}$$

VIII. Generalized Beta Conditionals

Gordy (1998b) proposed the "compound confluent hypergeometric" (CCH) model with probability density function:

$$f(x; p, q, r, s, \nu, \theta) = \frac{x^{p-1}(1 - \nu x)^{q-1} \{\theta + (1 - \theta)\nu x\}^{-r}}{B(p, q)H(p, q, r, s, \nu, \theta)}$$
$$\times \exp(-sx)I(0 < x < 1/\nu) \qquad \text{(VIII.1)}$$

with $p, q > 0$, $r, s \in \mathbb{R}$, $0 \leq \nu \leq 1$ and $\theta > 0$. The function H is given by:

$$H(p, q, r, s, \nu, \theta) = \nu^{-p} \exp(-s/\nu)\Phi_1(q, r, p + q, s/\nu, 1 - \theta)$$

where Φ_1 is the confluent hypergeometric function of two variables defined by:

$$\Phi_1(\alpha, \beta, \gamma, x, y) = \sum_{m=0}^{\infty} \sum_{n=0}^{\infty} \frac{(\alpha)_{m+n}(\beta)_n}{(\gamma)_{m+n}m!n!} x^m y^n,$$

where $(a)_k$ is the Pochhammer symbol defined by: $(a)_0 = 1$, $(a)_1 = a$ and $(a)_k = (a)_{k-1}(a + k - 1)$. If X has the density (VIII.1), we will write:

$$X \sim CCH(p, q, r, s, \nu, \theta).$$

In this section the most general bivariate densities with Gordy generalized beta conditionals are derived. The model (VIII.1), in addition to including, as a particular case, the classic beta, also includes three important existing models:

- Generalized Beta (GB) distribution, proposed by McDonald and Xu (1995). When $s = 0$, $r = p + q$, $\nu = (1 - c)/b$ and $\theta = 1 - c$ the CCH simplifies to the GB:

$$f(x; a, b, c, p, q) = \frac{|a| x^{ap-1} \{1 - (1 - c)(x/b)^a\}^{q-1}}{b^{ap}B(p, q) \{1 + c(x/b)^a\}^{p+q}}$$
$$\times I\{0 < x^a < b^a/(1 - c)\}$$

with $a = 1$. The CCH distribution can easily be extended to accommodate a peakedness parameter a as well: if $X \sim$

$GB(a, b, c, p, q)$ then

$$X^a \sim CCH\left(p, q, r = p + q, s = 0, \nu = (1 - c)/b, \theta = 1 - c\right).$$

- Gauss Hypergeometric (Armero and Bayarri, 1994). Making $s = 0$ and $\nu = 1$ the $CCH(p, q, r, s = 0, \nu = 1, \theta)$ simplifies to the $GH(p, q, r, \lambda)$ with pdf:

$$f(x; p, q, r, \lambda) = \frac{x^{p-1}(1 - x)^{q-1}(1 + \lambda x)^{-r}}{B(p, q)\, _2F_1(r, p, p + q; -\lambda)} I(0 < x < 1),$$

where $\lambda = (1 - \theta)/\theta$ and $_2F_1$ denotes the Gauss hypergeometric function. The GH gives to the ordinary beta if either $r = 0$ or $\lambda = 0$.

- Confluent Hypergeometric (Gordy, 1998a). Making $\nu = 1$ and $\theta = 1$, the $CCH(p, q, r, s, \nu = 1, \theta = 1)$ simplifies to the confluent distribution $CH(p, q, s)$ with pdf:

$$f(x; p, q, s) = \frac{x^{p-1}(1 - x)^{q-1}\exp(-sx)}{B(p, q)\, _1F_1(p, p + q, -s)} I(0 < x < 1).$$

We seek the most general bivariate density of (X, Y) such that the associated conditionals satisfy:

$$X|Y = y \sim CCH\left(p_1(y), q_1(y), r_1(y), s_1(y), \nu_1, \theta_1\right) \qquad \text{(VIII.2)}$$
$$Y|X = x \sim CCH\left(p_2(x), q_2(x), r_2(x), s_2(x), \nu_2, \theta_2\right) \qquad \text{(VIII.3)}$$

where $p_i(x)$, $q_i(x)$, $r_i(x)$ and $s_i(x)$, $i = 1, 2$ are unknown functions. Note that the parameters ν_i and θ_i, $i = 1, 2$ are fixed and known. If we write (VIII.1) in the form:

$$f(x; p, q, r, s) \propto x^{-1}(1 - \nu x)^{-1} \exp\left[p \log x + q \log(1 - \nu x)\right.$$
$$\left. -r \log\left\{\theta + (1 - \theta)\nu x\right\} - sx\right]$$

a 4-parameter exponential family is obtained, where ν and θ are fixed and known). Thus, Theorem 1 can be applied. Defining the vector:

$$v_{\nu, \theta}(x) = (1, \log(x), \log(1 - \nu x), -\log(\theta + (1 - \theta)\nu x), -x)$$

the most general density that satisfies (VIII.2) and (VIII.3) is given by:

$$f(x, y) = x^{-1} (1 - \nu_1 x)^{-1} y^{-1} (1 - \nu_2 y)^{-1}$$
$$\times \exp\{v_{\nu_1, \theta_1}(x) M v_{\nu_2, \theta_2}(y)^T\} \qquad \text{(VIII.4)}$$

where $0 < x < 1/\nu_1$ and $0 < y < 1/\nu_2$ and $M = \{m_{ij}\}$ is a 5×5 matrix of parameters. The resulting density depends on $(25 - 1) + 4 = 28$ parameters. The m_{ij}, ν_i and θ_i parameters must be selected in such a way that one of the marginal densities (VIII.4) will be integrable, in accord with Theorem 1.

IX. Other Conditioning Schemes

In the context of bivariate survival modeling, we may consider joint densities specified by conditioning on events such as $\{X > x\}$ and $\{Y > y\}$. For example, let (X, Y) be a two dimensional random variable with support $[0, 1]^2$ such that for each $y \in (0, 1)$,

$$\Pr(X > x | Y > y) = (1 - x)^{\alpha_1(y)} \qquad \text{(IX.1)}$$

and, for each $x \in (0, 1)$,

$$\Pr(Y > y | X > x) = (1 - y)^{\alpha_2(x)} \qquad \text{(IX.2)}$$

These are beta conditional densities. It is not difficult (using the Stephanos-Levi-Civita-Suto theorem again) to verify that the corresponding joint survival function must be of the form:

$$\bar{F}(x, y) = \Pr(X > x, Y > y)$$
$$= (1 - x)^a (1 - y)^b \exp\left[-c \log(1 - x) \log(1 - y)\right] \quad \text{(IX.3)}$$

for $0 < x < 1$, $0 < y < 1$ where $a, b, c > 0$. Note that these models could also be viewed as having proportional conditional hazard functions.

Beta conditionals can also be encountered by conditioning on events of the form $\{X \leq x\}$ and $\{Y \leq y\}$. In this case the joint distribution will be of the form

$$F(x, y) = \Pr(X \leq x, Y \leq y) = x^a y^b \exp\left[-c \log x \log y\right] \qquad \text{(IX.4)}$$

Arnold, Castillo, and Sarabia

for $0 < x < 1$, $0 < y < 1$ where again $a, b, c > 0$.

More details on models such as these may be found in Arnold, Castillo and Sarabia (1999, Chapter 11).

Acknowledgment

The authors of this paper want to thank to the Ministerio de Ciencia y Tecnología (project BEC2000-1186) for the partial support of this research.

References

1. Armero, C.; Bayarri, M.J. Prior assessments for prediction in queues, The Statistician **1994**, *43*, 139-153.

2. Arnold, B.C.; Castillo, E.; Sarabia, J.M. *Conditional Specification of Statistical Models*; Springer Series in Statistics, Springer Verlag: New York, 1999.

3. Arnold, B.C.; Castillo, E.; Sarabia, J.M. A multivariate version of Stein's identity with applications to moment calculations and estimation of conditionally specified distributions, Communications in Statistics–Theory and Methods **2001**, *30*, 2517-2542.

4. Arnold, B.C.; Strauss, D. Bivariate distributions with conditionals in prescribed exponential families, Journal of the Royal Statistical Society, B **1991**, *53*, 365-375.

5. Gordy, M.B. Computationally convenient distributional assumptions for common-value auctions, Computational Economics **1998a**, *12*, 61-78.

6. Gordy, M.B. A generalization of generalized beta distributions, Federal Reserve Board **1998b**, FEDS 1998-18.

7. James, I.R. Multivariate distributions which have beta conditional distributions, Journal of the American Statistical Association **1975**, *70*, 681-684.

8. McDonald, J.B.; Xu, Y.J. A generalization of the beta distribution with applications, Journal of Econometrics **1995**, *66*, 133-152.

Parameter Specification of the Beta Distribution and its Dirichlet Extensions Utilizing Quantiles

J. René van Dorp and Thomas A. Mazzuchi

Department of Engineering Management and
Systems Engineering
1776 G Street, Suite 110
The George Washington University
Washington, DC 20052

I. Introduction

The time-honored 2 parameter Beta distribution

$$\frac{\Gamma(a + b)}{\Gamma(a)\Gamma(b)}\, x^{a-1}(1 - x)^{b-1}, a, b > 0, x \in [0, 1], \tag{I.1}$$

which is the main subject matter of this volume is well known in Bayesian methodology as a prior distribution on the success probability p of a binomial distribution (see, e.g. Carlin and Louis (2000)). Many authors (see, e.g. Gavaskar (1988)) have quoted the suitability of a Beta random variable X in different applications due to its flexibility. The transformation $Y = -\log(X)$ transforms the $[0, 1]$ support of X into the support $[0, \infty)$ of Y, while still inheriting the flexibility of X. Hence, the use of the Beta distribution as a prior distribution is by no means restricted to a bounded domain. For example, Van Dorp (1998) utilizes the above transformation to specify a prior distribution on the positive shape parameter of a Weibull distribution.

Amongst m-variate extensions of the Beta distribution (i.e. m - dimensional joint distributions with Beta marginals) the Dirichlet distribution (see, e.g., Kotz et al. (2000)),

$$\frac{\Gamma\left(\sum_{i=1}^{m+1}\theta_i\right)}{\prod_{i=1}^{m+1}\Gamma(\theta_i)}\left\{\prod_{i=1}^{m}x_i^{\theta_i-1}\right\}\left\{1-\sum_{i=1}^{m}x_i\right\}^{\theta_{m+1}-1}, \tag{I.2}$$

where $x_i \geq 0$, $i = 1,\ldots,m$, $\sum_{i=1}^{m}x_i \leq 1$, $\theta_i > 0$, $i = 1,\ldots,m+1$ has enjoyed wide popularity in Bayesian methodology (see, e.g., Cowell (1996), Johnson and Kokalis (1994) and Dennis (1998)). Application areas include Reliability Analysis (see, e.g., Kumar and Tiwari (1989), Coolen (1997) and Neath and Samaniego (1996)), Econometrics (see, e.g., Lancaster (1997)), and Forensics (see, e.g., Lang (1995)). The use of the m-variate Ordered Dirichlet distribution (see, e.g., Wilks (1962))

$$\frac{\Gamma\left(\sum_{i=1}^{m+1}\theta_i\right)}{\prod_{i=1}^{m+1}\Gamma(\theta_i)}\prod_{i=1}^{m+1}(x_i - x_{i-1})^{\theta_i-1}, \tag{I.3}$$

where $x_0 \equiv 1 - x_{m+1} \equiv 0$, $0 < x_{i-1} < x_i < 1$, $i = 1,\ldots,m+1$, $\theta_i > 0$, $i = 1,\ldots,m+1$ in Bayesian applications is less prevalent and the distribution is generally less well known. To the best of our knowledge, applications of (I.3) have been limited so far to reliability analysis problems (see, e.g., Van Dorp et al. (1996), Van Dorp et al. (1997), Erkanli et al. (1998), Van Dorp and Mazzuchi (2003)). A fundamental difference between the Dirichlet distribution (defined on a simplex) and the Ordered Dirichlet distribution (defined on the upper pyramidal cross section of the unit hyper cube) is their support. Figure 1 below illustrates this difference for the bivariate case. Both the

Dirichlet and Ordered Dirichlet random vector $X = (X_1, \ldots, X_m)$ inherit the flexibility of its Beta marginals X_i, $i = 1, \ldots, m$. Transforming these m-variate extensions by means of the transformation $Y_i = -\log(X_i)$, $i = 1, \ldots, m$ allow for flexible prior distributions not restricted to the unit hyper cube. Van Dorp and Mazzuchi (2003) used a similar transformation to define a prior distribution on a set of ordered failure rates on $[0, \infty)^m$ via an ordered Dirichlet distribution.

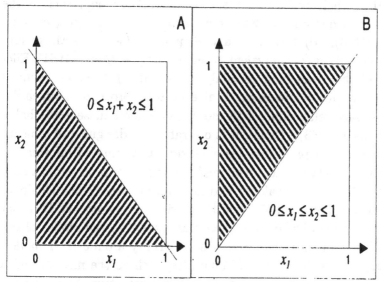

Figure 1. A: Support of a bivariate Dirichlet distribution, B: Support of a bivariate ordered Dirichlet distribution.

Practical implementation of subjective Bayesian methods involving the Beta distribution and its Dirichlet extensions evidently require the specification of their parameters. To avoid being incoherent in these Bayesian analyses, the specification of these prior parameters preferably should not rely on classical estimation techniques which use data, such as maximum likelihood estimation or the method of moments. The specification of the prior parameters ought to be based on expert judgment elicitation. To define the prior parameters, expert judgment about quantities of interest are elicited and equated to their theoretical expression for central tendency such as mean, median, or mode (see e.g. Chaloner and Duncan (1983)). In addition, some quantification of the quality of the expert judgment is often given by specifying a variance or a probability interval for the prior quantity. Solving these equations generally would lead to the required parameter estimates.

Methods for eliciting the parameters of a Beta distributions have focused on eliciting: (a) a measure of central tendency such as the mean and a measure of dispersion such as the variance (see, e.g., Press (1989)), (b) the mean and a quantile (see, e.g. Martz and Waller (1982)) or (c) equivalent observations (e.g. Cooke, 1991). Elicitation of the mean (and certainly the variance), however, requires a level of cognitive processing that elicitation procedures which demand it, may well produce little more than random noise (see, Chalone and Duncan (1983)). Hence, it is desirable, for designing of a meaningful elicitation procedure for engineers, that elicited information can be easily related (i.e. involving little cognitive processing) to observables (see, e.g., Chaloner and Duncan (1983)). While Chaloner and Duncan (1983), (1987) elicit Beta prior parameters and Dirichlet prior parameters by relating these parameters to the modes of observable random variables and non-uniformity around their modes, they also advocate the use of quantiles, such as the median and a lower quantile, for the elicitation of prior parameters. An additional advantage of eliciting quantiles is that it allows for the use of betting strategies in an indirect elicitation procedure (see e.g. Cooke (1991)).

This chapter addresses the problem of specification of prior parameters of a Beta distribution and its Dirichlet extensions above via quantile estimates. It is envisioned that these quantile estimates are elicited utilizing expert judgment techniques thereby allowing coherent and practical application of the Beta distribution and its Dirichlet extensions in Bayesian Analyses. Solving for the parameters of these prior parameters via quantile estimates involves using the incomplete Beta function $B(x|a,b)$ given by

$$B(x \,|a,b) = \frac{\Gamma(a+b)}{\Gamma(a)\Gamma(b)} \int_0^x p^{a-1}(1-p)^{b-1}dp, \qquad (I.4)$$

where $a > 0$, $b > 0$. The incomplete Beta function $B(x|a,b)$ has no closed-form (analytic) expression. Hence, Weiler (1965) resorted to solving graphically for the two parameters of the Beta distribution given the qth and $(1-q)$th quantile. This graphical approach, however, is limited to the number of graphs plotted. For intermediate solutions interpolation methods must be used, which are often subject to an interpolation error.

The adaptability of the Beta distribution will be reconfirmed in Section II by proving that a solution exists for the parameters of a Beta distribution for any combination of a lower quantile and upper quantile constraint. A numerical procedure will be described which solves for parameters a and b of a Beta distribution (cf. (I.1)) given these constraints. The contents of Section II is based on Van Dorp and Mazzuchi (2000). The numerical procedure derived in Section II can be easily adapted to the Weiler's (1965) methodology and improves on his graphical method. In addition, the numerical procedure can be adapted to the case where the median and an another quantile are specified as measures of central tendency and dispersion. In Sections III and IV the methods of Section II will be utilized to specify the parameters of the Dirichlet and Ordered Dirichlet distributions, respectively. In addition, some properties of the Dirichlet and Ordered Dirichlet distribution will be listed in Sections III and IV.

II.   Specification of Prior Beta Parameters

For reasons to become evident from the discussion below, we will reparameterize the Beta density given by (I.1) by setting $\beta = a + b$ and $\alpha = \frac{a}{a+b}$; This yields the following expression for the probability density function of a Beta random variable X

$$\frac{\Gamma(\beta)}{\Gamma(\beta \cdot \alpha)\Gamma(\beta \cdot (1-\alpha))} \, x^{\beta \cdot \alpha - 1}(1-x)^{\beta \cdot (1-\alpha)-1}, x \in [0,1], \quad (II.1)$$

where $0 \leq \alpha \leq 1, \beta > 0$. The reparameterization is a one-to-one transformation from (I.1) to (II.1) and vice versa. Note that the condition $\alpha \in [0, 1]$ is identical to the condition on the original random variable X. For the purpose of this chapter, a random variable X distributed following (II.1) will be denoted as $X \sim Beta(\alpha, \beta)$. The latter notation is somewhat unconventional as $Beta(a, b)$ usually refers to the structural form of the pdf provided by (I.1). Perhaps the consistent use of Greek notation α and β rather than Latin notation a and b may help alleviate this source of confusion.

A. Basic Properties of the Beta Distribution

It easily follows from (II.1) that

$$E[X|\alpha, \beta] = \alpha \tag{II.2}$$

and

$$Var[X|\alpha, \beta] = \frac{\alpha \cdot (1 - \alpha)}{(\beta + 1)}. \tag{II.3}$$

Hence, the reparameterization provided in (II.1) allows one to interpret α as a location parameter and β as a shape parameter that determines the uncertainty in X. The nth moment of X around zero in terms of α and β can be expressed utilizing (II.1) as

$$\begin{aligned}
E[X^n|\alpha, \beta] &= \frac{\prod_{i=1}^{n}(\beta \cdot \alpha + n - i)}{\prod_{i=1}^{n}(\beta + n - i)} \\
&= \alpha \cdot \frac{\prod_{i=1}^{n-1}(\beta \cdot \alpha + n - i)}{\prod_{i=1}^{n-1}(\beta + n - i)}, \quad n = 1, 2, 3, \ldots
\end{aligned} \tag{II.4}$$

with the usual convention that $\prod_{i=1}^{0}\{\cdot\} = 1$. Using the structure of (II.1), (II.2), (II.3) and (II.4) we can readily draw conclusions regarding the limiting distributions of a Beta random variable by letting $\beta \to \infty$ and $\beta \downarrow 0$ (for any fixed value of α). Consider the two different classes of degenerate distributions presented in Figure 2. It follows from (II.2) and (II.3) that the degenerate distribution in Class 1 of Figure 2 is the limiting distribution obtained by letting $\beta \to \infty$. From (II.4) it follows that the moments of the limiting distribution when letting $\beta \downarrow 0$ coincide with the moments of the degenerate distribution in Class 2 of Figure 2 (i.e. of a Bernoulli variable with a point mass of α at 1). As both the limiting distribution of X by letting $\beta \downarrow 0$ and the degenerate distribution of Class 2 have a bounded support, it follows from the agreement of their moments that the degenerate distribution in Class 2 is the limiting distribution by letting $\beta \downarrow 0$ (see e.g. Harris (1966), page 103). The limiting distributions of Class 1 and Class 2 (and how they arise from the limiting behavior of the parameter β) play a central role in deriving the theoretical result in the next section. An additional property of the Beta distribution utilized in this derivation is that for $b > 0$ (using the notation of (I.1))

$$0 < a_1 < a_2 \Rightarrow B\left(x \mid a_1, b\right) > B\left(x \mid a_2, b\right), \forall x \in (0,1), \quad \text{(II.5)}$$

and for $a > 0$

$$b_1 > b_2 > 0 \Rightarrow B\left(x \mid a, b_1\right) > B\left(x \mid a, b_2\right), \forall x \in (0,1), \quad \text{(II.6)}$$

(see e.g. Proschan and Singpurwalla (1979)). From (II.5) and (II.6), it follows (using the notation of (II.1)) that

$$\alpha_2 > \alpha_1 > 0, \beta > 0 \Rightarrow \Pr(X \leq x | \alpha_2, \beta) < \Pr(X \leq x | \alpha_1, \beta).$$
$$\text{(II.7)}$$

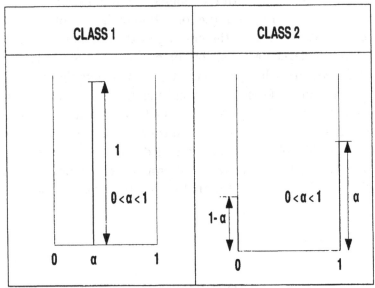

Figure 2. Two classes of degenerate beta distributions.

Finally, the quantile constraint concept defined in Definition 1 below will be used as well.

Definition 1: *Let* $0 < x_q < 1$, $0 < q < 1$. *A random variable* X *with support* $[0, 1]$ *satisfies quantile constraint* (x_q, q) *if and only if* $\Pr\{X \le x_q\} = q$.

B. Solving for the Prior Beta Parameters

To specify the prior Beta parameters based on quantile estimates we need to solve problem $\mathcal{P}_1$ below. Solving problem $\mathcal{P}_1$ involves the use of the incomplete Beta function given by (I.4) and therefore has no closed form (analytic) solution. Also, the quantile constraints in problem $\mathcal{P}_1$ can be considered a set of two nonlinear constraints in two unknowns, i.e. α and β, and may not necessarily have a feasible solution. To construct a numerical procedure with solves problem $\mathcal{P}_1$ in a finite number of iterations, it is necessary to verify that problem $\mathcal{P}_1$ has a solution for any combination of the two quantile constraints. This assertion will be proved in Theorem 1 by means of limiting arguments.

Problem $\mathcal{P}_1$:. *Solve α and β for $X \sim Beta(\alpha, \beta)$ (cf. (II.1)) under the two quantile constraints(x_{q_L}, q_L) and (x_{q_U}, q_U), where $q_L < q_U$.*

Theorem 1: *There exists a solution (α^*, β^*) of problem $\mathcal{P}_1$.*

Proof :The proof involves four steps. In the first step it will be proved, using the notation in (II.1), that for a given $\beta > 0$ and a quantile constraint (x_q, q), a unique α° exists such that $X \sim Beta$ (α°, β) (cf. (II.1)) satisfies this quantile constraint. In the second step it will be shown that for $\beta \downarrow 0$ the parameter $\alpha^\circ \rightarrow (1 - q)$. The third step validates that for $\beta \rightarrow \infty$ the parameter $\alpha^\circ \rightarrow x_q$. Finally, in the fourth step, the statement of this theorem will be verified.

Step 1: Let a quantile constraint (x_q, q) be specified for X. Assume that $\beta > 0$ is given and introduce the function $\xi(\alpha, \beta)$ such that

$$\xi(\alpha, \beta) = \Pr\{X \leq x_q | \alpha, \beta\} - q, \ 0 < \alpha < 1, \ \beta > 0. \qquad (II.8)$$

From the structure of (II.1) it follows that $\xi(\alpha, \beta)$ is a continuous differentiable function for $0 < \alpha < 1$, $\beta > 0$. Consider $\xi(\alpha, \beta)$ when $\alpha \downarrow 0$ and $\beta > 0$ fixed. From (II.2) and (II.3) it follows, respectively, that

$$\lim_{\alpha \downarrow 0} E[X | \alpha, \beta] = 0,$$
$$\lim_{\alpha \downarrow 0} Var[X | \alpha, \beta] = 0$$

respectively, for any fixed $\beta > 0$. Hence, when $\alpha \downarrow 0$, the distribution of X converges to a degenerate distribution with a single point mass concentrated at 0. With $0 < x_q < 1$, $0 < q < 1$ it thus follows from (II.8) that

$$\lim_{\alpha \downarrow 0} \xi(\alpha, \beta) = 1 - q > 0, \qquad (II.9)$$

for any fixed $\beta > 0$. Similarly, using (II.2), (II.3) and using the fact that the distribution of X converges to a degenerate distribution with a single point mass concentrated at 1 as $\alpha \uparrow 1$, we obtain

$$\lim_{\alpha\uparrow1}\xi(\alpha, \beta) = -q < 0 \qquad (II.10)$$

for any fixed $\beta > 0$. From (II.9), (II.10) with $\xi(\alpha, \beta)$ being a continuous function, it follows that

$$\exists\, \alpha° \in (0,1): \xi(\alpha°,\beta) = 0, \forall\, \beta > 0. \qquad (II.11)$$

Utilizing expression (II.7), it follows that $\xi(\alpha, \beta)$ is a strictly decreasing function in α for any fixed $\beta > 0$. Thus, given fixed $\beta > 0$, $\alpha°$ is the unique solution to $\xi(\alpha, \beta) = 0$ and $X \sim Beta(\alpha°, \beta)$ (cf. (II.1)) satisfies the quantile constraint (x_q, q) given fixed $\beta > 0$.

Before proceeding to Step 2, note that the solution $\alpha°$ depends on q, x_q (cf. (II.8)) and β (cf. (II.11)) motivating the following notation

$$\alpha° = \mathcal{G}_{x_q}(\beta), \qquad (II.12)$$

where $\mathcal{G}_{x_q}(\cdot): (0,\infty) \to (0,1)$, $0 < x_q < 1$, $0 < q < 1$, such that

$$\xi(\mathcal{G}_{x_q}(\beta), \beta) = 0, \forall \beta > 0. \qquad (II.13)$$

From the structure of (II.1), (II.13) and the implicit function theorem, it follows that $\mathcal{G}_{x_q}(\beta)$ is also a continuous function for $\beta > 0$. Using the definition of $\xi(\alpha, \beta)$ given by (II.8) and (II.13), it follows that

$$\Pr\{X \le x_q | \mathcal{G}_{x_q}(\beta), \beta\} = q, \forall \beta > 0. \qquad (II.14)$$

Step 2: Consider $X \sim Beta(\alpha°, \beta)$ (cf. (II.1)), where $\alpha° = \mathcal{G}_{x_q}(\beta)$, and let $\beta \downarrow 0$. From continuity of $\Pr\{X \le x_q | \mathcal{G}_{x_q}(\beta), \beta\}$ in β for fixed x_q, it follows from (II.14) that

$$\lim_{\beta\downarrow0}\Pr\{X \le x_q | \mathcal{G}_{x_q}(\beta), \beta\} = q. \qquad (II.15)$$

For the structure of the density (II.1) it has been shown above that as $\beta \downarrow 0$ the distribution of X converges to a degenerate distribution of Class 2 in Figure 2. The limiting expectation of X as $\beta \downarrow 0$ thus becomes the expectation of a Bernoulli random variable and from (II.15) it follows that

$$\lim_{\beta\downarrow0}E[\,X|\,\mathcal{G}_{x_q}(\beta), \beta\,] = 1 - q. \qquad (II.16)$$

However, from (II.2) we have

$$E[X|\mathcal{G}_{x_q}(\beta),\beta] = \mathcal{G}_{x_q}(\beta), \tag{II.17}$$

for any $\beta > 0$ and using (II.16) and (II.17) one concludes

$$\lim_{\beta \downarrow 0} \mathcal{G}_{x_q}(\beta) = 1 - q. \tag{II.18}$$

In other words, the parameter $\alpha° \to (1 - q)$ as $\beta \downarrow 0$.

Step 3: Consider $X \sim Beta(\alpha°, \beta)$ (cf. (II.1)), where $\alpha° = \mathcal{G}_{x_q}(\beta)$, and let $\beta \to \infty$. From (II.3), it follows that as $\beta \to \infty$ the distribution of X converges to a degenerate distribution of Class 1 in Figure 2 with a single point mass concentrated at some $x^* \in [0,1]$. From continuity of $\Pr\{X \leq x_q|\mathcal{G}_{x_q}(\beta),\beta\}$ in β for fixed x_q, it follows from (II.14) that $x^* = x_q$. This means that,

$$\lim_{\beta \to \infty} E[X|\mathcal{G}_{x_q}(\beta),\beta] = x_q.$$

Hence, from (II.2) we have

$$\lim_{\beta \to \infty} \mathcal{G}_{x_q}(\beta) = x_q. \tag{II.19}$$

In other words, the parameter $\alpha° \to x_q$ as $\beta \to \infty$.

Step 4: Let $X \sim Beta(\alpha, \beta)$ (cf. (II.1)). Let (x_{q_L}, q_L) and (x_{q_U}, q_U) be two quantile constraints specified for X, such that $q_L < q_U$. Consider the associated functions $\mathcal{G}_{x_{q_L}}(\beta)$ and $\mathcal{G}_{x_{q_U}}(\beta)$ each defined implicitly by (II.8) and (II.13), respectively. Introducing the function

$$H(\beta) = \mathcal{G}_{x_{q_L}}(\beta) - \mathcal{G}_{x_{q_U}}(\beta)$$

it follows from (II.18) that

$$\lim_{\beta \downarrow 0} H(\beta) = (1 - q_L) - (1 - q_U) = q_U - q_L > 0. \tag{II.20}$$

Similarly, from (II.19) it follows that

$$\lim_{\beta \to \infty} H(\beta) = x_{q_L} - x_{q_U} < 0. \tag{II.21}$$

From the continuity of $\mathcal{G}_{x_{q_L}}(\beta)$ and $\mathcal{G}_{x_{q_U}}(\beta)$, (II.20) and (II.21) it follows that

$$\exists \beta^* > 0 : H(\beta^*) = 0. \tag{II.22}$$

Denoting $\alpha^* = \mathcal{G}_{x_{q_L}}(\beta^*)$ (cf. (II.12)) it follows from (II.22) that

$$\alpha^* = \mathcal{G}_{x_{q_L}}(\beta^*) = \mathcal{G}_{x_{q_U}}(\beta^*).$$

In other words, $X \sim Beta(\alpha^*, \beta^*)$ (cf. (II.1)) satisfies both quantile constraints(x_{q_L}, q_L) and (x_{q_U}, q_U) and thus (α^*, β^*) is a solution to problem $\mathcal{P}_1$. $\square$

Theorem 1 proves the existence of a solution to problem $\mathcal{P}_1$. The uniqueness of the solution (α^*, β^*) to $\mathcal{P}_1$ would follow by showing that; (i) $H(\beta)$ has 0 or 1 stationary points for $\beta > 0$; (ii) if $H(\beta)$ has a stationary point for $\beta > 0$ this stationary point coincides with a global maximum. It is conjectured that the above assertions hold. Numerical analyses in the examples below support this conjecture (see Van Dorp and Mazzuchi (2000)). In case multiple solutions exist to problem $\mathcal{P}_1$, the numerical algorithm below is designed so that the selected solution coincides with the solution with the lowest value for β^*, and thus the highest level of uncertainty. The latter solution would be a preferred solution, given that x_{q_L} and x_{q_U} ought to be elicited through expert judgment in Bayesian Analysis.

C.   Design of a Numerical Procedure

Since problem $\mathcal{P}_1$ cannot be solved in a closed form, a numerical procedure that determines a solution to problem $\mathcal{P}_1$ with a prescribed level of accuracy, in a finite number of iterations, is desirable. Below, such a numerical procedure will be informally described. The numerical method uses a procedure for solving for the qth quantile of a Beta distribution. Such a procedure is described in the Appendix in Pseudo Pascal (denoted $BISECT1$).

From (II.2) and (II.3), it follows that α is a location parameter and β is an uncertainty parameter given the value of α and higher values of β coincide with lower uncertainty levels. These interpretations of the parameters α and β are used in the design of the numerical procedure to obtain a solution to $\mathcal{P}_1$. Assume for now that an interval

$[a_1, b_1]$ is obtained containing β^* which yields a solution (α^*, β^*) of $\mathcal{P}_1$, where $\alpha^* = \mathcal{G}_{x_q}(\beta^*)$. Let β_1 be the midpoint of this interval. The kth iteration of the numerical procedure will be described below.

To solve $(\alpha^\circ)_k$ satisfying the quantile constraint (x_{q_U}, q_U) of $\mathcal{P}_1$ given a value for β_k successive shrinking intervals $[d_n, e_n]$ are calculated containing the solution $(\alpha^\circ)_k$. From (II.1), follows that $(\alpha^\circ)_k \in [0, 1]$. Hence, $[d_1, e_1] = [0, 1]$. Next, α_n is set to the midpoint of $[d_n, e_n]$ and the probability mass $(q_U)_n = \Pr\{X \leq x_{q_U} | \alpha_n, \beta_k\}$ is calculated. In case $(q_U)_n \leq q_U$, the Beta distribution is skewed excessively towards 1. Therefore, it follows from (II.2) that the value of the location parameter α_n is too high. Hence, the next interval containing $(\alpha^\circ)_k$ is $[d_{n+1}, e_{n+1}] = [d_n, \alpha_n]$. On the other hand, when $(q_U)_n > q_U$, the Beta distribution is skewed excessively towards 0. Therefore it follows from (II.2) that the value of the location parameter α_n is too small. Hence, the next interval containing $(\alpha^\circ)_k$ can be set to $[d_{n+1}, e_{n+1}] = [\alpha_n, e_n]$. Finally, the next estimate α_{n+1} is set to be the midpoint of the interval $[d_{n+1}, e_{n+1}]$. The above procedure is repeated until $(q_U)_n$ is close to q_U with a pre-assigned level of accuracy. The quantile constraint (x_{q_U}, q_U) of $\mathcal{P}_1$ is satisfied once this accuracy has been reached and $(\alpha^\circ)_k$ is set equal to the α_n. The algorithm above is a bisection method (see, for example, Press et al., 1989) and is provided in the Appendix in Pseudo Pascal (denoted $BISECT\,2$). A specific example of the algorithm in $BISECT2$ is presented in Figure 3, where $(x_{q_U}, q_U) = (0.80, 0.70)$ and $\beta_1 = 3$. The starting interval for α_1 equals $[d_1, e_1] = [0, 1]$, hence $\alpha_1 = 0.5$. Thus, it follows that $(q_U)_1 > 0.7$, hence $[d_2, e_2] = [0.5, 1]$ and $\alpha_2 = 0.75$. Now we have $(q_U)_2 < 0.7$, hence $[d_3, e_3] = [0.5, 0.75]$ and $\alpha_3 = 0.625$. Consequently we have $(q_U)_3 \approx 0.7 = q_U$, $(\alpha^\circ)_1$ is set to $\alpha_3 = 0.625$, the algorithm terminates and $X \sim Beta((\alpha^\circ)_1, \beta_1)$ (cf. (II.1)) satisfies the quantile constraint $(x_{q_U}, q_u) = (0.80, 0.70)$.

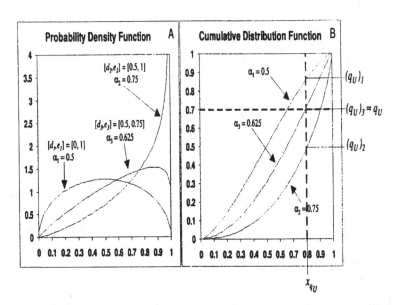

Figure 3. An example of bisection method $BISECT\,2$ (using $\beta_1 = 3$). A: Beta pdf's and shrinking bisection intervals $[d_n, e_n]$, B: Beta CDF's and sequence of $(q_U)_n$, $n = 1, \ldots, 3$.

After solving for $(\alpha^\circ)_k$ (utilizing $BISECT2$) the procedure calculates the q_Lth quantile $(x_{q_L})_k$ (utilizing $BISECT1$) of $Beta((\alpha^\circ)_k, \beta_k)$. When $(x_{q_L})_k < x_{q_L}$ the uncertainty in $Beta((\alpha^\circ)_k, \beta_k)$(cf. (II.1)) is too high. Therefore, the current estimate of the uncertainty parameter β_k should be too low. Hence, the next interval which contains β^* can be set to $[a_{k+1}, b_{k+1}] = [\beta_k, b_k]$. On the other hand, if $(x_{q_L})_k > x_{q_L}$ the uncertainty in $Beta((\alpha^\circ)_k, \beta_k)$ is too low. Therefore, the current estimate of the uncertainty parameter β_k is too high. Hence, the next interval which contains β^* can be set to $[a_{k+1}, b_{k+1}] = [a_k, \beta_k]$. Finally, the next estimate β_{k+1} is taken to be the midpoint of the interval $[a_{k+1}, b_{k+1}]$. The above procedure is then repeated until the current estimate $(x_{q_L})_k$ is close to x_{q_L} with the pre-assigned desired level of accuracy. The quantile constraint (x_{q_L}, q_L) of $\mathcal{P}_1$ is met once this accuracy has been reached. The parameters (α^*, β^*)that solve $\mathcal{P}_1$ are set equal to the pair $((\alpha^\circ)_k, \beta_k)$. The algorithm above is a bisection method and is provided in the Appendix in Pseudo Pascal (denoted $BISECT\,3$). A specific example of the algorithm $BISECT\,3$ is presented in Figure 4, where $(x_{q_L}, q_L) = (0.20, 0.10)$, $(x_{q_U}, q_U) = (0.80, 0.70)$, $\beta_1 = 3$ and $(\alpha^\circ)_1 = 0.625$.

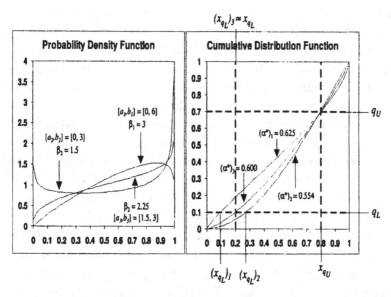

Figure 4. An example of bisection method $BISECT3$. A: Beta pdf's and shrinking bisection intervals $[a_k, b_k]$, B: Beta CDF's and a sequence of $(x_{q_L})_k$, $k = 1, \ldots, 3$.

The starting interval for β_1 equals $[a_1, b_1] = [0, 6]$. It follows that $(x_{q_L})_1 < 0.2$, hence $[a_2, b_2] = [0, 3]$ and $\beta_2 = 1.5$, $(\alpha^\circ)_2 = 0.554$ (determined using $BISECT2$). It follows that $(x_{q_L})_2 > 0.2$, hence $[a_3, b_3] = [1.5, 3]$ and $\beta_3 = 2.25$, $(\alpha^\circ)_2 = 0.600$ (determined using $BISECT2$) It now follows that $(x_{q_L})_1 \approx 0.2 = q_L$, the algorithm terminates, β^* is set $\beta_3 = 2.25$, α^* is set to $(\alpha^\circ)_3 = 0.600$ and we have $X \sim Beta(\alpha^*, \beta^*)$ which satisfies the quantile constraints $(x_{q_L}, q_L) = (0.20, 0.10)$, $(x_{q_U},$

$q_U) = (0.80, 0.70)$.

To determine a starting interval $[a_1, b_1]$ containing β^* the following steps could be adopted in the procedure. Set the lower bound $a_1 = 0$. To obtain the upper bound b_1 set $\beta_{1,k} = 1$, where $k = 1$, and solve for $(\alpha^\circ)_{1,k}$ satisfying the quantile constraint x_{q_U} of problem $\mathcal{P}_1$ utilizing $BISECT2$. Next, solve for the q_Lth quantile $(x_{q_L})_{1,k}$ of $Beta((\alpha^\circ)_{1,k}, \beta_{1,k})$ (utilizing $BISECT1$). In case $(x_{q_L})_{1,k} < x_{q_L}$ the uncertainty in $Beta((\alpha^\circ)_{1,k}, \beta_{1,k})$ is too high. Therefore, $\beta_{1,k} < \beta^*$. In that case, set $\beta_{1,k+1} = 2\beta_{1,k}$ and repeat the above procedure. Conversely, in the case $(x_{q_L})_{1,k} > x_{q_L}$ the uncertainty in $Beta((\alpha^\circ)_{1,k}, \beta_{1,k})$ is too low. Therefore, $\beta_{1,k} > \beta^*$. In that case, set $b_1 = \beta_{1,k}$ and the starting interval $[a_1, b_1]$ has been determined. Note that if multiple solutions exist to problem $\mathcal{P}_1$, the starting interval is chosen in such a manner that the selected solution for $\mathcal{P}_1$ by means of the algorithm coincides with the solution with the lowest value for β^*, and consequently the highest level of uncertainty.

The three different bisection methods $BISECT1$, $BISECT2$ and $BISECT3$ were implemented in a PC-based program BETA-CALCULATOR. Figure 5 displays a screen capture of BETA-CALCULATOR. The accuracy for δ in the bisection methods $BISECT1$ and $BISECT2$ was set to be 10^{-8}. The accuracy in the bisection method $BISECT3$ was set to be 10^{-4}. Table 1 contains solutions to problem $\mathcal{P}$ for 4 different combinations of a lower quantile and upper quantile constraint calculated using BETA-CALCULATOR. In addition, Table 1 provides the maximum number of iterations in each bisection method to yield the solutions with the above settings of error-tolerances. Figure 5 contains the results for Example 1 in Table 1. Example 4 in Table 1 coincides with the setup of the Weiler's (1965) graphical method. Finally, Figure 6 depicts the probability density functions and cumulative distribution functions associated with the examples in Table 1. Note that the U-Shaped, J-Shaped and Unimodal forms of the Beta distribution are represented in Figure 6.

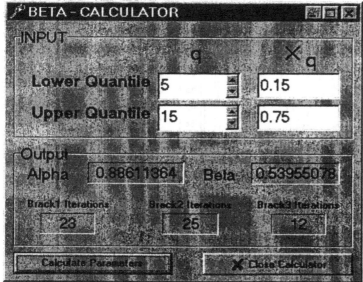

Figure 5. Screen capture of BETA-CALCULATOR with calculation results for the first row in Table 1.

Table 1. Some calculation examples of beta parameters given an upper and a lower quantile constraints.

		q	x_q	α^*	β^*	# 1	# 2	# 3
Example 1	L	0.05	0.15	0.8861	0.5396	23	25	12
	U	0.15	0.75					
Example 2	L	0.49	0.25	0.2672	11.3906	71	24	11
	U	0.99	0.60					
Example 3	L	0.20	0.10	0.3437	2.6328	26	26	10
	U	0.50	0.30					
Example 4	L	0.05	0.45	0.6330	19.5625	44	25	10
	U	0.95	0.80					

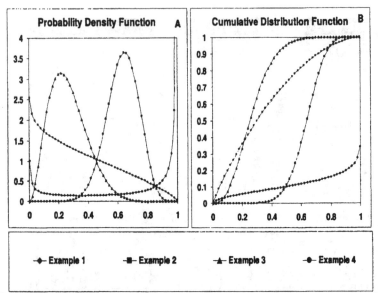

Figure 6. A: Beta probability density functions for the examples in Table 1, B: Beta cumulative distribution functions for the examples in Table 1.

III.   Specification of Prior Dirichlet Parameters

Analogously to (II.1) and for a straightforward application of the numerical procedures derived in Section II (and provided in the Appendix), we reparameterize the Dirichlet distribution given by (I.2) by introducing the new parameters $\beta = \sum_{i=1}^{m+1} \theta_i$ and $\alpha_i = \frac{\theta_i}{\beta}$, $i = 1, \ldots, m$, yielding the probability density

$$\frac{\Gamma(\beta)}{\left\{ \prod_{i=1}^{m} \Gamma(\beta \cdot \alpha_i) \right\} \Gamma\left(\beta \cdot (1 - \sum_{i=1}^{m} \alpha_i) \right)} \left\{ \prod_{i=1}^{m} x_i^{\beta \cdot \alpha_i - 1} \right\}$$

$$\times \left\{ 1 - \sum_{i=1}^{m} x_i \right\}^{\beta \cdot (1 - \sum_{i=1}^{m} \alpha_i) - 1}, \qquad (III.1)$$

where $x_i \geq 0$, $\sum_{i=1}^{m} x_i \leq 1$, $\alpha_i \geq 0$, $i = 1, \ldots, m$, $\sum_{i=1}^{m} \alpha_i \leq 1$ and $\beta > 0$. A random vector $X = (X_1, \ldots, X_m)$ distributed according to the reparameterized Dirichlet distribution (III.1) will be denoted by $Dirichlet(\alpha, \beta)$. Note that, as in the case of (II.1), the condition on the parameters $\alpha = (\alpha_1, \ldots, \alpha_m)$ is identical to the conditions on the variables X_i, $i = 1, \ldots, m$.

A. Basic Properties of the Dirichlet Distribution

Let a random vector $X = (X_1, \ldots, X_m) \sim Dirichlet(\underline{\alpha}, \beta)$. It may be derived from (III.1) that marginals distribution of X_i are given by $X_i \sim Beta(\alpha_i, \beta)$ (in parameterization of (II.1)), $i = 1, \ldots, m$. The moments $E[X_i^n | \alpha_i, \beta]$ follow by substituting α_i for α in (II.4). Analogously, the mean and the variance of X_i follow by substituting α_i for α in (II.2) and (II.3), respectively. Hence, the parameter β of the $Dirichlet(\alpha, \beta)$ distribution may be interpreted as the common shape parameter amongst $X = (X_1, \ldots, X_m)$, whereas the vector $\underline{\alpha} = (\alpha_1, \ldots, \alpha_m)$ may be interpreted as a location parameter of X. Such an interpretation was not valid for the original parameterization given by (I.2) involving the parameters θ_i, $i = 1, \ldots, m+1$. Similar to the analysis in Section A it follows that we may draw conclusions regarding the limiting distributions of the $Dirichlet(\underline{\alpha}, \beta)$ based solely on the limiting behavior of the parameter β. Letting $\beta \to \infty$ we observe that a $Dirichlet(\underline{\alpha}, \beta)$ distribution converges to a degenerate distribution with a single point mass concentrated at α. Letting $\beta \downarrow 0$, we deduce that the $Dirichlet(\underline{\alpha}, \beta)$ distribution converges to an m-variate Bernoulli distribution with marginal parameters α_i in Class 2 of Figure 2, $i = 1, \ldots, m$. The dependence structure in the limiting m-variate Bernoulli distribution is obtained by studying the limiting behavior of the pairwise correlation coefficients in a $Dirichlet(\underline{\alpha}, \beta)$ distribution as $\beta \downarrow 0$. Utilizing the reparameterization in (III.1) it follows that

$$Cov\,(X_i, X_j) = -\frac{\alpha_i \alpha_j}{\beta + 1} \qquad \text{(III.2)}$$

and with (III.2) and (II.3)

$$Cor\,(X_i, X_j) = \frac{Cov\,(X_i, X_j)}{\sqrt{Var\,(X_i)\,Var\,(X_j)}}$$

$$= -\sqrt{\frac{\alpha_i \alpha_j}{(1-\alpha_i)\,(1-\alpha_i)}}. \tag{III.3}$$

Apparently, the correlation structure in a $Dirichlet(\underline{\alpha}, \beta)$ distribution does not depend on the common scale parameter β, and (III.3) describes the dependence structure of the limiting m-variate Bernoulli distribution when $\beta \downarrow 0$. Relation (III.3) is consistent with the well-known result (see Kotz et al. (2000)) that the correlations in a "classical" Dirichlet distribution are negative.

We now present several basic properties of the $Dirichlet(\underline{\alpha}, \beta)$ distribution below utilizing reparameterization (III.1). It would appear (similar to the result in (III.3)) that some further transparency may be achieved by expressing these properties in terms of $(\underline{\alpha}, \beta)$. Firstly, for any index set $A \subset \{1, \ldots, m\}$,

$$X^A \sim Dirichlet(\underline{\alpha}^A, \beta), \tag{III.4}$$

where $X^A = \{X_i | i \in A\}$ and $\underline{\alpha}^A = \{\alpha_i | i \in A\}$. Next,

$$\sum_{i \in A} X_i \sim Beta(\sum_{i \in A} \alpha_i, \beta),$$

$$\frac{X_j}{\sum_{i \in A} X_i} \sim Beta(\frac{\alpha_j}{\sum_{i \in A} \alpha_i}, \beta \sum_{i \in A} \alpha_i),$$

where $j \in A$. Finally, utilizing (III.4) we may derive the conditional probability density function of $(X^A | X^{A^c})$, where A^c denotes the complement of A, i.e. $A^c = \{1, \ldots, m\} \setminus A$, yielding

$$\left\{\frac{1}{\xi}\right\}^{\beta^\bullet - 1} \frac{\Gamma(\beta^\bullet)}{\left\{\prod_{i \in A} \Gamma(\beta^\bullet \cdot \alpha_i^\bullet)\right\} \Gamma\left(\beta \cdot (1 - \sum_{i \in A} \alpha_i^\bullet)\right)}$$

$$\times \left\{\prod_{i \in A} y_i^{\beta^\bullet \cdot \alpha_i^\bullet - 1}\right\} \left\{\xi - \sum_{i \in A} y_i\right\}^{\beta^\bullet \cdot (1 - \sum_{i \in A} \alpha_i^\bullet) - 1}, \tag{III.5}$$

where

$$\xi = 1 - \sum_{j \in A^c} x_j,$$

$$\beta^\bullet = \beta \cdot \left(1 - \sum_{i \in A^c} \alpha_i\right), \qquad \text{(III.6)}$$

$$\alpha_i^\bullet = \frac{\alpha_i}{1 - \sum_{i \in A^c} \alpha_i}.$$

The distribution in (III.5) may be recognized as that of an $|A|$- dimensional vector Y, where $|A|$ indicates the cardinality of the index set A, and

$$Y = \xi\, Z,$$
$$Z \sim Dirichlet(\underline{\alpha}^\bullet, \beta^\bullet),$$

where $\underline{\alpha}^\bullet = \{\alpha_i^\bullet | i \in A\}$ and $\alpha_i^\bullet$ and $\beta^\bullet$ are given by (III.6). Setting $A = \{i\}$ in (III.5) and (III.6) yields what is called the full conditional distribution of X_i as a transformed $Beta(\alpha_i^\bullet, \beta^\bullet)$ with the support

$$\left[0, 1 - \sum_{j=1, j \neq i} x_j\right].$$

The latter result is relevant to the application of the Markov Chain Monte Carlo (MCMC) methods utilizing a $Dirichlet(\underline{\alpha}, \beta)$ (see, e.g., Casella and George (1992)). The MCMC methods have spurted an emergence of numerous Bayesian applications (see, e.g. Gilks et al. (1995)) as these methods allow for sampling from a posterior distribution by successively sampling from posterior full conditional distributions, without having a closed form of the posterior distribution.

B.   Solving for the Dirichlet Prior Parameters

In order to solve for the common shape parameter β and location parameter $\underline{\alpha} = (\alpha_1, \ldots, \alpha_m)$ of an m dimensional random vector $X \sim$ Dirichlet $(\underline{\alpha}, \beta)$ distribution using quantile estimates, it is required to solve problem $\mathcal{P}_2$ below.

Problem $\mathcal{P}_2$: *Solve α and β for $X \sim Dirichlet(\alpha, \beta)$, $X = (X_1,$ ..., $X_m)$ under the two quantile constraints $(x_{q_L}^i, q_L^i)$ and $(x_{q_U}^i, q_U^i)$ for X_i where $q_L^i < q_U^i$, and $m - 1$ single quantile constraints (x_q^j, q^j) for X_j, $j = 1, \ldots, m$, $j \neq i$.*

Note that the quantile levels q^j and q_U^i (or q_L^i) may differ amongst $X_j, j = 1, \ldots, m$. Since $X_i \sim Beta(\alpha_i, \beta)$, it follows immediately from Theorem 1 that a solution to problem $\mathcal{P}_2$ exists. When multiple solutions are available to problem $\mathcal{P}_2$, the numerical algorithms in the Appendix are designed such that the selected solution coincides with the solution with the lowest value for β, and thus the highest level of uncertainty. The latter solution would be a preferred solution given that the quantile constraints in $\mathcal{P}_2$ ought to be elicited through expert judgment in Bayesian Analysis. The algorithm to solve $\mathcal{P}_2$ is provided below in Pseudo Pascal using the bisection methods described in the Appendix.

$STEP1 : BISECT3(\alpha_i, \beta, x_{q_L}^i, x_{q_U}^i, q_L^i, q_U^i) : j = 1;$
$STEP2 : If\, j \neq i\, then\, BISECT2(\alpha_j, x_q^j, \beta, q^j);$
$STEP3 : If\, j < m\, then\, j := j + 1;\, Goto\, STEP2;\, Else\, Stop;$

Table 2 below describes two instances of problem $\mathcal{P}_2$ and their solutions using the algorithm above for $X = (X_1, X_2)$, where $X \sim Dirichlet(\underline{\alpha}, \beta)$ and $\underline{\alpha} = (\alpha_1, \alpha_2)$. Note that, $(x_{q_L}^1, q_L^1)$ and $(x_{q_U}^1, q_U^1)$ in Table 2 coincide with the third row of Table 1. Hence, α_1 and β also coincide in Tables 1 and 2 resulting in a J-shaped marginal form for the pdf of X_1 given in Figure 6A. The resulting marginal form in Example 1(2) of Table 2 for X_2 is J-shaped (uni-modal). Figure 7 displays the resulting Dirichlet densities for the examples in Table 2. Note that, the marginal density of X_1 in both Figures 7A and 7B is identical and J-Shaped, whereas the marginal form of X_2 in Figure 7B is uni-modal. As a result, the joint pdf in Figure 7B has a single mode at $(x_1, x_2) = (0, 1)$. Figure 7A displays three modes at each corner of the unit simplex.

		X_1	X_2			
		(q, x_q)	(q, x_q)	α_1	α_2	β
$Example\,1$	L	$(0.20, 0.10)$	$(0.20, 0.10)$	0.3437	0.3437	2.6328
	U	$(0.50, 0.30)$				
$Example\,2$	L	$(0.20, 0.10)$	$(0.40, 0.50)$	0.3437	0.5665	2.6328
	U	$(0.50, 0.30)$				

IV. Specification of Ordered Dirichlet Parameters

Analogously to (II.1) and (III.1) we reparameterize the Ordered Dirichlet distribution given by (I.3) by introducing $\beta = \sum_{i=1}^{m+1} \theta_i$ and $\alpha_i = \dfrac{\theta_i}{\beta}$, $i = 1, \ldots, m$ yielding the probability density

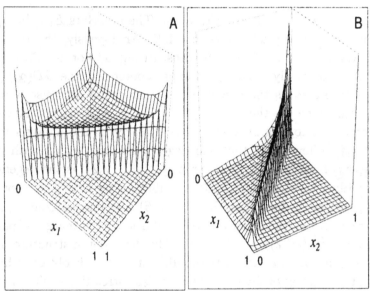

Figure 7. A: Dirichlet PDF of Example 1 in Table 2, B: Dirichlet PDF of Example 2 in Table 2.

$$\frac{\Gamma(\beta)}{\left\{ \prod_{i=1}^{m} \Gamma\left(\beta \cdot \alpha_i\right) \right\} \Gamma\left(\beta \cdot \left(1 - \sum_{i=1}^{m} \alpha_i\right)\right)} x_1^{\beta \cdot \alpha_i - 1}$$

$$\times \left\{ \prod_{i=2}^{m} (x_i - x_{i-1})^{\beta \cdot \alpha_i - 1} \right\} (1 - x_m)^{\beta \cdot (1 - \sum_{i=1}^{m} \alpha_i) - 1}, \quad \text{(IV.1)}$$

where $0 \leq x_{i-1} < x_i \leq 1$, $\alpha_i \geq 0$, $i = 1, \ldots, m$, $\sum_{i=1}^{m} \alpha_i \leq 1$ and $\beta > 0$. The distribution given by (IV.1) will be denoted by $OD(\underline{\alpha}, \beta)$, where $\underline{\alpha} = (\alpha_1, \ldots, \alpha_m)$.

A.   Some Properties of the Ordered Dirichlet Distribution

Let a random vector $X = (X_1, \ldots, X_m) \sim OD(\underline{\alpha}, \beta)$. It easily follows from (IV.1) that marginals distribution of X_i are given by

$$X_i \sim Beta(\alpha_i^+, \beta), \quad \alpha_i^+ = \sum_{k=1}^{i} \alpha_k \qquad \text{(IV.2)}$$

(in parameterization (II.1)), $i = 1, \ldots, m$. The moments $E[X_i^n | \underline{\alpha}, \beta]$ follow by substituting α_i^+ for α in (II.4). Analogously, the mean and the variance of X_i follow by substituting α_i^+ for α in (II.2) and (II.3), respectively. As above, the parameter β of the $OD(\underline{\alpha}, \beta)$ may be interpreted as the common scale parameter amongst $X = (X_1, \ldots, X_m)$, whereas the vector $\underline{\alpha}^+ = (\alpha_1^+, \ldots, \alpha_m^+)$ may be interpreted as a location parameter of X. Similarly to the analysis in Section A it follows that that the degenerate distribution with a point mass concentrated at parameter $\underline{\alpha}^+$ (cf. (IV.2)) is the degenerate distribution of an $OD(\underline{\alpha}, \beta)$ distribution by letting $\beta \to \infty$. Letting $\beta \downarrow 0$ we deduce that the $OD(\underline{\alpha}, \beta)$ distribution converges to an ordered m-variate Bernoulli distribution with marginal parameters $\underline{\alpha}_i^+$ (cf. (IV.2)), $i = 1, \ldots, m$. The dependence structure in the limiting ordered m-variate Bernoulli distribution is obtained by studying the limiting behavior of the pairwise correlation coefficients in a $OD(\underline{\alpha}, \beta)$ distribution as $\beta \downarrow 0$. Utilizing the reparameterization in (IV.1) it follows that

$$Cov(X_i, X_j) = \frac{Cov(X_i, X_j)}{\sqrt{Var(X_i) Var(X_j)}} = \frac{\alpha_i^+(1 - \alpha_j^+)}{\beta + 1} \qquad \text{(IV.3)}$$

and with (IV.3) and (II.4) we have

$$Cor(X_i, X_j) = \sqrt{\frac{\alpha_i^+(1 - \alpha_j^+)}{(1 - \alpha_i^+)\alpha_j^+}}. \qquad \text{(IV.4)}$$

Utilizing the pdf reparameterization in (IV.1), it follows from (IV.4) that the correlation structure in a $OD(\underline{\alpha}, \beta)$ does not depend on the common scale parameter β as it is in the case of the

$Dirichlet(\underline{\alpha}, \beta)$ distribution. Note that, unlike the case of $Dirichlet(\underline{\alpha},$ $\beta)$ distribution the correlations are positive. The difference between the signs of the correlations in the $OD(\underline{\alpha}, \beta)$ and $Dirichlet(\underline{\alpha}, \beta)$ distributions may in part be explained by the differences in their support (see, Figure 1). An additional useful property for the $OD(\underline{\alpha}, \beta)$ is that for any index set $A \subset \{1, \ldots, m\}$,

$$X^A \sim OD(\underline{\alpha}^A, \beta), \tag{IV.5}$$

where $X^A = \{X_i | i \in A\}$, $\underline{\alpha}^A = \{\alpha_l^A | l = 1, \ldots, A\}$ and

$$\alpha_l^A = \begin{cases} \alpha_{A_{(l)}}^+ & l = 1 \\ \alpha_{A_{(l)}}^+ - \alpha_{A_{(l-1)}}^+ & l = 2, \ldots, |A| \end{cases}$$

where, as above, $|A|$ indicates the number of elements in the index set A and $A_{(l)}$ indicates the lth element in A, such that $A_{(l)} > A_{(l-1)}$, $k = 2, \ldots, |A|$. Furthermore,

$$(X_j - X_i) \sim Beta(\alpha_j^+ - \alpha_i^+, \beta),$$

$$\frac{X_i}{X_j} \sim Beta(\frac{\alpha_i^+}{\alpha_j^+}, \beta\alpha_j^+),$$

where $j > i$ and α_i^+ are defined in (IV.5). Finally, utilizing (III.4) we may derive the conditional probability density function of $(X^A | X^{A^c})$, where A^c denotes the complement of A, i.e. $A^c = \{1, \ldots, m\} \setminus A$ and $A = \{i\}$, $i \in \{1, \ldots, m\}$ yielding

$$\frac{\Gamma\big(\beta(\alpha_i + \alpha_{+1})\big)}{\Gamma(\beta\alpha_i)\Gamma(\beta\alpha_{+1}))} \frac{(x_i - x_{i-1})^{\beta\alpha_i - 1}(x_{i+1} - x_i)^{\beta\alpha_{i+1} - 1}}{(x_{i+1} - x_{i-1})^{\beta(\alpha_{+1} + \alpha_i) - 1}} \tag{IV.6}$$

The distribution in (IV.6) may be recognized as that of a random variable Y where

$$Y = (x_{i+1} - x_{i-1})Z + x_{i-1}$$

$$Z \sim Beta\left(\frac{\alpha_i}{\alpha_{i+1}^+ - \alpha_{i-1}^+}, \beta(\alpha_{i+1}^+ - \alpha_{i-1}^+)\right), \tag{IV.7}$$

where α_{i+1}^+ is defined by (IV.2). Hence, the distribution of $(X^A | X^{A^c})$ where $A = \{i\}$ (referred to as the full conditional distribution of X_i)

is a transformed Beta distribution with support

$$\begin{cases} [0, x_{i+1}] & i = 1 \\ [x_{i-1}, x_{i+1}] & i = 1, \dots, m-1 \\ [x_{i-1}, 1] & i = m \end{cases}$$

As above, the result in (IV.6) and (IV.7) is relevant to the application of Markov Chain Monte Carlo (MCMC) methods (see, e.g. Casella and George (1992)) utilizing an $OD(\underline{\alpha}, \beta)$ distribution.

B. Solving for the Ordered Dirichlet Prior Parameters

To solve for the common shape parameter β and location parameter $\underline{\alpha} = (\alpha_1, \dots, \alpha_m)$ of an m dimensional random vector $X \sim OD(\underline{\alpha}, \beta)$ distribution using quantile estimates, we are required to solve problem $\mathcal{P}_3$ below.

Problem $\mathcal{P}_3$: *Solve $\underline{\alpha}$ and β for $X \sim OD(\alpha, \beta)$, $X = (X_1, \dots, X_m)$ under the two quantile constraints$(x_{q_L}^i, q_L^i)$ and $(x_{q_U}^i, q_U^i)$ for X_i, where $q_L^i < q_U^i$, and $m-1$ single quantile constraints (x_q^j, q^j) for X_j, $j = 1, \dots, m$, $j \neq i$ such that*

$$x_q^1 < x_q^2 \cdots < x_q^{i-1} < x_{q_L}^i < x_{q_U}^i < x_q^{i+1} < \cdots < x_q^m \qquad (IV.8)$$

Note that the quantile levels q^j and q_U^i (or q_L^i) may differ amongst X_j, $j = 1, \dots, m$. Since $X_i \sim Beta(\alpha_i^+, \beta)$, it follows immediately from Theorem 1 that a solution to problem $\mathcal{P}_3$ exists. In case multiple solutions exist to problem $\mathcal{P}_3$, the numerical algorithms in the Appendix are designed in such a manner that the selected solution coincides with the solution with the lowest value for β, and thus the highest level of obtained uncertainty. As mentioned above, the latter solution would be a preferred solution. The algorithm for solving $\mathcal{P}_3$ is provided below in Pseudo Pascal using the bisection methods described in the appendix.

$STEP\,1 : BISECT3(\alpha_i^+, \beta, , x_{q_L}^i, x_{q_U}^i, q_L^i, q_U^i) : j = 1;$
$STEP\,2 : If\,j \neq i\,then\,BISECT2(\alpha_j^+, x_q^j, \beta, q^j);$
$STEP\,3 : If\,j < m\,then\,j := j + 1;\,Goto\,STEP\,2;\,Else\,Goto\,STEP\,4$
$STEP\,4 : \alpha_1 := \alpha_1^+$
$STEP\,5 : For\,i := 2\,to\,m\,do\,\alpha_i := \alpha_i^+ - \alpha_{i-1}^+$

Note that Steps 1 to 3 are identical to those in the algorithm to solve problem $\mathcal{P}_2$ associated with a $Dirichlet(\underline{\alpha}, \beta)$ distribution. As an example of the procedure above, note that the Example 2 in the second row of Table 2 satisfies the order restriction (IV.8) since $x_{q_L}^1 = 0.10$, $x_{q_U}^1 = 0.30$ and $x_q^2 = 0.50$. Note that the corresponding quantile levels $q_L^1 = 0.20$, $q_U^1 = 0.50$ and $q^2 = 0.40$ differ. From Table 2 it follows that

$$\alpha_1^+ = 0.3437; \alpha_2^+ = 0.5665; \beta = 2.6328$$

Executing Steps 4 and 5 in the algorithm above it follows that for this example

$$\begin{aligned} \beta &= 2.6328; \\ \alpha_1 = \alpha_1^+ &= 0.3437; \\ \alpha_2 = \alpha_2^+ - \alpha_1^+ &= 0.5665 - 0.3437 = 0.2228. \end{aligned} \qquad \text{(IV.9)}$$

The probability density function of the Ordered Dirichlet distribution associated with (IV.9) is presented in Figure 8.

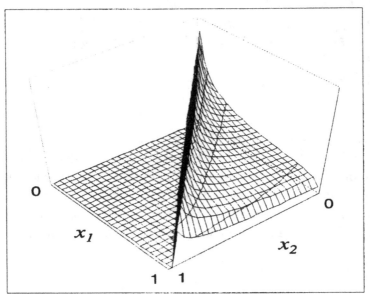

Figure 8. Ordered Dirichlet PDF associated with (IV.9).

C. Transforming the Ordered Dirichlet Distribution and Numerical Stability

The use of an m-variate Ordered Dirichlet distribution to specify a prior distribution may be extended to the unbounded domain $(\mathbb{R}^+)^m$ by transforming to $Y_i = -\log(X_i)$ or $X_i = \exp(-Y_i)$, where $X = OD(\underline{\alpha}, \beta)$, $X = (X_1, \ldots, X_m)$. Although the quantile levels q^j and q_U^i (or q_L^i), $j = 1, \ldots, m$, may differ in the specification of problem $\mathcal{P}_3$ and in its solution, it is perhaps more practical from an expert judgment elicitation point of view to set e.g. $q^j = q_L^i = 0.50$ and $q_U^i = 0.95$ (which are quantile levels widely used in practice). Hence, measures of location for every Y_i are established by eliciting their median values and one could utilize the median (50% quantile) of Y_i and the 95% quantile for Y_i to determine the common shape parameter β. Table 3 below contains such median estimates $y_1^{0.50}$ for failure rates of an exponential life time distribution in different stress environments (see, Van Dorp and Mazzuchi (2003)). In addition, the 95%

quantile $y_1^{0.95}$ of the failure rate at Environment 1 (which is typically the use-stress environment in an accelerated life testing set-up) is provided as well.

Table 3. Environments, prior failure rates and transformations.

Environment	Temp	Volt	Prior $y_i^{0.50}$	Prior $x_i^{0.50}$	Prior $x_i^{0.50}$
	$(^\circ F)$	(VDC)	$(hours)^{-1}$	$c = 1$	$c = 841.61$
1	100	10.0	$5.036 \cdot 10^{-5}$	0.99995	0.9585
2	125	13.0	$1.100 \cdot 10^{-4}$	0.99998	0.9116
3	160	15.0	$5.732 \ 10^{-4}$	0.99943	0.6173
4	200	17.0	$1.429 \cdot 10^{-3}$	0.99857	0.3004
5	250	19.0	$3.781 \cdot 10^{-3}$	0.99623	0.0415
$y_1^{0.95}$ at use stress			$1.315 \cdot 10^{-3}$	0.99869	0.3306

Note that the median failure rates $y_i^{0.50}, i = 1, \ldots, 5$, are very small (common to the accelerated life testing area) and increase rapidly when the stress in an environment increases (see, e.g., Van Dorp and Mazzuchi (2002)). The resulting transformed medians $x_i^{0.50}$ follow utilizing the transformation $x_i^{0.50} = \exp(-y_i^{0.50})$ and are provided in fourth column in Table 3. Note that these values for $x_i^{0.50}, i = 1, \ldots, 5$ are very close to 1.

Instead of the transformation $Y_i = -\log(X_i)$ we may utilize a more general transformation

$$Y_i = \frac{-\log(X_i)}{c} \Leftrightarrow X_i = \exp(-cY_i) \qquad \text{(IV.10)}$$

to define a prior distribution on the domain $[0, \infty)^m$ via the Ordered Dirichlet distribution, where c is a preset transformation factor. As an example, the fifth column in Table 3 contains the transformed median values $x_i^{0.50} = \exp(-cy_i^{0.50}), i = 1, \ldots, 5$, where $c = 841.61$. Note that these median values are now spread over the entire support $[0, 1]$ rather than being in the vicinity at 1. The motivation for utilizing (IV.10) follows from (i) the value of the medians $y_i^{0.50}$, (ii) the Beta marginal distributions of X_i and (iii) the fact that no closed-form expression is available for the incomplete Beta function given $B(x \mid a, b)$ given by (I.4).

Several numerical algorithms exist to approximate the incomplete Beta function given by (I.4) (see, e.g., Press et al. (1989)). These approximations are well-behaved for parameter values $a \geq 1$, $b \geq 1$. However, in case $a < 1$, the Beta density explodes at $x = 0$, resulting in numerical instability for the approximation of $B(u|a, b)$ for the values close to $u = 0$. Vice versa, in case $b < 1$, the Beta density explodes at $x = 1$, resulting in numerical instability of $B(u|a, b)$ for values in the vicinity of $u = 1$. Figure 9A depicts the situation in Table 3 with $c = 1$ and medians $x_i^{0.50}$ close to 1. As the probability mass in the interval $[x_i^{0.50}, 1]$ has to account for 50% of the total probability mass, the associated Beta marginal densities will have an infinite mode at 1, resulting in numerically unstable behavior of evaluation of $B(x \mid a, b)$ (cf. (I.4)) when setting $c = 1$ into (IV.10). The closer these median values $x_i^{0.50}$ are to the boundaries 1 (or 0), the larger will the numerical instability be in evaluating $B(x \mid a, b)$ (cf. (I.4)).

To reduce instability in numerical evaluation of $B(x \mid a, b)$ (cf. (I.4)) one may select a preset transformation factor c in (IV.10) such that distance between median estimates $x_i^{0.50}$, $i = 1, \ldots, 5$, and the boundaries of the support $[0, 1]$ be as large as possible. Figure 9B depicts the motivation behind using (IV.10) with a value of $c > 1$. The idea in Figure 9B is to choose a preset transformation parameter c such that the incomplete Beta function $B(x \mid a, b)$ corresponding to $x_5^{0.50}$ (the median at the highest stress level) is as well-behaved at $x = 0$ as $B(x|a, b)$ corresponding to $x_1^{0.50}$ (the median at the lowest stress level) at $x = 1$.

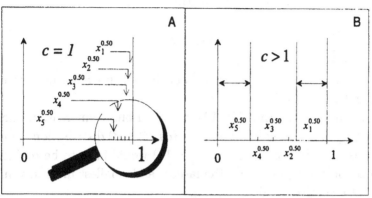

Figure 9. Transformation to medians $x_i^{0.50}$, $i = 1, \ldots, 5$ with transformation factor c.

The suggestion in Figure 9B is to select c such that $x_5^{0.50} = 1 - x_1^{0.50}$ or

$$\exp\left(-c \cdot y_5^{0.50}\right) = 1 - \exp\left(-c \cdot y_1^{0.50}\right). \tag{IV.11}$$

Unfortunately, (IV.11) cannot be solved in closed form. General root finding algorithms may be used to solve for the transformation constant c numerically up to a pre-assigned level of accuracy. The resulting value for the transformation factor c, utilizing $y_1^{0.50}$ and $y_5^{0.50}$ in Table 3 and (IV.11) equals 841.61 as indicated in Table 3. Note that, indeed in the fifth column in Table 3, $x_5^{0.50} = 1 - x_1^{0.50}$. Utilizing the median $x_i^{0.50}$ in the fifth column of Table 3 the parameters of the $OD(\underline{\alpha}, \beta)$ may be solved by means of the algorithm described in the previous section. The resulting parameter values are:

$$\beta = 1.6656; \, \alpha_1 = 0.1522; \alpha_2 = 0.0481; \tag{IV.12}$$
$$\alpha_3 = 0.2198; \alpha_4 = 0.2168; \alpha_5 = 0.2109;$$

It follows from (IV.2) and (IV.12) that

$$1 - \sum_{i=1}^{5} \alpha_i = 1 - \alpha_5^+ = 0.1552 = \alpha_1 = \alpha_1^+$$

indicating an identical numerical behavior when evaluating the cdf of X_5 and X_1.

It should be noted that (IV.10) may be useful to increase numerical stability when transforming a $Dirichlet(\underline{\alpha}, \beta)$ distribution. In the latter case, the indices 1 and 5 in (IV.11) would have to be replaced by those indices representing the largest and smallest median values in the $Dirichlet(\underline{\alpha}, \beta)$ distribution.

V. Conclusions

Algorithms have been developed for solving the prior parameters of the Beta, Dirichlet and Ordered Dirichlet distributions based on quantile constraints. The development of these algorithms involved a reparameterization which allows interpretation of these prior parameters in terms of a location parameter and a shape parameter $\beta > 0$. Limiting distributions of the Beta, Dirichlet and Ordered Dirichlet distributions follow in a transparent manner from the limiting behavior of the common shape parameter β. Existence of parameter solutions for the Beta, Dirichlet and Ordered Dirichlet distributions was proved utilizing their limiting distributions. We note in passing that the reparameterization advocated in this paper may be related to the orthoganality of parameters. We have not checked as yet whether indeed the condition for orthoganality as described for example in Cox and Reid (1987), is valid in our case.

Appendix

Let $X \sim Beta(\alpha, \beta)$ using the reparameterization given by (II.1). The bisection methods below use the numerical algorithm given in Press et al. (1989) to evaluate the incomplete Beta function $B(\cdot | a, b)$ given by (I.4). $BISECT1(x_q, \alpha, \beta, q)$ solves for the qth quantile x_q of

X. Output parameters are indicated in bold. $BISECT2(\alpha^\circ, x_q, \beta, q)$ solves for the parameter α°, satisfying the quantile constraint (x_q, q). $BISECT3(\alpha^*, \beta^*, x_{q_L}, x_{q_U}, q_L, q_U)$ solve for the required solution (α^*, β^*). A method to determine a starting interval $[a_1, b_1]$, containing β^* is given in $STEP\,1$, $STEP\,2$, and $STEP\,3$ of $BISECT\,3$.

$BISECT1(x_q, \alpha, \beta, q)$:

$STEP\,1$	$m := 1;\ Set\,[d_1, e_1] = [0, 1]\,;$
$STEP\,2$	$x_{q,m} := \frac{d_m + e_m}{2}\,;\ q_m := B(x_{q,m}\mid \alpha, \beta);$
$STEP\,3$	$If\ q_m \leq q\ then\ d_{m+1} := x_{q,m}\,;\ e_{m+1} := e_m;$
	$Else\ e_{m+1} := x_{q,m}\,;\ d_{m+1} := d_m;$
$STEP\,4$	$If\,\mid q_m - q\mid\ < \delta\ then\ x_q := x_{q,m};\ Stop;$
	$Else\ m := m + 1;\ Goto\,STEP\,2;$

$BISECT2(\alpha^\circ, x_q, \beta, q)$:

$STEP\,1$	$n := 1;\ Set\,[d_1, e_1] = [0, 1]\,;$
$STEP\,2$	$\alpha_{n+1} := \frac{d_n + e_n}{2};\ q_n := B(x_q\mid \alpha_{n+1}, \beta);$
$STEP\,3$	$If\ q_n \leq q\ then\ e_{n+1} := \alpha_{n+1};\ d_{n+1} := d_n;$
	$Else\ d_{n+1} = \alpha_{n+1}\,;\ e_{n+1} := e_n;$
$STEP\,4$	$If\,\mid q_n - q\mid\ < \delta\ then\ Stop;$
	$Else\ n := n + 1;\ Goto\,STEP\,2;$

$BISECT3(\alpha^*, \beta^*, x_{q_L}, x_{q_U}, q_L, q_U)$:

$STEP\,1$	$k := 1;\ \beta_{1,k} := 1;$
$STEP\,2$	$(\alpha^\circ)_{1,k} := BISECT2(x_{q_U}, \beta_{1,k}, q_U);$
	$(x_{q_L})_{1,k} := BISECT1((\alpha^\circ)_{1,k}, \beta_{1,k}, q_L);$
$STEP\,3$	$If\ (x_{q_L})_{1,k} < x_{q_L}\ then\ \beta_{1,k+1} := 2 \cdot \beta_{1,k};\ Goto\ STEP\,2;$
	$Else\ [a_1, b_1] := [0, \beta_{1,k}];$
$STEP\,4$	$k := 1;$
$STEP\,5$	$\beta_k := \frac{a_k + b_k}{2};\ (\alpha^\circ)_k := BISECT2(x_{q_U}, \beta_k, q_U);$
	$(x_{q_L})_k := BISECT1((\alpha^\circ)_k, \beta_k, q_L);$
$STEP\,6$	$If\ (x_{q_L})_k < x_{q_L}\ then\ a_{k+1} := \beta_k;\ b_{k+1} := b_k;$
	$Else\ a_{k+1} := a_k;\ b_{k+1} := \beta_k;$
$STEP\,7$	$If\,\mid (x_{q_L})_k - x_{q_L}\mid\ < \delta\ then\ \alpha^* := (\alpha^\circ)_k;\ \beta^* := \beta_k;\ Stop;$
	$Else\ k := k + 1;\ Goto\,STEP\,5;$

The authors would like to extend their gratitude to Samuel Kotz for his most valuable comments in developing this chapter. We are indepted to Saralees Nadarajah for his editorial support.

References

1. Carlin, B.P.; Louis, T.A. *Bayes and Empirical Bayes Methods for Data Analysis*; Chapman and Hall/CRC: New York, 2000.
2. Casella, G.; George, E.I. Explaining the Gibbs sampler, The American Statistician **1992**, *46*, 167-174.
3. Chaloner, K.M.; Duncan, G.T. Assessment of a beta prior distribution: PM elicitation, The Statistician **1983**, *32*, 174-181.
4. Chaloner, K.M.; Duncan, G.T. Some properties of the Dirichlet-multinomial distribution and its use in prior elicitation, Communications in Statistics–Theory and Methods **1987**, *16*, 511-523.
5. Cooke, R.M. *Experts in Uncertainty: Opinions and Subjective Probability in Science*; Oxford University Press: Oxford, 1991.
6. Coolen, F.P.A. An imprecise Dirichlet model for Bayesian analysis of failure data including right-censored observations, Reliability Engineering and Systems Safety **1997**, *56*, 61-68.
7. Cowell, R.G. On compatible priors for Bayesian networks, IEEE Transactions on Pattern Analysis and Machine Intelligence **1996**, *18*, 901-911.
8. Cox, D.R.; Reid, N. Parameter orthogonality and approximate conditional inference, Journal of Royal Statistical Society, B **1987**, *49*, 1-39.
9. Dennis III, S.Y. A Bayesian analysis of ratios of proportions in tree-structured experiments, Journal of Statistical Planning and Inference **1998**, *74*, 293-322.
10. Erkanli, A.; Mazzuchi A.; Soyer, R. Bayesian computations for a class of reliability growth models, Technometrics **1998**, *40*, 14-23.
11. Gavaskar, U. A comparison of two elicitation methods for a

prior distribution for a binomial parameter, Management Science **1988**, *34*, 784-790.

12. Gilks, W.R.; Richardson S.; Spiegelhalter, D.J. *Markov Chain Monte Carlo in Practice*; CRC Press, 1995.

13. Harris, B. *Theory of Probability*; Addision-Wesley: Reading, 1966.

14. Johnson, W.O.; Kokolasis, G.E. Bayesian classification based on multivariate binary data, Journal of Statistical Planning and Inference **1994**, *41*, 21-35.

15. Lancaster, T. Bayes WESML posterior inference from choice-based samples, Journal of Econometrics **1997**, *79*, 291-303.

16. Lange, K. Application of the Dirichlet distribution to forensic match probabilities, Genetica **1995**, *96*, 107-117.

17. Kotz, S.; Balakrishnan, N; Johnson, N.L. *Continuous Multivariate Distributions, Volume 1: Models and Applications* (second edition); John Wiley and Sons: New York, 2000.

18. Kumar, S.; Tiwari, R.C. Bayesian estimation of reliability under a random environment governed by a Dirichlet prior, IEEE Transactions of Reliability **1989**, *38*, 218-223.

19. Martz, H.F.; Waller, R.A. *Bayesian Reliability Analysis*; John Wiley and Sons: New York, 1982.

20. Neath A.N.; Samaniego, F.J. On Bayesian estimation of the multiple decrement function in the competing risks problem, Statistics and Probability Letters **1996**, *31*, 75-83.

21. Press, S.J. *Bayesian Statistics, Principles, Models and Applications*; John Wiley and Sons, New York, 1989.

22. Press, W.H.; Flannery, B.P.; Teukolsky S.A.; Vettering W.T. *Numerical Recipes in Pascal*; Cambridge University Press: Cambridge, 1989.

23. Proschan, F.; Singpurwalla, N.D. Accelerated life testing - a pragmatic Bayesian approach, in: *Optimizing Methods in Statistics* (editor J. Rustagi) **1979**, pp. 385 -401, Academic Press: New York.

24. Van Dorp, J.R. A General Bayesian Inference Model for Accelerated Life Testing, Dissertation, The George Washington University, Washington, D.C., 1998.

25. Van Dorp, J.R.; Mazzuchi, T.A. Solving for the parameters of a beta Distribution under two quantile constraints, Journal of Statistical Computation and Simulation **2000**, *67*, 189-201.

26. Van Dorp, J.R.; Mazzuchi, T.A. A general Bayes exponential inference model for accelerated life testing, to appear in Journal of Statistical Planning and Inference, 2003.

27. Van Dorp, J.R.; Mazzuchi T.A.; Fornell, G.E.; Pollock L.R. A Bayes approach to step-stress accelerated life testing, IEEE Transactions on Reliability **1996**, *45*, 491-498.

28. Van Dorp, J.R.; Mazzuchi, T.A.; Soyer, R. Sequential inference and decision making for single mission systems development, Journal of Statistical Planning and Inference **1997**, *62*, 207-218.

29. Weiler, H. The use of incomplete beta functions for prior distributions in binomial sampling, Technometrics **1965**, *7*, 335-346.

30. Wilks, S.S. *Mathematical Statistics*; John Wiley and Sons: New York, 1962.

Extensions of Dirichlet Integrals: their Computation and Probability Applications

Milton Sobel[1] and Krzysztof S. Frankowski[2]

[1]Department of Statistics and Applied Probability
University of California
Santa Barbara, California 93106
[2]Department of Computer Science
University of Minnesota
Minneapolis, Minnesota 55455

I.   Definitions of I, J and IJ-Functions and Their Properties.

In Sobel et al. (1977) the Dirichlet I-function was studied, developed and tabulated; on page 40 a related function, the Dirichlet J, was also introduced. Both of these are used in a multinomial setting with a given sample size and given cell probabilities and neither of these should be used in the case of a finite population size such as a card deck, unless we are sampling with replacement. We also introduce the Dirichlet IJ-function which is quite useful when we have two different restrictions on different cells that we wish to apply. We have shown in Sobel et al. (1977) that cumulative multinomial probabilities can be written as multiple Dirichlet integrals in addition to

319

the usual manner of writing them as sums. These integral expressions led to highly efficient exact algorithms for their calculation without any approximate integration methods. We illustrate our methods by writing programs in *Mathematica*. After each of our short programs we give a numerical examples. All answers with fewer then 8 decimals are understood to be exact(unless the approximation symbol $\approx$ is used). Tabled values are given to only 6 decimal figures.

For n, b positive integers and vector arguments $\bar{r} = (r_1, r_2, \ldots, r_b)$, $\bar{p} = (p_1, p_2, \ldots, p_b)$ with $G = \prod_{i=1}^{b} \Gamma(r_i)$, $n \geq R$ (defined as $r_1 + r_2 + \cdots + r_b)$ and $\sum_{i=1}^{b} p_i \leq 1$, the incomplete Dirichlet integral of Type-1 is defined by the b-fold integral

$$I_{\bar{p}}^{(b)}(\bar{r}, n) = \frac{n!}{(n-R)!G} \int_0^{p_1} \int_0^{p_2} \cdots \int_0^{p_b} \left(1 - \sum_{i=1}^{b} x_i\right)^{n-R} \prod_{i=1}^{b} x_i^{r_i-1} dx_i$$

(I.1)

Illustration I.1. *Compute $I_p^{(b)}(\bar{r}, n)$ using the integral representation (I.1)*

We write a short program in *Mathematica* and use it for numerica illustration:

```
iInt[rvec_List, n_,p_] :=
(* Compute I-function for vector r and scalar p *)
 Module[{b, rr, r=rvec, tt,i},
    b = Length[r]; rr = Sum[r[[i]], {i, b}];
    tt = (n)!/(n-rr)!/Product[(r[[i]] - 1)!, {i, b}];
    tt = tt*(Product[x[i]^(r[[i]] - 1), {i, b}]*
    (1 - Sum[x[i],{i, b}])^(n- rr));
    Do[tt = Integrate[tt, {x[i], 0, p}], {i, b}]; tt]
```

A short description of *Mathematica* is given in the Appendix to this paper; here we will only indicate briefly what we are doing: The first line defines a function iInt with 3 parameters: *rvec*, which is a list of *r*-values, *n*, and *p*, which are scalars. The '_ ' symbol after the name signifies 'arbitrary' value. Without it, the function would work for values with only the name of the parameter.

The second line is a comment; the third line (Module), defines local variables (protecting the variables with the same names outside the function). The fourth line computes b-number of integrals, which has to be the same as the length of the vector $\bar{r}$ and then computes $rr = \sum_{i=1}^{b} r_i$. The last three lines compute the constant outside the integrand and the integrand itself; finally the Do statement does the integration.

The careful reader will notice that the parentheses $(\cdots)$ are used for computation only, the brackets $[\cdots]$ separate the functional name from parameters, the braces $\{\cdots\}$ are used to denote list, like in the invocation of the function iInt $[\{1,2\},3,.1]$ and double brackets $[[\cdots]]$ denote a subscript.

As a numerical example, we compute iInt $[\{1,2\},3,.1]$=0.003.

Finally we notice that the program is 'informal', not suitable for library, because it does not check impossible parameters and can be used only for small values of b. We also note that iInt $[\{1,2\},3,p]$ gives $3p^2$, the algebraic result, which holds for any p.

The formula (I.1) has been shown to be identical to the joint probability that an independent sample of size n will contain at least r_i from the category i(with cell probability p_i) $(i = 1, 2, \ldots, b)$. Readers acquainted with the relation of the usual incomplete beta distribution to either 'tail' of a binomial series will recognize that the result (I.1) is a direct generalization of that well-known result.

The special case in which all p_i are equal and all r_i $(i = 1, 2, \ldots, b)$ are equal is the most important application and we have (with $G = \Gamma^b(r)$)

$$I_p^{(b)}(r,n) = \frac{n!}{(n-br)!G} \int_0^p \int_0^p \cdots \int_0^p \left(1 - \sum_{i=1}^{b} x_i\right)^{n-br} \prod_{i=1}^{b} x_i^{r-1} dx_i.$$

(I.2)

Illustration I.2. *Compute $I_p^{(b)}(r,n)$ using (I.2)*

We will write the informal program as before in **Illustration I.1.** Since the program is shorter and simpler we do not discuss it here.

```
iInt[b_,r_,n_,p_] :=
(* Compute I-function by integration for scalars r and p *)
Module[{tt,i}, tt = (n)!/(n-b r)!/((r - 1)!)^b;
    tt = tt*(Product[x[i]^(r - 1), {i, b}]*
```

```
(1 - Sum[x[i], {i, b}])^(n-b r));
   Do[tt = Integrate[tt, {x[i], 0, p}], {i, b}]; tt]
```

We use the above to compute $I_p^{(2)}(2,4)$: iInt[2,2,4,p]=6p^4.
The dual of the I-function is the J-function, which enables us to work
with the condition "freq.$< r$" instead of "freq.$\geq r$". The J-function
does not have a simple integral form as the I has in (I.1); the exact
upper limits of integration may be difficult to write in the general
case. Instead we define it for (I.2) by inclusion-exclusion in terms of
the corresponding I-function (and conversely for I in terms of J) by
writing for scalar p and r

$$J_p^{(b)}(r,n) = \sum_{\alpha=0}^{b}(-1)^\alpha \binom{b}{\alpha} I_p^{(\alpha)}(r,n),$$

$$I_p^{(b)}(r,n) = \sum_{\alpha=0}^{b}(-1)^\alpha \binom{b}{\alpha} J_p^{(\alpha)}(r,n).$$

(I.3)

Illustration I.3. *Compute $J_p^{(b)}(r,n)$ by (I.3).*

```
jInc[b_,r_,n_,p_] :=
(* Compute J-function by inclusion-exclusion for scalars r and p *)
  Module[{i},1+Sum[(-1)^i Binomial[b,i]iInt[i,r,n,p],{i,b}]];
jInc[2,2,4,.1]=0.896
```

For the scalar case, we can write the integral representation for J-
function, (with $G = \Gamma^b(r)$ as before), namely

$$J_p^{(b)}(r,n) = \frac{n!}{(n-br)!G} \int_p^{u_1} \int_p^{u_2} \cdots \int_p^{u_b} \left(1 - \sum_{i=1}^{b} x_i\right)^{n-br} \prod_{i=1}^{b} x_i^{r-1}dx_i.$$

(I.4)

Here the upper limits are;
$u_1 = 1 - (b-1)p,\ u_2 = 1 - (b-2)p - x_1, \ldots, u_i = 1 - (b-i)p - x_1$
$-x_2 \cdots - x_{i-1}, \ldots, u_b = 1 - \sum_{\alpha=1}^{b-1} x_\alpha$. These limits are much more
complicated and hence the integral representation is less useful. We
have to assume also that $n - br \geq 0$; otherwise the integral representation is not defined, even if the probability interpretation and (I.3)
is well defined.

Illustration I.4. *Compute $J_p^{(b)}(r,n)$ from (I.4).*

```
jInt[b_,r_,n_,p_] :=
(* Compute J-function by integration for scalar r and  p *)
 Module[{tt,i,j,u}, u=Table[1-(b-i)p-Sum[x[j],{j,i-1}],{i,b}];
    tt = (n)!/(n-b r)!/((r - 1)!)^b;
    tt = tt*(Product[x[i]^(r - 1), {i, b}]*(1 - Sum[x[i], {i, b}])^
    (n- b r)); Do[tt = Integrate[tt, {x[i], p,u[[i]]}], {i, b,1,-1}];
    tt];   jInt[2,2,4,.1]=0.896
```

The function Table[ ] computes the list of the upper limits $u[[i]]$, which are used in the integration in the Do-statement.

For the vector case we associate each subset of r-components of size α with the corresponding subset of p-components; the binomial coefficient $\binom{b}{\alpha}$ in (I.3) must then be replaced by a sum over all $\binom{b}{\alpha}$ possible subsets of α components from the r-vector. Since we usually use (I.3) with scalar arguments we shall not write out this result explicitly.

For fast computations and in some applications we generalize (I.2) and introduce an additional parameter j as follows:

$$I_p^{(b,j)}(r,n) = \frac{n!\,(p^r/r)^j}{(n-br)!\Gamma^b(r)} \underbrace{\int_0^p \int_0^p \cdots \int_0^p}_{(b-j)\text{--}fold}$$

$$\left(1 - jp - \sum_{i=j+1}^{b} x_i\right)^{n-br} \prod_{i=j+1}^{b} x_i^{r-1} dx_i. \tag{I.5}$$

The expression (I.5) has a simple probability interpretation which is often quite useful:

It represents the joint probability that a sample of size n will contain $b - j$ cells with frequency $\geq r$ and j cells with frequency exactly r. For $j = b$ all integrals in (I.5) vanish and we are left with

$$I_p^{(b,b)}(r,n) = \frac{n!}{(n-br)!(r!)^b} p^{br} (1 - bp)^{n-br}. \tag{I.6}$$

When $n = br$, the integral in (I.5) is easy to compute and we obtain:

$$I_p^{(b,j)}(r, br) = \frac{(br)!p^{br}}{(r!)^b}. \tag{I.7}$$

Finally, integrating by parts in (I.5), we obtain (with the help of the new parameter j) a beautiful recurrence relation, which together with (I.6) and (I.7) is a basis of our fast and accurate computation of $I_p^{(b)}(r, n)$.

$$(n - jr) I_p^{(b,j)}(r, n) = n(1 - jp) I_p^{(b,j)}(r, n - 1) + r(b - j) I_p^{(b,j+1)}(r, n). \tag{I.8}$$

Looking at this relation, we notice that for fixed b, r and p, the only changing 2 parameters are j and n. Using (I.6) and (I.7) as boundary conditions, we obtain a very fast program, which is a basis for our fast and accurate computations.

Illustration I.5. *Compute $I_p^{(b)}(r, n)$ from (I.8).*

```
dI[(b_Integer)?NonNegative, (r_Integer)?Positive,
(n_Integer)?Positive, pp_] :=
(* Compute I-function from recurrence relation *)
  Module[{j, p = N[pp], l, m, v = Table[0, {n}], res = 0},
  If[b==0,Return[1.]];
   If[n - b*r < 0 || 1 - b*p < 0, Return[0],
     v[[b*r]] = Product[(p*(l*r - m + 1))/m, {l, 1, b}, {m, 1, r}];
  Do[v[[m]] = (r*(b - j)*v[[m]] + m*(1 - j*p)*v[[m - 1]])/(m - j*r),
     {j, b, 0, -1}, {m, b*r + 1, n}]; res = v[[n]]; ]; N[res,10]];
  dI[2,2,4,1/10]=0.0006
```

Starting with *Mathematica* 4.0, the statement N[res,10] (which we used to print exactly 10 digits of res) stopped working and we replaced it with different function SetPrecision[res,10], which in turn does not work for earlier versions.

For the I-function with vector arguments, we can eliminate one cell at-a-time, obtaining:

$$I_{\tilde{p}}^{(b)}(\tilde{r}; n) = \sum_{\alpha = r_1}^{n - \sum_{i=2}^{b} r_i} \binom{n}{\alpha} p_1^{\alpha} q_1^{n-\alpha} I_{\tilde{p}_1/q_1}^{(b-1)}(\tilde{r}_1; n - \alpha), \tag{I.9}$$

where $\tilde{r}_1 = \tilde{r}$ with the first component removed. Similarly $\tilde{p}_1 = \tilde{p}$ with the first component removed and $q_1 = 1 - p_1$.

Illustration I.6. *Compute* $I_{1/6}^{(2)}(2,4;7)$ *from (I.9).*
We have

$$I_{1/6}^{(2)}(2,4;7) = \sum_{\alpha=2}^{3} \binom{7}{\alpha} \left(\frac{1}{6}\right)^{\alpha} \left(\frac{5}{6}\right)^{7-\alpha} I_{1/5}^{(1)}(4,7-\alpha) \approx .00170039.$$

The computation of the corresponding J-function is similar:
The formula (I.6) for $j = b$ is still valid for $J_p^{(b,b)}(r,n)$, but the condition for $n = br$ has to be changed to a different condition at the point $n = jr$, namely

$$J_p^{(b,j)}(r,jr) = \frac{(jr)!}{(r!)^j} p^{jr}. \tag{I.10}$$

As with the I-function the recurrence relation for J is nearly identical to (I.8) with only a sign change:

$$(n - jr) J_p^{(b,j)}(r,n) = n(1 - jp) J_p^{(b,j)}(r,n-1) - r(b-j) J_p^{(b,j+1)}(r,n). \tag{I.11}$$

As before, using this new recurrence relation with (I.10) and the same boundary conditions (I.6) (with I replaced by J) and (I.10), we have again a very fast and useful computational program:

Illustration I.7. *Compute* $J_p^{(b)}(r,n)$ *from (I.11).*

```
dJ[(b_Integer)?NonNegative, (r_Integer)?Positive,
(n_Integer)?Positive, pp_] :=
(* Compute J-function from recurrence relation *)
Module[{j, p = N[pp], l, m, v = Table[0, {n + 1}], res = 0, t},
  If[n<r || b==0,Return[1.]];
    If[1 - b*p < 0 , Return[0], j = Floor[n/r] + 1;
      While[j > 0, j = j - 1; If[j == 0, t = 1,
        Do[t = Product[(p*(l*r - m + 1))/m, {l, 1, j}, {m, 1, r}]]];
        v[[j*r + 1]] = t; Do[v[[m + 1]] = (-(r*(b - j)*v[[m + 1]]) +
          m*(1 - j*p)*v[[m]])/(m - j*r), {m, j*r + 1, n}]];
      res = v[[n + 1]]; ]; If[res < 0, 0, N[res,10]]]];
dJ[2,2,4,1/10]=0.896
```

For J-function with vector arguments, we have to change the limits of summation in the formula (I.9), namely

$$J_{\tilde{p}}^{(b)}(\tilde{r};n) = \sum_{\alpha=0}^{r_1-1} \binom{n}{\alpha} p_1^{\alpha} q_1^{n-\alpha} J_{\tilde{p}_1/q_1}^{(b-1)}(\tilde{r}_1;n-\alpha) \tag{I.12}$$

with $\bar{p}_1$, $\tilde{q}_1$ and q_1 defined as in (I.9). In some cases we wish to use both types of conditions: "freq.$\geq r$ and freq.$< r$" on different cells but in the same problem. The combined function **IJ** enables us to do this.

For computational reasons it is often useful to express **IJ** in terms of **I** or **J**. To do this we use the inclusion-exclusion formula (I.3) and obtain 2 different results, namely:

$$
IJ_{(p_1,p_2)}^{(g;h)}(r,s;n) = \sum_{\alpha=0}^{g}(-1)^\alpha \binom{g}{\alpha} J_{(p_1,p_2)}^{(\alpha;h)}(r,s;n)
$$
$$
= \sum_{\alpha=0}^{h}(-1)^\alpha \binom{h}{\alpha} I_{(p_1,p_2)}^{(g;\alpha)}(r,s;n).
$$

$$(I.13)$$

Here in (I.13) and in (I.14) the semicolon in the superscript indicates two separate conditions and is not to be confused with the comma used in the superscript (I.5) above and in (I.15) below.

If $p_1 = p_2 (= p$, say) and $r = s$, then the functions **I** and **J** do not have vector arguments and we can write $J_{p_1}^{(h+\alpha)}(r,n)$ in the middle expression of (I.13) and $I_{p_1}^{(g+\alpha)}(r,n)$ in the right hand expression of (I.13), obtaining

$$
IJ_p^{(g;h)}(r,n) = \sum_{\alpha=0}^{g}(-1)^\alpha \binom{g}{\alpha} J_p^{(h+\alpha)}(r,n) = \sum_{\alpha=0}^{h}(-1)^\alpha \binom{h}{\alpha} I_p^{(g+\alpha)}(r,n).
$$

$$(I.14)$$

Illustration I.8. *Compute $IJ_p^{(g;h)}(r,n)$ from (I.14).*

```
dIJ[i_Integer,j_Integer,(r_Integer)?Positive,(n_Integer)?Positive,pp_]
   := Module[{p, t, k}, p = N[pp];
   t = If[i <= j, Sum[(-1)^k*Binomial[i, k]*dJ[k + j, r, n, p],
   {k, 0, i}], Sum[(-1)^k*Binomial[j, k]*dI[k + i, r, n, p],
   {k, 0, j}]]; N[t]];   dIJ[1,2,3,10,.1]=.065791312
```

Note, that in this version of the program we try to compute the shortest sum, i.e. the one in which the upper limit is the minimum of g and h.

Illustration I.9. *Compute* $IJ_{1/6}^{(1;1)}(2,3;5)$ *using (I.13).*

$$IJ_{1/6}^{(1;1)}(2,3;5) = I_{1/6}^{(1)}(2,5) - I_{1/6}^{(1;1)}(2,3;5) \approx .194959.$$

In some applications, we have to compute, or transform a formula with the j-value (i.e. that part of the second superscript which is after a comma) not equal to zero. In such cases we use the so called "reduction formula", which enables us to reduce the superscript j to 0 and express (b,j) by $(b-j,0)$, for which we have fast programs.

$$I_p^{(b,i+j)}(r,n) = \frac{n!}{(r!)^j (n-jr)!} p^{rj} (1-jp)^{n-jr} I_{p/(1-jp)}^{(b-j,i)}(r, n-jr).$$

$$(\text{I.15})$$

We use this mostly with $i = 0$. Exactly the same formula holds with J or IJ functions on both sides of (I.15).

Illustration I.10. *Compute* $I_p^{(b,j)}(r,n)$ *from (I.15) with* $i = 0$

```
djI[(b_Integer)?Positive, (j_Integer)?Positive, (r_Integer)?Positive,
    (n_Integer)?Positive, pp_] :=
(* Compute I[b,j,r,n,p] from reduction formula *)
Module[{p, q, t}, p = N[pp]; q = 1 - j*p;
  t = If[n - j*r == 0, 1, q^(n - j*r)*dI[b - j, r, n - j*r, p/q]];
  N[(n!/(n - j*r)!/r!^j)*p^(j*r)*t]];
djI[6,1,1,10,1/6]=0.1379481977
```

In some applications we need further generalization of (I.5), by using the same parameter j and changing both p and r to vectors. Letting R denote $\sum_{i=1}^b r_i$ as in (I.1), we have

$$I_{\tilde{p}}^{(b,j)}(\tilde{r};n) = \frac{n!}{(n-R)!} \frac{\prod_{i=1}^j (p_i^{r_i}/r_i)}{\prod_{i=1}^b \Gamma(r_i)} \int_0^{p_{j+1}} \cdots \int_0^{p_b}$$

$$\left(1 - \sum_{i=1}^j p_i - \sum_{\alpha=j+1}^b x_\alpha\right)^{n-R} \prod_{i=j+1}^b x_i^{r_i-1} dx_i.$$

$$(\text{I.16})$$

This has the probability interpretation that for j specified cells the ith one has frequency exactly r_i $(i = 1, 2, \ldots, j)$ and for the remaining $b - j$ cells the αth cell has frequency at least $r_{j+\alpha}$ $(\alpha = 1, 2, \ldots, b - j)$. One immediate advantage of this more general form is seen by differentiating with respect to p_{j+1}. For $\alpha = j + 1$ we obtain the simple result

$$\frac{\partial I_{\tilde{p}}^{(b,j)}(\tilde{r}; n)}{\partial p_\alpha} = \frac{r_\alpha}{p_\alpha} I_{\tilde{p}}^{(b,j+1)}(\tilde{r}; n). \tag{I.17}$$

To write this for any $\alpha \geq j + 1$, we must remove p_α from the $(b - j)$ set of p-values and add it to the j set of p-values. Note that this result holds only for $\alpha \geq j + 1$, but since our main application is for for $j = 0$, we are not concerned with this condition.

A generalization of (I.15) for vectors $\tilde{p}$ and $\tilde{r}$ gives

$$I_{\tilde{p}}^{(b,j)}(\tilde{r}; n) = \begin{bmatrix} n \\ r_1, \ldots, r_j \end{bmatrix} \left(\prod_{i=1}^{j} p_i^{r_i} \right) \left(1 - \sum_{i=1}^{j} p_i \right)^{n-R}$$
$$I_{\tilde{p}_j/\left(1-\sum_{i=1}^{j} p_i\right)}^{(b-j)} (\tilde{r}_j, n - R), \tag{I.18}$$

where $R = r_1 + r_2 + \cdots + r_j$ and $\tilde{p}_j = \tilde{p}$, $\tilde{r}_j = \tilde{r}$ both with the first j components removed. Of course, $(n - R)!$ is understood to be part of the multinomial coefficient in (I.18). Here again we have a combination of equalities for j cells and inequalities for $b - j$ cells so that a comma is appropriate in the superscript. It was assumed only that the two b-vectors $\tilde{r}$ and $\tilde{p}$ were similarly partitioned.

The case of scalar p and vector $\tilde{r}$ seems to come up in many applications. If the r-values are not widely dispersed then the following recurrence relation is very useful

$$I_p^{(b,j)}(\tilde{r}; n) = I_p^{(b,j)}(\tilde{r}^{j+1}; n) + I_p^{(b,j+1)}(\tilde{r}; n), \tag{I.19}$$

where $\tilde{r}^{j+1} = \tilde{r}$ with $(j + 1)$st component increased by one.

In the case of $j = 0$, using (I.18) and (I.19), we obtain a formula for evaluating $I_p^{(b)}(\tilde{r}; n)$

$$I_p^{(b)}(\tilde{r}; n) = I_p^{(b)}(\tilde{r}^1; n) + \binom{n}{r_1} p^{r_1} (1 - p)^{n-r_1} I_{p/(1-p)}^{(b-1)}(\tilde{r}_1; n - r_1), \tag{I.20}$$

where $\tilde{r}^1 = \tilde{r}$ with first component incremented by one and $\tilde{r}_1 = \tilde{r}$ with the first component dropped.

Illustration I.11. *Compute $I_p^{(b)}(\tilde{r}, n)$ from (I.20).*

```
iRulUp[rt_List, nn_, pp_] :=
(* Compute I[{r1,...,rb},n,p] using  vector r and scalar p. *)
Module[{b, n = nn, r, t, rul1, rul2, rul3, tt, tta, p = pp},
    r = rt; b = Length[r]; f[r_, 0, m_, a_] := 1;
  rul1 = f[r_List, b_, m_, p_] :> (tt = 1 - p; tta = p/(1 - p);
  f[Prepend[Rest[r], r[[1]] + 1], b, m, p] + Binomial[n, r[[1]]]*
tt^(n - r[[1]])*p^r[[1]]*f[Rest[r], b - 1, m - r[[1]], tta]) /;
                r[[1]] < r[[b]] && b > 0;
    rul2 = f[r_, b_, m_, a_] :> f[Sort[r], b, m, a];
    rul3 = f[r_, b_, m_, a_] :> dI[b, r[[1]], m, a];
    t = f[r, b, n, p]; Do[t = t //. rul2;
    t = Expand[t //. rul1], {b - 1}]; t = t //. rul3; t];
    iRulUp[{2,4},7,1/6]=0.00170038866
```

A few words of explanation are in order for the above illustration:
Since p is a common scalar and $j = 0$, the order of the $\tilde{r}$-components
does not influence results. We sort the vector $\tilde{r}$ (rul2) and incre-
ment r_1 using (I.20) repeatedly $b - 1$ times until all components of
the vectors are equal to r_b (rul1), which has to be the largest after
sorting. Then we substitute I-functions for the temporary f-functions
used in the program (rul3).

Previously (see equation (I.17)) we have written formula for a deriva-
tive of the I-function with vectors $\tilde{r}$ and $\tilde{p}$ with respect to p_i component. Now we want to write a few first terms of the Taylor series. Our
main interest is the case of $j = 0$. In order to maximize the rate of
convergence of the Taylor expansion we choose to expand $I_{\tilde{p}}^{(b)}(\tilde{r}, n)$
about the weighted average $\bar{\bar{p}} = (\sum_{\alpha=1}^{b} p_\alpha r_\alpha)/\sum_{\alpha=1}^{b} r_\alpha$, which reduces to the usual average $\bar{p}$ for scalar r.

The first derivatives have combined coefficient $\sum(p_\alpha - \bar{\bar{p}}) = 0$, so we
have to calculate higher derivatives. Since (I.17) is good only for
$j + 1 \leq \alpha \leq b$ we calculate the other formulas needed. These are:

$$\frac{\partial}{\partial p_\alpha} I_{\tilde{p}}^{(b,j)}(\tilde{r}, n) = \frac{r_\alpha}{p_\alpha} I_{\tilde{p}}^{(b,j)}(\tilde{r}, n) - n I_{\tilde{p}}^{(b,j)}(\tilde{r}, n - 1) \quad (1 \leq \alpha \leq j)$$

$$(I.21)$$

and second derivatives for $j + 1 \leq \alpha, \beta \leq b$

$$\frac{\partial^2}{\partial p_\alpha \partial p_\beta} I_{\tilde{p}}^{(b,j)}(\tilde{r}, n) = \frac{r_\alpha}{p_\alpha} \frac{r_\beta}{p_\beta} I_{\tilde{p}}^{(b,j+2)}(\tilde{r}, n) \quad (\alpha \neq \beta) \qquad (I.22)$$

and

$$\frac{\partial^2}{\partial p_\alpha^2} I_{\bar{p}}^{(b,j)}(\tilde{r},n) = -\frac{r_\alpha n}{p_\alpha} I_{\bar{p}}^{(b,j+1)}(\tilde{r},n-1) + \frac{r_\alpha(r_\alpha-1)}{p_\alpha^2} I_{\bar{p}}^{(b,j+1)}(\tilde{r},n).$$

$$(I.23)$$

Using (I.22) and (I.23) we obtain the Taylor expansion for scalar r and vector $\bar{p}$ good to $\mathcal{O}(p_\alpha - \bar{p})^3$.

$$I_{\bar{p}}^{(b)}(r,n) = I_{\bar{p}}^{(b)}(r,n) + \frac{W_1}{2}\left[\frac{(r-1)}{\bar{p}} I_{\bar{p}}^{(b,1)}(r,n) - n I_{\bar{p}}^{(b,1)}(r,n-1)\right]$$
$$+ W_2\left[\frac{r^2}{\bar{p}^2} I_{\bar{p}}^{(b,2)}(r,n)\right],$$

$$(I.24)$$

where

$$W_1 = \frac{r}{\bar{p}}\sum_{\alpha=1}^{b}(p_\alpha - \bar{p})^2, \quad W_2 = \sum_{\alpha<\beta=1}^{b}(p_\alpha - \bar{p})(p_\beta - \bar{p}) \quad (I.25)$$

Illustration I.12. *Compute* $I_{\bar{p}}^{(b)}(r,n)$ *from (I.24).*

```
taylorI[r_, n_, vecp_] :=
  (* Compute I[b,r,n,vecp] using Taylor series around p average *)
      Module[{b, p, pbar, i, w1, w2, dv, t, ii},
      p = vecp; b = Length[p]; pbar = Sum[p[[i]], {i, b}]/b;
      dv = Table[p[[i]] - pbar, {i, b}];
      w1 = r/pbar*Sum[dv[[i]]^2, {i, b}];
      w2 = Sum[dv[[i]]*dv[[ii]], {i, b - 1},{ii, i + 1, b}];
       t = n * djI[b, 1, r, n-1, pbar];
      Return[dI[b, r, n, pbar] +w1/ 2*((r-1)/pbar *
      djI[b, 1, r, n, pbar] -t)+
            w2*(r/pbar)^2 djI[b, 2, r, n, pbar] ]]
```

The function djI[b, j, r, n, pbar] used above was programmed in *Illustration*(I.8).
To check taylorI[r_, n_, vecp_] numerically we compute $I_{p_1,p_2}^{(2)}(2,5)$ by integration obtaining for $p_1 = .38$, $p_2 = .42$

$$I_{p_1,p_2}^{(2)}(2,5) = 10 p_1^2 p_2^2 (3 - 2p_1 - 2p_2) = .35661024 \quad (I.26)$$

and from taylorI[2, 5, {.38, .42}] $\approx .356608$.
The formula (I.17) for partial derivative can be used to prove the

following interesting theorem:

Theorem:

Given a multinomial setting with b cells and $\sum_{i=1}^{b} p_i = 1$, the probability of getting at least r observations in the same cell in n trials achieves its maximum when the cell probabilities are all equal, i.e. when $p_1 = p_2 = \cdots = p_b = 1/b$.

We prove it using the integral representation (I.16) of $f(\tilde{p}) = I_{\tilde{p}}^{(b)}(r, n)$. Let $p_b = 1 - p_1 - p_2 - \cdots - p_{b-1}$. Then from the definition (I.16) and from (I.17) we obtain for $j = 0$:

$$\frac{\partial f}{\partial p_\alpha} = \frac{\partial I_{\tilde{p}}^{(b)}(r, n)}{\partial p_\alpha} = \frac{r}{p_\alpha} I_{\tilde{p}}^{(b,1)}(r, n) - \frac{r}{p_b} I_{\tilde{p}}^{(b,1)}(r, n) \qquad (I.27)$$

for $\alpha = 1, 2, \ldots, b - 1$.

The necessary condition for maximum of $f(\tilde{p})$ is the vanishing of all derivatives $\frac{\partial f}{\partial p_\alpha}$ for $\alpha = 1, 2, \ldots, b - 1$. For $p_\alpha = 1/b$ using (I.27) this condition is trivially satisfied.

To complete the proof we study the function $f(\tilde{p})$ on the boundary of the b-dimensional simplex

$$p_1 \geq 0, \quad p_2 \geq 0, \ldots, p_{b-1} \geq 0, \quad p_1 + p_2 + \cdots + p_b \leq 1. \qquad (I.28)$$

From the integral representation (I.16), it follows that the function f for each $p_i = 0$ has to be zero and since $f \geq 0$ (being a probability) the function f goes toward its maximum QED.

We illustrate this behavior with example of $I_{p1,p2,p3}^{(3)}(2, 7)$. To compute this function we modify the program from Illustration (I.2)

Illustration I.13. *Compute* $I_{\tilde{p}}^{(b)}(r, n)$ *from integral representation (I.1).*

```
iIntVecp[r_,n_,vecp_List] :=
(* Compute I-function by integration for scalar r and vector p *)
 Module[{b,tt,i,p},
    p=vecp; b=Length[p];
    tt = (n)!/(n-b r)!/((r - 1)!)^b;
    tt = tt*(Product[x[i]^(r - 1), {i, b}]*
    (1 - Sum[x[i], {i, b}])^(n-b r));
    Do[tt = Integrate[tt, {x[i], 0, p[[i]] }], {i, b}]; tt];
```

Using this function $\texttt{iIntVecp}[2,7,\{p1,p2,p3\}] = 210p1^2p2^2p3^2(3 - 2p1 - 2p2 - 2p3)$. With $p1 + p2 + p3 = 1$ we look for maximum of $f = 210p1^2p2^2p3^2$. This happens at $p1 = p2 = p3 = 1/3$ giving $f \approx .288066$.

Remark:

The condition $\sum_{i=1}^{b} p_i = const$ is sufficient for the proof. Without placing any restrictions on p_i, we get the maximum also for $p_1 = p_2 = \cdots = p_b$, but outside the simplex (I.28).

The Taylor expansion in p (for all p_i-values equal) is of special interest. By straightforward differentiation of (I.2) we obtain:

$$\frac{d}{dp}I_p^{(b)}(r,n) = \frac{n}{p} \triangle I_p^{(b)}(r,n-1) = \frac{rb}{p}I_p^{(b,1)}(r,n). \qquad (I.29)$$

Here, $\triangle f(x) = f(x+1) - f(x)$ denotes the finite difference operator.

To obtain the sth derivative we use a Leibnitz-type theorem, namely

$$\triangle^s\left[F\left(x\right)G\left(x\right)\right] = \sum_{i=0}^{s} \binom{s}{i} \left[\triangle^i F\left(x\right)\right]\left[\triangle^{s-i}G\left(x+i\right)\right]. \qquad (I.30)$$

We use (I.30) to generalize (I.29), namely

$$D_p^s I_p^{(b)}(r,n) = \frac{n^{[s]}}{p^s}\triangle^s I_p^{(b)}(r,n-s), \qquad (I.31)$$

where D_p^s denotes the sth derivative and $n^{[s]} = n\cdots(n-s+1)$.

Using (I.30) with (I.31) we obtain the desired Taylor expansion(about $p = p_0$)

$$I_p^{(b)}(r,n) = \sum_{\alpha=0}^{n-br} \left(\frac{p-p_0}{p_0}\right)^\alpha \binom{n}{\alpha}\triangle^\alpha I_{p_0}^{(b)}(r,n-\alpha). \qquad (I.32)$$

Here the sum is exact since $I_p^{(b)}(r,n)$ is a polynomial in p.

The expansion (I.32) can also be written without differences:

$$I_p^{(b)}(r,n) = \begin{cases} \sum_{\beta=0}^{n-br} B_{p'}(n,\beta) I_{p_0}^{(b)}(r,n-\beta) & \text{for } p \le p_0, \\[2mm] \left(\dfrac{2p-p_0}{p_0}\right)^n \sum_{\beta=0}^{n-br}(-1)^\beta B_{p''}(n,\beta) I_{p_0}^{(b)}(r,n-\beta) & \text{for } p \ge p_0, \end{cases}$$

$$(I.33)$$

where $p' = (p_0 - p)/p_0$, $p'' = (p - p_0)/(2p - p_0)$ and $B_p(n,r) = \binom{n}{r} p^r$ $(1-p)^{n-r}$ is the usual binomial probability of r successes in n trials when the common probability of success on a single trial is p. For derivation of this result we send the interested reader to Sobel et al. (1977).

II. Probability and Combinatorial Problems

Using straightforward probability interpretation of Dirichlet I, J and IJ-functions we can solve many interesting probability and combinatorial problems.

Problem **II.1.** _(Die Problem)_
A 6-sided fair die is tossed 10 times. What is the probability that (exactly) one face will appear at least t $(3 \le t \le 7)$ times and all the others at most twice ?

Answer : Using the definition of the IJ-function with the binomial distribution for the one face we get

$$A = \binom{6}{1} IJ_{1/6}^{(1;5)}(t,3;10) = 6 \sum_{i=t}^{10} \binom{10}{i}\left(\frac{1}{6}\right)^i \left(\frac{5}{6}\right)^{10-i} J_{1/5}^{(5)}(3,10-i)$$

$$(II.1)$$

or in the _Mathematica_ language:

```
6 Table[Sum[Binomial[10,i](1/6)^i(5/6)^(10-i) dJ[5,3,10-i,1/5],
    {i,t,10}],{t,3,7}]
```

The numerical values are given in the following table

$t \to$	3	4	5	6	7
$A \to$	5.372E-1	2.371E-1	6.831E-2	1.280E-2	1.546E-3

Table 2.1

Problem **II.2.** *(Wine Problem)*
A company produces 5 different kinds of wine in similar amounts
and in similar size bottles. If the bottles are put randomly into a
carton that holds 10 bottles, what is the chance A that the carton
will contain (exactly) j different kinds of wine ($j=5,4,3,2,1$)?

Answer :

$$A = \binom{5}{j} I J_{1/5}^{(j;5-j)}(1,10) = \quad \text{for } j = 5,4,3,2,1 \ . \tag{II.2}$$

Another expression for A is $\binom{5}{j}(j/5)^{10} I_{1/j}^{(j)}(1,10)$.

`Table[Binomial[5,j](j/5)^10*dI[j,1,10,1/j],{j,5,1,-1}]`

The numerical answers are arranged in the following table:

$j \rightarrow$	5	4	3	2	1
$A \rightarrow$	.52255	.41908	.05732	.00105	.00000

Table 2.2

We see as a check that the probabilities add up to 1.
If we want the probability that the carton contains exactly 2 of each
kind, the answer is $I_{1/5}^{(5)}(2,10) \approx .0116122$, since for $n = 10$ the events
"at least 2 of each kind " and "exactly 2 of each kind " are identical.

Problem **II.3.** *(Birthmonth Problem)*
In a classroom sample of 30 students, what is the probability of finding for each of the 12 months at least one person with that birthmonth ? Assume that the 12 month are all equally long.
   Answer :

$$I_{1/12}^{(12)}(1,30) \approx .359145 \ . \tag{II.3}$$

Problem **II.4.** *(Bus Problem)*
   For a bus ride n people independently and randomly select seat
assignments. If the bus has total of 100 seats, how large can n be if
we wish to keep the probability of any overlapping seat assignments
below .10 ?

Answer :

$$J_{1/100}^{(100)}(2,n) \geq .90 \qquad \text{(II.4)}$$

For $n=2,3,4,5,6$ the numerical answers for $J = J_{1/100}^{(100)}(2,n)$ are in tabular form:

$n \rightarrow$	2	3	4	5	6
$J \rightarrow$	.990000	.970200	.941094	.903450	.858278

Table 2.3

Hence the answer to the question is $n = 5$.

In some applications we have vector arguments. One of the methods to deal with such problems is to use (I.9) and (I.12) of the previous section.

Problem **II.5.** _Compute the probability of at least 2 sixes and at least 4 threes in 7 tosses of a fair six-sided die_ (see **Illustration(I.6)**).

$$I_{1/6}^{(1;1)}(2,4;7) = \sum_{\alpha=2}^{3} \binom{7}{\alpha} \left(\frac{1}{6}\right)^{\alpha} \left(\frac{5}{6}\right)^{7-\alpha} I_{1/5}^{(1)}(4,7-\alpha) \approx .00170039$$

$$\text{(II.5)}$$

As another illustration of unequal r-values (see **Illustration(I.9)**):

Problem **II.6.** _The probability of getting at least 2 sixes and fewer than 3 threes in 5 independent tosses of a fair 6-sided die is (using (I.13))_

$$IJ_{1/6}^{(1;1)}(2,3;5) = I_{1/6}^{(1)}(2,5) - I_{1/6}^{(2)}(2,3;5) \approx .194959 \qquad \text{(II.6)}$$

Problem **II.7.** _(Quality Control Problem)_ Suppose an item has 3 categories of quality called good, mediocre and trash. We wish to accept the lot if a random sample of 10 items contains at most 1 trash and at least 5 good items. What is the probability of accepting the lot if the population drawn from has 50% good, 10% trash and 40% mediocre;

Answer :

$$IJ_{.5,.1}^{(1;1)}(5,2;10) = I_{.5}^{(1)}(5,10) - I_{.5,.1}^{(2)}(5,2;10)$$
$$= (.9)^{10} I_{5/9}^{(1)}(5,10) + 10\,(.1)\,(.9)^9\, I_{5/9}^{(1)}(5,9) \quad \text{(II.7)}$$
$$\approx .261554 + .245816 = .507370$$

Problem **II.8.** *With 12 tosses of a fair 6-sided die, what is the probability of getting each of faces #1 and #2 at least 3 times and the remaining 4 faces each at most 3 times.*

Answer : To satisfy our conditions the possible partitions between seeing the first two faces (#1 and #2) vs. the remaining are (6,6), (7,5), (8,4), (9,3), (10,2), (11,1) and (12,0). Hence a simple method of getting our answer is

$$Ans. = \sum_{j=6}^{12} \binom{12}{j} \left(\frac{1}{3}\right)^j \left(\frac{2}{3}\right)^{12-j} \left[I_{1/2}^{(2)}(3,j) \cdot J_{1/4}^{(4)}(4,12-j) \right]$$

$$\approx .0676264 \ .$$

$$\text{(II.8)}$$

We could also use the expression $IJ_{1/6}^{(2;4)}(3,4;12)$.

Problem **II.9.** *If 5 ordinary fair dice are used in each of t tosses, what is the probability of seeing each of the six faces at least 3 times*

(A) not necessarily on the same toss

(B) for each face the three repeats are necessarily on the same toss.

Answer : Ans.$(A) = I_{1/6}^{(6)}(3,5t)$. Doing some calculations it is also easy to solve an inverse problem of finding the smallest value of t such that Ans.$(A) \geq .90$. Since for $t = 8$, Ans.$(A) \approx .8403$ and for $t = 9$, Ans. $\approx .9195$, the smallest value of $t = 9$. For the problem (B) the answer is Ans.$(B) = 6! \left[I_{1/6}^{(1)}(3,5) \right]^6 \approx 1.43963E - 6$

Problem **II.10.** *In a toss of 5 ordinary fair dice we use poker terminology and poker ranking (e.g., 2 pair < 3 of a kind < straight < full house < 4 of a kind < 5 of a kind). What is the probability in a single toss of tieing or beating someone who has (only) a pair of sixes.*

Answer :

$$A1 = IJ_{1/6}^{(1;5)}(2,2;5) - IJ_{1/6}^{(1;5)}(3,2;5)$$

$$= \binom{5}{2}\left(\frac{1}{6}\right)^2 \left(\frac{5}{6}\right)^3 J_{1/5}^{(5)}(2,3)$$

$$= \frac{600}{6^5} \text{ for a pair of sixes only.}$$

$$A2 = \binom{6}{2} I_{1/6}^{(2)}(2,5) = \frac{2100}{6^5} \text{ for two pair or a full house}$$

$$A3 = 2I_{1/6}^{(5)}(1,5) = \frac{240}{6^5} \text{ for a straight (1 to 5 or 2 to 6)}$$

$$A4 = \binom{6}{1} I_{1/6}^{(1)}(3,5) = \frac{1656}{6^5} \text{ for 3 or 4 or 5 of a kind or a full house}$$

$$A5 = \binom{6}{1}\binom{5}{1} I_{1/6}^{(2)}(3,2;5) = \frac{300}{6^5} \text{ for a full house.}$$

Doing the arithmetic, we get

$$Ans = \sum_{i=1}^{4} A_i - A_5 = \frac{4296}{6^5} = \frac{179}{324} \approx .552469$$

Since there are no suits or faces of different color on a single die, there is no need to bring into consideration the flush, straight flush or royal flush.

Problem **II.11.** *(Rat Problem.)*
In some medical experiments rats are treated with different chemicals and observed for a fixed time T. These rats can die of $b = 5$ different causes with a common probability $p \leq 1/b$ (these probabilities need not add to one). How many rats N do we have to buy in order to have probability $P^* = .95$(say) that at the end of this time period T there will be at least r rats in each of the b categories?

Answer :
We have to satisfy the condition by solving for the smallest value of N such that

$$I_p^{(b)}(r,N) \geq P^* . \tag{II.9}$$

The following table gives the results for $b = 5$, $P^* = .95$, $p = .01$, .05,

.10, .15, .20 and $r = 1, 2, 3, 4, 5, 10$.

r	$p=.01$	$p=.05$	$p=.10$	$p=.15$	$p=.20$
1	457	90	44	29	21
2	659	130	64	42	31
3	835	165	81	53	39
4	999	197	97	64	47
5	1154	228	113	74	55
6	1304	258	127	84	62
7	1450	287	142	93	69
8	1593	315	156	102	76
9	1733	343	170	112	83
10	1870	371	183	121	89

Table 2.3

Remark: If the five p-values are not equal then using the smallest of the five as a common value will give an upper bound on the value of N needed to satisfy the same condition. We can also get a good estimate of N by using the average of the $b = 5$ p-values. It is interesting to note that the ratios of N-values are nearly inversely proportional to the ratios of the corresponding p-values.

Problem **II.12.** *(Basic Birthday Problem).*

The basic birthday problem as well as variations on the same theme are all "grist for the Dirichlet mill". The most common form is to find the "median" number of people n needed to get $r = 2$ people with the same birthday, i.e., the number needed for this event to have probability $P^* = 1/2$. The table below extends it in at least two directions:

(1) by letting r(number of people with the same birthday) range from 2 to 10 and

(2) by using the five different P^*-values .50, .75, .90, .95 and .99. In each of the resulting 9*5=45 cells we give the two successive n-values that enclose the specified P^* and the probability P for each of the two n-values that the event will occur. The calculations are all based on 365 days in the year (without considering the leap years). The result 23 in the very first cell is well known in probability circles.

For given r and P^* the second result in each cell is the smallest value of n such that

$$1 - J_{1/365}^{(365)}(r, n) \geq P^* \qquad \text{(II.10)}$$

and the exact P-values for both n and $n - 1$ in each cell confirm the result.

Nr. of People needed and Probabilities of the same Birthday.

r	$P^* = .50$ $n \quad P$	$P^* = .75$ $n \quad P$	$P^* = .90$ $n \quad P$	$P^* = .95$ $n \quad P$	$P^* = .99$ $n \quad P$
2	22→.4757	31→.7305	40→.8912	46→.9483	56→.9883
	23→.5073	32→.7533	41→.9032	47→.9548	57→.9901
3	87→.4995	109→.7363	131→.8963	194→.9490	167→.9895
	88→.5111	110→.7546	132→.9063	195→.9519	168→.9902
4	186→.4958	225→.7476	259→.8990	279→.9500	314→.9896
	187→.5027	226→.7532	260→.9022	280→.9519	315→.9901
5	312→.4962	366→.7472	412→.8982	439→.9497	486→.9898
	313→.5011	367→.7513	413→.9006	440→.9512	487→.9902
6	459→.4986	527→.7476	585→.8993	618→.9499	676→.9899
	460→.5024	528→.7509	586→.9012	619→.9510	677→.9902
7	622→.4998	704→.7493	772→.8987	811→.9495	879→.9897
	623→.5029	705→.7521	773→.9003	812→.9505	880→.9900
8	797→.4976	893→.7495	972→.8998	1016→.9495	1094→.9898
	798→.5003	894→.7518	973→.9012	1017→.9504	1095→.9901
9	984→.4986	1092→.7489	1181→.8995	1231→.9499	1318→.9899
	985→.5009	1093→.7510	1182→.9008	1232→.9506	1319→.9901
10	1180→.4988	1300→.7489	1398→.8990	1453→.9494	1549→.9898
	1181→.5009	1301→.7508	1399→.9001	1454→.9501	1550→.9900

Table 2.5

Several authors have given asymptotic approximations (ct e.g. mathworld.wolfram.com) for the value of n needed, but these are generally useful only for P^* near $1/2$. Thus using the best approximation from the above noted source with $r = 10$ and $P^* = .95$, the result was 1130 instead of the value 1454 in the table, i.e., the result based on the approximation was off by more than 300.

Their Properties

Up to now we dealt with **I**, **J** and **IJ**-functions, or with Type 1 Dirichlet functions. In this section we introduce Type 2 Dirichlet functions, which include the **C**, the **D** and the **CD**-functions; The functions of Type 2 constitute the main tool used in the next section.

The Type 1 Dirichlet deals primarily with problems involving a fixed number n of observations, the Type 2 Dirichlet deals with more interesting and challenging (probability and expectation) problems that have some sequential aspects.

First there is a large class of waiting time (WT) problems, where we usually ask for the expected WT (or number of observations) until a specified random event occurs. A second important set of problems asks for the probability of a set of events occurring in categories 1,2,..., b at a random time when some specified event occurs in another (disjoint) category S, called the sink (or counting cell). Thus, if we deal out the 52 cards (one-by-one) from the usual deck, we might ask for the probability that at least 2 aces remain when we deplete all 4 kings. 'Sooner' and 'later' problems or problems dealing with the jth ordered of a sequence of events, like birthday problems, die problems, even (random) graph theory problems are all grist for the Type 2 Dirichlet mill.

The Type 2 Dirichlet functions are defined, studied, tabulated with many illustrations of usage in Sobel et al. (1985) and we repeat them briefly.

For any positive integers b and m with vectors $\bar{a} = (a_1, \ldots, a_b)$ (real) and $\bar{r} = (r_1, \ldots, r_b)$ (natural) we define

$$C_{\bar{a}}^{(b)}(\bar{r}, m) = \frac{\Gamma(m + R)}{\Gamma(m) \prod_{i=1}^{b} \Gamma(r_i)} \int_0^{a_1} \cdots \int_0^{a_b} \frac{\prod_{i=1}^{b} x_i^{r_i - 1} dx_i}{\left(1 + \sum_{i=1}^{b} x_i\right)^{m+R}},$$

(III.1)

where $R = r_1 + \cdots + r_b$, $a_i = p_i/p_0$ $(i = 1, 2, \ldots, b)$, p_i are the cell probabilities and $p_0 = 1 - \sum_{i=1}^{b} p_i < 1$.

Illustration III.1. *Compute $C_{\bar{a}}^{(b)}(\bar{r}, m)$ using the integral represen-*

tation (III.1)

We write again a short program in *Mathematica* and use it for numerical illustration:

```
cInt[rvec_List, m_,a_] :=
(* Compute C-function for vector r and scalar a *)
 Module[{b, rr, r=rvec, tt,i},
    b = Length[r]; rr = Sum[r[[i]], {i, b}];
    tt = (m+rr-1)!/(m-1)!/Product[(r[[i]] - 1)!, {i, b}];
    tt = tt*(Product[x[i]^(r[[i]] - 1), {i, b}]/
    (1 + Sum[x[i],{i, b}])^(m + rr));
    Do[tt = Integrate[tt, {x[i], 0, a}], {i, b}]; tt];
    cInt[{4,2},3,1.0]=0.29873971
```

For the probability interpretation of (III.1) we consider a multinomial setting with b blue cells, where the ith cell has cell probability p_i and quota r_i. In addition we have a counting cell with cell probability p_0. We sample from this multinomial until the counting cell reaches frequency $m \geq 1$ for the first time, referring to this as "at stopping time" or ast. Then (III.1) is the probability that for each of the b blue cells, the ith cell has its quota of at least r_i ast.

The dual probability that for each of the b cells, the ith cell has less than r_i ($i=1,\ldots,b$) ast is given by:

$$D_{\bar{a}}^{(b)}(\check{r}, m) = \frac{\Gamma(m + R)}{\Gamma(m) \prod_{i=1}^{b} \Gamma(r_i)} \int_{a_1}^{\infty} \cdots \int_{a_b}^{\infty} \frac{\prod_{i=1}^{b} x_i^{r_i - 1} dx_i}{\left(1 + \sum_{i=1}^{b} x_i\right)^{m+R}},$$

(III.2)

with the same notation as in (III.1).

Illustration III.2. *Compute $D_{\bar{a}}^{(b)}(r, m)$ using the integral representation (III.2)*

Here is *Mathematica* program:

```
dIntavec[r_, m_,avec_List] :=
(* Compute D-function for scalar r and vector a *)
 Module[{b, a=avec, tt,i},
    b = Length[a]; tt = (m+b*r-1)!/(m-1)!/(r-1)!^b;
    tt = tt*(Product[x[i]^(r - 1), {i, b}]/
    (1 + Sum[x[i],{i, b}])^(m + b*r));
    Do[tt = Integrate[tt, {x[i], a[[i]], Infinity}], {i, b}]; tt];
```

dIntavec$[2, 3, \{a1, a2\}] = 12a1\ a2\ s^5 + s^4 - 4s^3 + 4s^2$,

where $s = 1/(1 + a1 + a2)$

We need also the mixed integral, where $0 \le c \le b$ integrals are from 0 to a_i or of the C-type while the remaining $b - c$ integrals are from a_i to ∞ or of the D-type, namely

$$CD_{\tilde{a}}^{(c;b-c)}(\tilde{r}, m) = \frac{\Gamma(m + R)}{\Gamma(m) \prod_{i=1}^{b} \Gamma(r_i)} \int_0^{a_1} \cdots \int_0^{a_c} \int_{a_{c+1}}^{\infty} \cdots \int_{a_b}^{\infty}$$

$$\frac{\prod_{i=1}^{b} x_i^{r_i - 1} dx_i}{\left(1 + \sum_{i=1}^{b} x_i\right)^{m+R}}.$$

(III.3)

The probability interpretation is given in the same multinomial setting, with the obvious difference that for each of the c cells the ith cell $(i = 1, \ldots, c)$ has its quota of at least r_i and for the rest of the $b - c$ cells the ith cell $(i = c + 1, \ldots, b)$ has less than r_i. For $b = 0$ we define $C^{(0)} = D^{(0)} = CD^{(0,0)} = 1$.

In the homogeneous case we have all $p_i = p$ $(i = 1, 2, \ldots, b)$, $p_0 = 1 - bp$ and we use the common scalar $a = p/p_0$ instead of $\tilde{a}$. Whenever we use more than one r (or s) value in a Dirichlet function, we separate them from the last argument with a semicolon.

As with **I, J**-functions we need a slight generalization of (III.1). For the present $\tilde{r} = (r_1, r_2, \ldots, r_b)$ and $\tilde{a} = (a_1, a_2, \ldots, r_b)$ are each b vectors. Let X_i denote the frequency in the ith blue cell $(i = 1, 2, \ldots, b)$ and $R = r_1 + r_2 + \cdots + r_b$ as before. Let $P = C_{\tilde{a}}^{(b,j)}(\tilde{r}; m)$ denote the probability, when the counting cell reaches m (for the first time), that the joint event $X_i = r_i$ $(i = 1, 2, \ldots, j)$ and $X_\alpha \ge r_\alpha$ $(\alpha = j + 1, j + 2, \ldots, b)$ for any given j with $0 \le j < b$ has occurred. Then we have

$$C_{\tilde{a}}^{(b,j)}(\tilde{r}; m) = \frac{\Gamma(m + R) \prod_{i=1}^{j} (a_i^{r_i}/r_i)}{\Gamma(m) \prod_{i=1}^{b} \Gamma(r_i)} \int_0^{a_{j+1}} \cdots \int_0^{a_b}$$

$$\frac{\prod_{\alpha=j+1}^{b} x_\alpha^{r_\alpha - 1} dx_\alpha}{\left(1 + \sum_{i=1}^{j} a_i + \sum_{\alpha=j+1}^{b} x_\alpha\right)^{m+R}}.$$

(III.4)

Clearly the j cells with equality are taken to be the first j cells only for notational convenience. For $j = 0$ we drop the second superscript and for $j = b$ we get multinomial probabilities .

The same result (III.4) holds for $D_{\tilde{a}}^{(b,j)}(\tilde{r}; m)$, where the event changes to $X_\alpha < r_\alpha$ $(\alpha = j+1, \ldots, b)$, except that all the limits of integration are from a_α to ∞.

In many applications all r-values and a-values are equal and we have

$$C_a^{(b,j)}(r,m) = M \int_0^a \cdots \int_0^a \frac{\prod_{\alpha=j+1}^b x_\alpha^{r-1} dx_\alpha}{\left(1 + ja + \sum_{\alpha=j+1}^b x_\alpha\right)^{m+br}}, \quad \text{(III.5a)}$$

$$D_a^{(b,j)}(r,m) = M \int_a^\infty \cdots \int_a^\infty \frac{\prod_{\alpha=j+1}^b x_\alpha^{r-1} dx_\alpha}{\left(1 + ja + \sum_{\alpha=j+1}^b x_\alpha\right)^{m+br}}, \quad \text{(III.5b)}$$

where

$$M = \frac{\Gamma(m+br)}{\Gamma(m)\Gamma^b(r)} \left(\frac{a^r}{r}\right)^j.$$

For $j = b$ we have the same expressions for C and D, namely

$$C_a^{(b,b)}(r,m) = \frac{\Gamma(m+br)}{(r!)^b \Gamma(m)} \left(\frac{a}{1+ba}\right)^{br} \left(\frac{1}{1+ba}\right)^m = D_a^{(b,b)}(r,m).$$
$$\text{(III.6)}$$

The next property we can use as a second boundary condition for computation of **D** integral in case of $m > r$

$$D_a^{(b,j)}(r,m) = \frac{1}{\binom{m-1}{r}} \sum_{\alpha=1}^r \frac{\binom{m-\alpha-1}{r-\alpha}}{a^\alpha} D_a^{(b,j+1)}(r, m-\alpha), \quad \text{(III.7)}$$

which is especially simple for $m = r+1$. The corresponding equation for the **C** integral is more complicated (see page 20 in Sobel et al. (1985)) and not so useful.

Sometimes it is convenient and interesting to consider the alternative (script) generalization. With

$$Mj = \frac{(1+ja)^m \Gamma(m+(b-j)r)}{\Gamma(m)\Gamma^{b-j}(r)},$$

we have

$$C_a^{(b,j)}(r,m) = Mj \int_0^a \cdots \int_0^a \frac{\prod_{\alpha=j+1}^b x_\alpha^{r-1} dx_\alpha}{\left(1 + ja + \sum_{\alpha=j+1}^b x_\alpha\right)^{m+(b-j)r}}$$

(III.8)

and

$$\mathcal{D}_a^{(b,j)}(r,m) = Mj \int_a^\infty \cdots \int_a^\infty \frac{\prod_{\alpha=j+1}^b x_\alpha^{r-1} dx_\alpha}{\left(1 + ja + \sum_{\alpha=j+1}^b x_\alpha\right)^{m+(b-j)r}}.$$

(III.9)

These functions also have a probability interpretation. Using a similar idea of stopping when the counting cell first obtains m observations except that at the outset the j specified cells are combined with the counting cell and we wait until the first time this union contains m observations. Then the interpretation is the probability that the $b - j$ remaining cells have frequency at least r for (III.8) and less than r for (III.9). We list a few properties for C and $\mathcal{D}$ functions below:

$$C_a^{(b,j)}(r,m) = C_{a/(1+ja)}^{(b-j,0)}(r,m) \quad \mathcal{D}_a^{(b,j)}(r,m) = \mathcal{D}_{a/(1+ja)}^{(b-j,0)}(r,m);$$

(III.10)

$$C_a^{(b,b)}(r,m) = 1 = \mathcal{D}_a^{(b,b)}(r,m);$$

(III.11)

$$C_a^{(b,j)}(r,1) = (b-j)\left[\frac{a}{1+(j+1)a}\right]^r C_a^{(b,j+1)}(r,r);$$

$$\mathcal{D}_a^{(b,j)}(r,1) = 1 - (b-j)\left[\frac{a}{1+(j+1)a}\right]^r \mathcal{D}_a^{(b,j+1)}(r,r).$$

(III.12)

The main purpose of introducing the additional parameter j (as in the case of I and J in section (I)) is to develop simple recurrence relations, which enable us to calculate with very high accuracy all multiple integrals without use of numerical quadratures. Below we write all four recurrence relations and illustrate them with working *Mathematica* programs:

For any values of r, m, b, j, a with $0 \le j \le b$ both integers

$$m\,(1 + ja)\,E_a^{(b,j)}(r, m + 1) = (1 + jr)\,E_a^{(b,j)}(r, m)$$
$$\pm\, r\,(b - j)\,E_a^{(b,j+1)}(r, m). \tag{III.13}$$

Here E is C function if we use $+$ sign, or D if we take $-$ sign; similarly, we have for $\mathcal{E}$:

$$\mathcal{E}_a^{(b,j)}(r, m + 1) = \mathcal{E}_a^{(b,j)}(r, m) \pm (b - j) \binom{m + r - 1}{m}$$
$$\frac{a^r\,(1 + ja)^m}{[1 + (j + 1)a]^{m+r}}\mathcal{E}_a^{(b,j+1)}(r, m + r). \tag{III.14}$$

Illustration III.3. *Compute $C_a^{(b)}(r, m)$ using the recurrence relation (III.14) and boundary conditions (III.12) and (III.11)*

```
dC[(b_Integer)?Nonnegative,(r_Integer)?Positive,
(m_Integer)?Positive, aa_:1] :=
(* Compute dC[b,r,m,a] from recurrence relation of script C;
  if a not given put a=1 *)
  Module[{j = b, a = N[aa], coe, mm, t1, t2, tj, tn, tt,
    c = Table[1., {m + b*r}]},
If[b==0, Return[1.0]];
  While[j > 0, j--; tj = 1 + j*a; coe = 1/(tj + a); t1 = (a*coe)^r;
  tn = t1; t2 = tj*coe; c[[1]] = tt = (b - j)*tn*c[[r]];
    Do[tn = tn*t2*((mm + r - 1)/mm); c[[mm + 1]] =
      tt = tt + (b - j)*tn*c[[mm + r]], {mm, 1, m + j*r - 1}]
      (*of Do*) ](*of While*); Return[N[c[[m]] ] ] ];
dC[2,2,2,0.1]=0.00144263
```

For a change we will compute $D_a^{(b)}(r, m)$, using regular (not script) definition.

Illustration III.4. *Compute $D_a^{(b)}(r, m)$ using the recurrence relation (III.5) and boundary conditions (III.6) and (III.7)*

```
dD[(b_Integer)?Nonnegative, (r_Integer)?Positive,
(m_Integer)?Positive,aa_:1.] :=
Module[{j, a, oba, mm, al, t1, t2, pa, ia, v, res = 1},
  a = N[aa]; oba= 1/(1 + a*b); ia= 1/a; v= Table[0, {Max[r+1, m]}];
    If[m <= r + 1, v[[1]] = ((b*r)!/r!^b)*(a*oba)^(r*b)*oba;
    Do[v[[mm + 1]] = (mm + b*r)*v[[mm]]*(oba/mm),{mm, 1, r}];j = b;
      While[j > 0, j = j - 1; pa = ia; t1 = 0;
        Do[t1 = t1 + v[[r + 1 - al]]*pa; pa = pa*ia, {al, 1, r}];
        Do[t2 = ((b - j)*r*v[[mm]] + mm*(1 + j*a)*t1)/(mm + j*r);
```

```
    v[[mm + 1]] = t1; t1 = t2, {mm, r, 1, -1}]; v[[1]] = t1];
res = v[[m]]; Return[N[res,10]],
v[[m]] =(N[(m + b*r - 1)!]/((m-1)!*r!^b))*(a*oba)^(r*b)*oba^m;
Do[v[[mm]]=(mm/oba/(mm+b*r))*v[[mm+1]], {mm, m-1, m-r, -1}];
j = b; While[j > 0, j = j - 1; pa = ia*(r/(m - 1)); t1 = 0;
Do[t1 =t1 + v[[m - al]]*pa; pa =pa*ia*((r - al)/(m - al - 1)),
    {al, 1, r}]; Do[t2 = ((b - j)*r*v[[mm]] + mm*(1 + j*a)*t1)/
    (mm + j*r); v[[mm+1]] = t1; t1=t2, {mm, m-1, m - r, -1}];
v[[m-r]]=t1]; res = v[[m]]; Return[N[res,10]]; ](* of If *)];
```

Since the boundary condition (III.7) is good only for $m > r$, we have two cases to consider either increasing m, or decreasing m.

Returning to the case of vectors $\tilde{r}$ and $\tilde{a}$, we get a useful pair of relations obtained by a simple integration by parts on the variable x_{j+1} in both $\mathbf{D}$ and $\mathbf{C}$ in (III.4) gives

$$D_{\tilde{a}}^{(b,j)}(\tilde{r};m) = D_{\tilde{a}}^{(b,j)}(\tilde{r}^{(j+1)};m) - D_{\tilde{a}}^{(b,j+1)}(\tilde{r};m), \qquad \text{(III.15a)}$$

$$C_{\tilde{a}}^{(b,j)}(\tilde{r};m) = C_{\tilde{a}}^{(b,j)}(\tilde{r}^{(j+1)};m) + C_{\tilde{a}}^{(b,j+1)}(\tilde{r};m), \qquad \text{(III.15b)}$$

where $\tilde{r}^{(j+1)} = (r_1, \ldots, r_j, r_{j+1} + 1, r_{j+2}, \ldots, r_b)$, i.e. we add one to the $(j+1)^{th}$ component of $\tilde{r}$, but leave $\tilde{a}$ unchanged.

The use of the above recurrence relations gives rise to Type 2 integrals with double subscript (b,j) so we develop a reduction formula, as for **I**, **J**-functions, which replaces (b,j) by $(b-j,0)$. The transformation $y_i = x_i/(1 + A_j)$, $(i = j + 1, \ldots, b)$, where $A_j = a_1 + a_2 + \cdots + a_j$ leads to the result

$$C_{\tilde{a}}^{(b,j)}(\tilde{r};m) = N\left(\frac{1}{1 + A_j}\right)^m \prod_{i=1}^{j}\left(\frac{a_i}{1 + A_j}\right)^{r_i} C_{\frac{\tilde{a}'}{1+A_j}}^{(b-j)}(\tilde{r}', m + R_j),$$

$$\text{(III.16)}$$

where $R_j = r_1 + r_2 + \cdots + r_j$, $N = \begin{bmatrix} m - 1 + R_j \\ m - 1, r_1, \ldots, r_j \end{bmatrix}$, $\tilde{r}' = (r_{j+1}, \ldots, r_b)$

and $\tilde{a}'$ is defined similarly. Exactly the same formula holds for D and CD functions.

In the special case of a and r scalars (III.16) becomes:

$$C_a^{(b,j)}(r,m) = \frac{(m - 1 + jr)!}{(m - 1)!(r!)^j}\left(\frac{1}{1 + ja}\right)^m\left(\frac{a}{1 + ja}\right)^{jr} C_{a/(1+ja)}^{(b-j)}(r, m + jr).$$

$$\text{(III.17)}$$

Illustration III.5. Compute $D_a^{(b,j)}(r,m)$ using the reduction formula (III.17)

```
djD[(b_Integer)?Positive, (j_Integer)?Positive,
 (r_Integer)?Positive,(m_Integer)?Positive,aa_:1.]:=
(* Compute D[b,j,r,m,a] using reduction formula *)
  Module[{tm, tr, a},
    a = N[aa]; tm = 1/(1 + j*a); tr = a*tm;
    N[((m + j*r - 1)!/(m - 1)!/r!^j)*tr^(j*r)*tm^m*dD[b - j, r,
    m + j*r, tr]]];
djD[2,1,2,3]=0.0658436214
```

We use (III.16) together with (III.15) to obtain a formula, which when used repeatedly enables us to compute **C** and **D**-functions with vector $\tilde{r}$ namely:

$$E_a^{(b)}(\tilde{r};m) = E_a^{(b)}(\tilde{r}^1;m) \pm \binom{m-1+r_1}{r_1}\left(\frac{1}{1+a}\right)^m$$
$$* \left(\frac{a}{1+a}\right)^{r_1} E_{\frac{a}{1+a}}^{(b-1)}(\tilde{r}_1;m+r_1). \tag{III.18}$$

Here E is C function if we use the $+$ sign, or D if we take the $-$ sign, $\tilde{r}^1$ is $\tilde{r}$ with first component incremented by one and $\tilde{r}_1$ is $\tilde{r}$ with first component removed.

Illustration III.6. Compute $D_a^{(b)}(\tilde{r};m)$ using the formula (III.18)

```
dRulUp[rt_List, mm_, aa_:1] :=
Module[{b, m = mm, r, t, rul1, rul2, rul3, tt, tta, a = aa},
  r = rt; b = Length[r]; f[r_, 0, m_, a_] := 1;
rul1 = f[r_List, b_, m_, a_] :> (tt = 1/(1 + a); tta = a*tt;
f[Prepend[Rest[r], r[[1]] + 1], b, m, a] - Binomial[m - 1 + r[[1]],
  m - 1]* tt^m*tta^r[[1]]*f[Rest[r], b - 1, m + r[[1]], tta]) /;
  r[[1]] < r[[b]] && b > 0;
  rul2 = f[r_, b_, m_, a_] :> f[Sort[r], b, m, a];
  rul3 = f[r_, b_, m_, a_] :> dD[b, r[[1]], m, a];
  t = f[r, b, m, a]; Do[t = t //. rul2;
  t = Expand[t //. rul1], {b - 1}]; t = t //. rul3; t]
dRulUp[{2,4},3,1.0]=0.26748971
```

To check (III.16) we take $(b,j) = (2,1)$ with $\tilde{a} = (a_1,a_2)$. Using (III.16

we have

$$D_{a_1,a_2}^{(2,1)}(r,s;m) = \binom{m-1+r}{r}\left(\frac{1}{1+a_1}\right)^m$$
$$\left(\frac{a_1}{1+a_1}\right)^r D_{\frac{a_2}{1+a_1}}^{(1)}(s,m+r). \tag{III.19}$$

To illustrate these formulas we evaluate $D_1^{(2)}(4,2;3)$ using (III.15b) twice in reverse order and (III.19), we have

$$D_1^{(2)}(4,2;3) = D_1^{(2)}(3,2;3) + D_1^{(2,1)}(3,2;3) = D_1^{(2)}(2,2;3)$$
$$+ D_1^{(2,1)}(2,2;3) + D_1^{(2,1)}(3,2;3) = D_1^{(2)}(2,3)$$
$$+ \binom{4}{2}\left(\frac{1}{2}\right)^5 D_{1/2}^{(1)}(2,5) + \binom{5}{2}\left(\frac{1}{2}\right)^6 D_{1/2}^{(1)}(2,6)$$
$$\approx .160494 + .187500(.351166) + .156250(.263374)$$
$$\approx .267490.$$

$$\tag{III.20}$$

Similar calculations can be performed using the formula

$$D_{\tilde{a}}^{(b,j)}(\tilde{r};m) = \sum_{\alpha=0}^{r_{j+1}-1} D_{\tilde{a}}^{(b,j+1)}(\tilde{r}^{(\alpha)};m), \tag{III.21}$$

where $\tilde{r}^{(\alpha)}$ is the same as $\tilde{r}$ except that α is in the $j+1^{st}$ position instead of r_{j+1}. This result follows from the probability interpretation of D.

We illustrate this with the same example (III.20)

$$D_1^{(2)}(2,4;3) = D_1^{(2,1)}(0,4;3) + D_1^{(2,1)}(1,4;3)$$
$$= \left(\frac{1}{2}\right)^3 D_{1/2}^{(1)}(4,3) + 3\left(\frac{1}{2}\right)^4 D_{1/2}^{(1)}(4,4)$$

with the same result as in (III.20).

In some cases it is convenient to express $\mathbf{D}$ (and similarly $\mathbf{C}$ functions) as $D_a^{(b1;b2)}(r1,r2;m)$ for vector r, in which the vector is expressed as having $b1$ $r1$ components followed by $b2$ $r2$ components. In such case if $b = b1 + b2$ is not too large we can use (III.6) dRulUp function for evaluation.

Illustration III.7. *Compute* $D_a^{(b1;b2)}(\tilde{r};m)$ *using the formula (III.18)*

```
dDvec[b1_,b2_,r1_,r2_,m_,a_:1]:=
  (* Compute D-function for vector r with b1 r1-values, b2
   r2-values and scalars m, a *)
  Module[{rr},
  rr = Join[Table[r1, {b1}], Table[r2, {b2}]];
    Return[dRulUp[rr, m, a]]]
```

We also can use the multiple multinomial sum expression for the Dirichlet D integral as follows:

$$D_{\tilde{a}}^{(b)}(\tilde{r},m) = \frac{1}{\left(1+\sum_{i=1}^{b}a_i\right)^m} \sum_{x_1<r_1}\cdots\sum_{x_b<r_b}\begin{bmatrix} m-1+\sum_{\alpha=1}^{b}x_\alpha \\ m-1, x_1,\ldots,x_b \end{bmatrix}$$

$$\prod_{i=1}^{b}\left(\frac{a_i}{1+\sum_{\alpha=1}^{b}a_\alpha}\right)^{x_i},$$

(III.22)

where the square bracket denotes the multinomial coefficient; for the case of common a and r, this result appears as (2.15) in Sobel et al. (1985).

In some problems we have to compute the CD-function. The following identity is very useful:

$$CD_a^{(c;b-c)}(r,m) = \sum_{\alpha=0}^{c}(-1)^\alpha\binom{c}{\alpha}D_a^{(\alpha+b-c)}(r,m)$$

$$= \sum_{\alpha=0}^{b-c}(-1)^\alpha\binom{b-c}{\alpha}C_a^{(\alpha+c)}(r,m).$$

(III.23)

Illustration III.8. *Compute* $CD_a^{(;)}(r,m)$ *using the above formula (III.23)*

```
dCD[ii_Integer, jj_Integer, (r_Integer)?Positive,
 (m_Integer)?Positive, aa_] :=
(* Compute dCD[c,d,r,m,a] using inclusion-exclusion *)
  Module[{a, t, k}, a = N[aa];
    t = If[ii <= jj, Sum[(-1)^k*Binomial[ii, k]*dD[k + jj, r, m, a],
    {k, 0, ii}], Sum[(-1)^k*Binomial[jj, k]
    * dI[k + ii, r, m, a], {k, 0, jj}]]; N[t,10]];
dCD[1,1,2,3,1]=0.15200617
```

One way to evaluate D-functions with unequal a_i is to use the Taylor series. The development of this series is done in Sobel et al. (1985), here we only give the ready formulas for the case of equal r-values.

Let $\bar{a}$ be the average of its components: $\bar{a} = \sum_{i=1}^{b} a_i/b$, then the Taylor series good to $\mathcal{O}(a_i - \bar{a})^3$ is:

$$D_{\bar{a}}^{(b)}(r,m) = D_{\bar{a}}^{(b)}(r,m) + \frac{mw_1}{2\bar{a}} D_{\bar{a}}^{(b,1)}(r,m+1)$$

$$- \frac{w_1(r-1)}{2\bar{a}^2} D_{\bar{a}}^{(b,1)}(r,m) + \frac{w_2}{2\bar{a}^2} D_{\bar{a}}^{(b,2)}(r,m), \qquad \text{(III.24)}$$

where

$$w_1 = r \sum_{i=1}^{b} (a_i - \bar{a})^2, \quad w_2 = \sum_{\alpha=1}^{b-1} \sum_{\beta=\alpha+1}^{b} (a_\alpha - \bar{a})(a_\beta - \bar{a}). \quad \text{(III.25)}$$

The program for the series with scalar r follows:

Illustration III.9. *Compute $D_{\bar{a}}^{(b)}(r,m)$ using the Taylor series (III.2*

```
taylorD[r_, m_, va_]:=
(* Compute D[r,m,{a1,...,ab}] using Taylor series *)
  Module[{b, a, abar, i, w1, w2, w3, ii},
    ddj1[bb_, rr_, mm_, aa_] := c[mm - 1 + rr, rr]*(aa/(1 + aa))^rr
    *(1/(1 + aa))^mm*dD[bb - 1, rr, mm + rr, aa/(1 + aa)]; a = va;
    b = Length[a]; abar = Sum[a[[i]], {i, 1, b}]/b;
    dv = Table[a[[i]] - abar, {i, 1, b}]; w1 = r*Sum[dv[[i]]^2,
    {i, 1, b}]; w2 = w1*(r - 1); w3 = r^2*Sum[dv[[i]]*dv[[ii]],
    {i, 1, b - 1}, {ii, i + 1, b}]; Return[dD[b, r, m, abar] +
      m*(w1/(2*abar))*ddj1[b, r, m + 1, abar] - (w2/(2*abar^2))*
    ddj1[b, r, m, abar] + (w3/abar^2)*((m - 1 + 2*r)!/(m - 1)!/r!^2
    /(1 + 2*abar)^m)*(abar/(1 + 2*abar))^(2*r)*dD[b - 2, r, m + 2*r,
      abar/(1 + 2*abar)]]];
taylorD[2,3,{.8,1.2}=.1585185
```

This can be compared with $107/675 \approx .15851851$ obtained from the exact formula:

dIntavec$[2, 3, \{4/5, 6/5\}] = \frac{107}{675}$.

One of the remarkable properties of **C**, **D** and **CD**-functions is possibility of expressing them as a single integral.

Let

$$G_r(x) = \frac{1}{\Gamma(r)} \int_0^x t^{r-1} \exp(-t) dt \qquad \text{(III.26)}$$

denote the usual incomplete gamma function with parameter $r > 0$. Then for any valid scalars of b, m, c and vectors $\tilde{r}$, $\tilde{a}$

$$C_{\tilde{a}}^{(b)}(\tilde{r}; m) = \int_0^\infty \left[\prod_{i=1}^b G_{r_i}(a_i x) \right] dG_m(x),$$

$$D_{\tilde{a}}^{(b)}(\tilde{r}; m) = \int_0^\infty \left\{ \prod_{i=1}^b \left[1 - G_{r_i}(a_i x) \right] \right\} dG_m(x),$$

$$CD_{\tilde{a}}^{(c;b-c)}(\tilde{r}; m) = \int_0^\infty \left\{ \prod_{i=1}^c G_{r_i}(a_i x) \prod_{i=c+1}^b \left[1 - G_{r_i}(a_i x) \right] \right\} dG_m(x).$$

$$(III.27)$$

Using these formulas, we can easily prove that

$$CD_1^{(c,b-c)}(r, r) = \frac{c!\,(b-c)!}{(b+1)!} \qquad (III.28)$$

and from it for $c=0$, $C_1^{(b)}(r, r) = D_1^{(b)}(r, r) = 1/(b+1)$.

IV. C, D and CD Probability Problems

Dirichlet C and D-functions are mostly useful for sequential problems in which the stopping rule usually depends on the frequency of some specified cell (and we are interested in the distribution in the remaining cells). In several of the problems below we extend this application to problems in which the stopping rule depends on maximum (or minimum) frequency or on the (earliest) time when you have at least r observations in every cell.

Problem IV.1. _(Gum Problem)_
Gum packages have either one baseball card enclosed or a blank card, not visible before making the purchase. There are b different baseball cards each with common probability p so that $1 - bp \geq 0$ is the probability of a blank card. If I stop buying when I have m blank cards, what are the probabilities
  (1) that I have at least 1 of each of the b cards
  (2) that I am still missing exactly 1 of the b cards (unspecified)

(3) that I am still missing at least 2 of the b cards.

Illustrate numerically for the case $b = 5$ baseball cards, with $p = .10$ and $m = 5, 10, 15$.

Answer :

(1) `Table[dC[5,1,m,.1/(1-.5)],{m,5,15,5}]`

$$C^{(b)}_{p/(1-bp)}(1, m) \approx .12964289, .46026595, .73179015. \qquad \text{(IV.1)}$$

(2) `Table[5dCD[4,1,1,m,.1/(1-.5)],{m,5,15,5}]`

$$\binom{b}{1} CD^{(b-1;1)}_{p/(1-bp)}(1, m) = b \left[C^{(b-1)}_{p/(1-bp)}(1, m) - C^{(b)}_{p/(1-bp)}(1, m)) \right]. \qquad \text{(IV.2)}$$

For the same parameter values as above the three answers for (2) are respectively .24952918, .33781180, .21917529.

(3) By adding the previous results and subtracting from 1 the answers we obtain for the same parameter values are .620828, .201921, .049035.

Problem **IV.2.** *(Quality Control Problem)*

Manufactured items are either good or defective but the defective items fail for one (or more) of b different reasons and the ith type of failure has probability p_i of occurring, so that the probability of a good item is $p_0 = 1 - \sum_{i=1}^{b} p_i$. Assume we have a quota for each type of failure (say, r_i for the ith type). If any one quota is reached the production line is stopped and a machine is repaired. If we plan ahead to get m good items, what is the probability that the production line

(1) is not halted?

(2) is halted for exactly 1 type of defective(unspecified)?

(3) is halted for at least 2 types of defectives?

Do it for $b = 5$, $m = 10$, $r_i = 5$ and $p_i = .10$ ($1 \le i \le 5$).

Answer :

(1) `dD[5,5,10,1/5]`

$$D^{(5)}_{1/5}(5, 10) \approx .73310099 \qquad \text{(IV.3)}$$

(the probability of no stoppage before 10 good items)

(2) For this problem we obtain

5dCD[1,4,5,10,1/5]

$$\binom{5}{1} CD_{1/5}^{(1;4)}(5, 10) = 5 \left[D_{1/5}^{(4)}(5, 10) - D_{1/5}^{(5)}(5, 10) \right] \approx .20392144.$$
(IV.4)

(3) With the same parameters by adding (1) and (2) and subtracting from unity, we obtain

$$1 - D_{1/5}^{(5)}(5, 10) - \binom{5}{1} DC_{1/5}^{(4;1)}(5, 5; 10) \approx .062978. \qquad \text{(IV.5)}$$

Problem **IV.3.** *(Die Problem)*
If I toss a fair 6-sided die until the face #6 is seen m times ($m = 1$, 2 and 3), what is the probability that

(1) I will see all the other faces each at least r times.

(2) I will see all the other faces each at most r times.

Answer :

(1) dC[5,r,m]

$$C_1^{(5)}(r, m), \qquad \text{(IV.6)}$$

(2) dD[5,r+1,m]

$$D_1^{(5)}(r + 1, m). \qquad \text{(IV.7)}$$

(3) What is probability that I will see the two other even faces (#2 and #4) each at least once and the 3 odd faces each at most 3 times.

$$Ans = CD_1^{(2;3)}(1, 4; m). \qquad \text{(IV.8)}$$

Using the integral representation and integrating out the two C-integrals we obtain

$$D_1^{(3)}(4, m) - \frac{2}{2^m} D_{1/2}^{(3)}(4, m) + \frac{1}{3^m} D_{1/3}^{(3)}(4, m). \qquad \text{(IV.9)}$$

For $m = 1$, 2 and 3 this gives respectively .228173, .306281 and .262925.

Problem IV.4. (Die Problem)

A single, fair $s = 6$-sided die is tossed until $m = 10$ sixes are obtained. This defines our random stopping time and the event that occurs is said to occur at stopping time(ast). What is the probability that exactly i of the 5 remaining cells have frequency $\geq r$ ast?

Answer : Using the probability interpretation of CD-function and the fact that the set of size i(as well as the value of i) are not specified, the formula for our result is

$$\binom{5}{i} CD_1^{(i;5-i)}(r, 10) = \binom{5}{i} \sum_{\alpha=0}^{5-i} (-1)^\alpha \binom{5-i}{\alpha} C_1^{(i+\alpha)}(r, 10).$$

(IV.10)

`Table[Binomial[5,i]dCD[i,5-i,r,10,1.0],{r,5,25,5},{i,0,5}]`

Numerical values for different i and r-values are given below.

i	$r = 5$	$r = 10$	$r = 15$	$r = 20$	$r = 25$
0	.003168	.166667	.606783	.895940	.981903
1	.010863	.166667	.189490	.071428	.014734
2	.027704	.166667	.097802	.020760	.002488
3	.066345	.166667	.056640	.007843	.000650
4	.173816	.166667	.032798	.003053	.000183
5	.718103	.166667	.016487	.000977	.000042

Table IV.1

Except for rounding, the sum in each column is clearly one.

Problem IV.5. (Minimum Frequency Die Problem)

Again we toss one fair die with $s=6$ faces, but the stopping time is when the smallest frequency reaches 10. What is the probability (ast) that exactly i faces have frequency ≥ 15, so that $s - 1 - i$ faces have frequency ≥ 10, but < 15.

Answer : Here we use inclusion-exclusion. For $r = 15$ the probability that exactly i $(0 \leq i \leq 5)$ faces will have frequency ≥ 15 and

$5 - i$ faces will have frequency ≥ 10 but < 15 is

$$P(ast) = 6\binom{5}{i} \sum_{\alpha=0}^{5-i} (-1)^\alpha \binom{5-b}{\alpha} C_1^{(i+\alpha;5-i-\alpha)}(15, 10; 10). \quad \text{(IV.11)}$$

We first calculate the 6 auxiliary quantities $Aux = C_1^{(i;5-i)}(15, 10; 10)$.

`Table[dCvec[i,5-i,15,10,10],{i,0,5}]`

Substituting these we obtain the table:

i $\rightarrow$	0	1	2	3	4	5
Aux $\rightarrow$	.16667	.08312	.04847	.03152	.02216	.01649
$P(ast)$ $\rightarrow$	.08887	.18712	.23344	.22156	.17009	.09892

Table IV.2

Problem **IV.6.** _(Maximum Frequency Die Problem)_
This problem is similar to problem(IV.5) with two minor changes:

(1) The stopping time is when the maximum frequency reaches 10,

(2) We want the probability (ast) that exactly i faces have frequency < 8, so that $s - 1 - i = 5 - i$ faces have frequency ≥ 8 and < 10. The auxiliary values for this problem are

$$Aux = D_1^{(i;5-i)}(8, 10; 10). \quad \text{(IV.12)}$$

with $i = 0, \ldots, 5$.

`Table[dDvec[i,5-i,8,10,10],{i,0,5}]`

and the final probability answer is

$$P(ast) = 6\binom{5}{i} \sum_{\alpha=0}^{5-i} (-1)^\alpha \binom{5-i}{\alpha} D_1^{(i+\alpha;5-b-\alpha)}(8, 10; 10). \quad \text{(IV.13)}$$

Again these results are arranged in a table with increasing i-values:

i $\rightarrow$	0	1	2	3	4	5
Aux $\rightarrow$	.16667	.12892	.10213	.08262	.06809	.05704
$P(ast)$ $\rightarrow$	.00324	.02452	.08983	.20852	.33165	.34223

Table IV.3

The fact that the $P(ast)$ values sum to one (except for rounding) is a check in both of the above problems. It should be pointed out that if we decrease the parameter 8 above (to say 6 or 7), the function $P(ast)$ would not be monotone in i, as it is above.

V.   Concluding Remarks

In this paper we gave a mathematical background and we have written programs in *Mathematica* for numerical evaluation of Dirichlet integrals and their various extensions as functions of Type 1(**I,J, IJ**) and Type 2(**C,D, CD**) with various generalizations. The intention of this paper was to make possible for many researchers to actually use the techniques proposed by the authors. We limited our examples to problems in probability only, so we can apply our techniques in the straightforward manner. The applications therefore touch only a tip of the iceberg of many other uses: Expected Waiting Time Problems, Coupon Collection Problems and many others.

As an example of other application, the Type 2 Dirichlet functions were used with the "Problem of Points" in (Sobel and Frankowski, 1994). At some unforeseen moment the host stops a multinomial game when each contestant has his own given number of wins but no one has the required number to win the prize. The problem is then to use the given information and properly share the prize money among the contestants; their probabilities of winning a single game are also given. This "problem of points" or "sharing problem" is the oldest problem in the theory of probability and is another direct application of Dirichlet integrals. As a result of publishing (Sobel and Frankowski, 1994), we received an unsolicited letter from Professor L. Takacs of Case-Western Reserve containing the following excerpt: "It seems the problem of points for several players did not get sufficient interest since the pioneering works of DeMoivre and Laplace. Subsequent publications simply repeated the old results. Only your recent paper with K. Frankowski made significant progress in solving various important problems".

It has been pointed out by a few people that some of the Dirichlet probabilities can be also obtained by ad hoc combinatorial methods, implying that if you are very clever with lots of time you may be able to calculate some of these answers without the help of the enclosed algorithms. We have never denied that this is so. In fact we often use ad hoc combinatorial methods as a check on the Dirichlet calculations. However we feel that there is a tremendous advantage in having a general methodology for a large number of applications in place of separate ad hoc solutions which differ from problem to problem.

Appendix

I. Short Introduction to Mathematica

All of our examples use language named *Mathematica*, created by S. Wolfram and is one of the richest Algebraic Manipulation systems. Here we give some rudimentary information about the system. For further information see for instance (Blachman, 1992).

In imperative computer languages like C the statement $x = a + b$, makes sense only if variables a and b are defined and initialized. In algebra systems, the value of x is $a + b$ itself.

The capabilities of *Mathematica* include: arbitrary precision arithmetic, calculus, expansion and factoring of expressions, variety of graphing techniques, linear algebra and many others.

Constants can be integers, fractions in the form int1/int2, reals with standard and arbitrary precision, complex in the form $a + I\, b$ and some named constants like Pi, E and many others. As the reader noticed the *Mathematica*'s predefined objects start with the capital letters. Constants like Log[2], Sqrt[5] and similar are unevaluated and the function $N[element]$, or $N[element, precision]$ are used to obtain given number of decimal figures.

Mathematica is usually used in the interactive mode and starts with the prompt: $In[1] :=$ waiting for the input of expression to be evaluated. It can be later referred to as %1. As the calculations progress *Mathematica* increments the consecutive numbers n in $In[n]$ and unless the expression was terminated with ";" mark, it produces immediately printed output.

Short Example:

Suppose we have the function $f(n) = n^2 + \sqrt{n}$.

To define this function in *Mathematica* we enter

```
f[n_]:= n ^ 2 + Sqrt[n]
```

Here the underscore _ n indicates arbitrary n within the scope of the definition, := denotes delayed replacement, when the expression is substituted for the parameter. With this definition of f we can evaluate f at $n = 0$ by f[0]. We could have used even n complex, as in f[1+2*I]

If we want to be sure that the function f is used only for integer n, we define it as

```
f[n_Integer]:= n ^ 2 + Sqrt[n]
```

This way when we write (by mistake)f[1+2*I] or f[1.0], the above definition is not used and f is not evaluated. If we want to use f for say, real x, we have to define it again. We can define f as many times (with the same name) as we wish.

We can further insure the safety of our definition by insisting that $n \geq 0$:

```
f[(n_Integer)?NonNegative]:= n ^ 2 + Sqrt[n]
```

We can also name the expression

```
expr = n ^ 2 + Sqrt[n],
```

but this is not a function, so evaluation expr[1] is not defined, but we can still use it for evaluation, as in t= expr /. n -> 2 , to substitute 2 for n in this expression to get the desired result in $t(4 + \sqrt{2})$. If we want to see the number, we may put 2.0 instead of 2 in the substitution. Then usually we get 6 digits. The function N[t,10] in most installations will give the same 6 digits, even if N[t,20] gives 20 digits. The function SetPrecision[t,10] will print 10 digits.

Below we list sample of commands used in our illustrations (with e denoting expression):

Command	Meaning	Example
$name = e$	Assign $name$ to e	$x = (a/b) \wedge 3 + Log[c]$
$(*\cdots*)$	Comment	$(*$ This is a comment$*)$
$f[x\_] := expr$	Define function f for any x	$f[x\_] := 2x - 1$
$D[e, v]$	Differentiate e vs v	$t = D[x - y/x, x]$
$D[e, \{v, n\}]$	Differentiate e vs v n times	$t = D[x - y/x, x, 2]$
$D[e, x, y]$	Differentiate e vs x, y	$t = D[x - y/x, x, y]$
$Do[e, \{n, end\}]$	Repeat e for $n = 1, \ldots, end$	$Do[Print[i*i], \{i, 3\}]$
$Expand[e]$		$Expand[(a + b)(a - b)]$
$Factor[e]$	Factor e	$Factor[x \wedge 2 - 2x + 4]$
$Floor[x]$		$Floor[26/5]$
$If[test, Tcase, Fcase]$	If conditional	$abs[x\_] := If[x < 0, -x,\ x]$
$Integrate[e, var]$	Integrate e vs var	$Integrate[xLog[x], x]$
$Integrate[e, \{v, vl, vu\}]$	Definite integral	$Integrate[x^2, \{x, 0, 1\}]$
$Join[list1, list2...]$	Concatenate lists	$Join[\{1, 2\}, \{a, b\}]$
$Length[list]$	Number of elements	$Length[\{a, b\}]$
$Module[\{x, y, \ldots\}, e]$	Defines $x, y...$ as local	
$N[e, n]$	Uses n digits in e	$N[1/3, 10]$
$Part[list, n]$	The same as $list[[n]]$	$v[[2]]$
$Prepend[list, elem]$	$List$ with $elem$ prepended	$Prepend[\{a, b\}, x]$
$Product[f, \{i, imax\}]$	Gives product of f_i	$Product[2n + 1, \{n, 3\}]$
$Rest[list]$	$List$ without first element	$Rest[\{1, a, b\}]$
$Return[e]$	Give result of e and return	$Return[x]$
$Sum[f, \{i, imax\}]$	Gives sum of f_i	$Sum[2n + 1, \{n, 3\}]$
$Table[f, \{i, imax\}]$	Gives list f_i	$Table[2n + 1, \{n, 3\}]$
$While[test, body]$	Evaluate $body$ as long as $test$ is true	$n = 4;$ $While[n > 1, Print[n];$ $n = n/2]$

Some commands use the standard *Mathematica* iteration. Those in our previous table include Do, Sum, Product, Table etc. These have the form: f[*expr*, {*iterator*}],where *iterator* can be:

- *number* - of times
- $i, imax$ - i goes from 1 to $imax$ in steps of 1
- $i, imin, imax$ - i goes from $imin$ to $imax$ in steps of 1
- $i, imin, imax, di$ - i goes from $imin$ to $imax$ in steps of di

We can also have a sequence of iterators but we will not deal with them here.

Obviously the information given here cannot make you a *Mathematica* programmer, but it might be enough to understand programs written here and to modify them as needed.

References

1. Blachman, N. *Mathematica: A Practical Approach*; Prentice Hall, 1992.

2. Sobel, M.; Frankowski, K. The 500th anniversary of the sharing problem (the oldest problem in the theory of probability), *American Mathematical Monthly* **1994**, 833-847 (see also pp. 896, 910).

3. Sobel, M.; Uppuluri, V.; Frankowski, K. Dirichlet distribution type-1, *Selected Tables in Mathematical Statistics* **1977**, *4*, American Mathematical Society: Providence, RI.

4. Sobel, M.; Uppuluri, V.; Frankowski, K. Dirichlet distribution type-2, *Selected Tables in Mathematical Statistics* **1985**, *9*, American Mathematical Society: Providence, RI.

Bayesian Inference

T. Pham-Gia

Département de Mathématique
Université de Moncton
Moncton, Canada E1A 3E9

I. Introduction

The beta distribution is very much used in Bayesian statistics, mainly as a prior distribution for either a proportion, or the probability of occurrence of an event, or the value of any random variable with variation between 0 and 1, such as the coefficient of determination or the reliability of a component. But the non-central beta distribution, defined on the unit interval, is less frequently encountered. Having a more general expression, and defined on a finite interval of R, is the general beta distribution, which can represent the distribution of a random variable with finite bounds. For this reason, it is encountered in several applications in Engineering and Management Science. However, the Bayesian approach to its study is much more elaborate, and has seen only a limited number of results. Extensions of the beta to vector and matrix variate distributions are receiving increasing attention, and, with the recent advances in computing technology, have found applications in several domains. Their Bayesian treatment seems quite complex at the present time. We will distinguish mainly between the beta type1 and beta type2.

361

- <u>Type1:</u> $X \sim beta(\alpha, \beta)$ if its density is defined on $[0, 1]$ by:

$$f(x; \alpha, \beta) = \frac{x^{\alpha-1}(1-x)^{\beta-1}}{B(\alpha, \beta)}, \tag{I.1}$$

with $\alpha, \beta > 0$, and

$$B(\alpha, \beta) = \Gamma(\alpha)\Gamma(\beta)/\Gamma(\alpha + \beta)$$

being the beta function. It has mean

$$\mu = \alpha/(\alpha + \beta),$$

variance

$$\sigma^2 = \alpha\beta/\left[(\alpha + \beta)^2(\alpha + \beta + 1)\right]$$

and mean absolute deviation (about its mean)

$$\delta_1 = 2\alpha^\alpha\beta^\beta/\left[B(\alpha, \beta)(\alpha + \beta)^{\alpha+\beta+1}\right].$$

Its mode (or antimode) is at $(\alpha - 1)/(\alpha + \beta - 2)$. A generalization of this distribution, defined on any finite interval (c, d), by:

$$f(t; \alpha, \beta; c, d) = \frac{(t - c)^{\alpha-1}(d - t)^{\beta-1}}{B(\alpha, \beta)(d - c)^{\alpha+\beta-1}}, \tag{I.2}$$

can be directly related to it by the linear transformation: $X = (T - c)/(d - c)$. This distribution is called the general beta in 4 parameters, denoted $gbeta(\alpha, \beta; c, d)$.

- <u>Type2:</u> $Y \sim betap(\alpha, \beta)$ if its density is defined on $[0, \infty)$ by:

$$f(y; \alpha, \beta) = \frac{y^{\alpha-1}}{B(\alpha, \beta)(1 + y)^{\alpha+\beta}}, \tag{I.3}$$

$\alpha, \beta > 0$. It has as mean

$$\alpha/(\beta - 1),$$

and as variance

$$\alpha(\alpha + \beta - 1)/[(\beta - 1)^2(\beta - 2)], \quad \beta > 2.$$

This distribution is also called *beta prime*, or *inverted beta*. Y can be obtained from $X \sim$ beta type1 by the odds-ratio transformation, $Y = X/(1 - X)$, and is also related to W, which follows the classical Fisher-Snedecor distribution with $v_1 = 2\alpha$ and $v_2 = 2\beta$ degrees of freedom, by the relation $\frac{\beta}{\alpha} Y = W$. A generalized form of this distribution, obtained by multiplication of Y by a positive parameter λ, is more frequently encountered. It is called the Generalized-F, has as density;

$$f(t; \alpha, \beta, \lambda) = \frac{\lambda^\alpha t^{\alpha-1}}{B(\alpha, \beta)(1 + \lambda t)^{\alpha+\beta}}, \tag{I.4}$$

$t \geq 0$ and $\alpha, \beta, \lambda \geq 0$, and is denoted $G3F(\alpha, \beta; \lambda)$. It is also called the Gamma-gamma distribution, denoted $Gg(\lambda, \alpha; 1/\beta)$ (Bernardo and Smith, 1994).

Several other derived distributions, obtained by performing elementary operations on X, or Y, or by using the beta type1 as a mixing weight, appear under various names. Similarly, generalizations of the beta, in one variable, or in several variables, are available in the literature. We will not treat them here, except those having a close relation with either beta, or those conjugate to either Bernoulli or Pascal sampling. In particular, we will consider the *ibetap*(α, β, b) distribution of the variable $W = b/X$, with $b > 0$, which has its density defined on $[b, \infty)$ by:

$$f(w; \alpha, \beta, \lambda) = \frac{(w - b)^{\beta-1} b^\alpha}{w^{\alpha+\beta} B(\alpha, \beta)}, \tag{I.5}$$

Its mean is

$$\mu = b(\alpha + \beta - 1)/(\alpha - 1)$$

and its variance

$$\sigma^2 = b^2 \beta(\alpha + \beta - 1)/\left[(\alpha - 1)^2(\alpha - 2)\right], \ \alpha > 2.$$

In this chapter, we will put more emphasis on the applied aspects of *Bayesian Statistics* and *Bayesian Decision Theory*, since these are the areas where the beta is more present, but will discuss theoretical results as well, with ample reference to specific sections of some excellent theoretical treatises on the topics concerned. Also, for convenience, we have regrouped the results under the four distinct parts of Bayesian analysis: the *prior*, the *likelihood (or sampling)*, the *posterior* and the *predictive*, which should be *consecutively* consulted regarding a particular model, to follow it through its complete evolution. The plan of this chapter is as follows: In Section II we present the general *Bayesian paradigm*. In Section III, various questions related to the prior beta distribution are discussed: *elicitation, non-informative, least informative* and *reference priors*. The sampling phase is treated in Section IV, with the *conjugacy* property highlighted. The *posterior distribution* is dealt with in Section V and the *predictive distribution* in Section VI. *Preposterior* analysis is discussed briefly in Section VII, while Section VIII treats *Bayesian Decision Theory*. The topics of *Sequential Bayes* and *Empirical Bayes* are presented in Section IX, while *mixtures of betas* are discussed in Section X. Section XI briefly surveys applications of the beta and of Bayesian methods in some fields of science, such as *Reliability, Pattern Recognition, Bayesian Quality Sampling Plans* and *Bayes Operator*. In Section XII, we deal with the four basic operations on independent beta variables and their Bayesian transforms. Distributions derived from the beta are discussed in Section XIII, and *generalized beta distributions* in one variable, in several variables, and in *matrix variates*, are discussed in Section XIV, where, again, they are treated in the four parts of Bayesian analysis. Finally, Section XV presents a problem that is attracting the interest of researchers, and where the beta plays an active role.

NOTE: Since the Bayesian approach is encountered in most statistical areas this chapter covers a wide variety of topics, and mathematical notations can become very complex if they are to conform to the notations used in each topic. We will try to make them as simple, and as clear as possible, and for this reason, they do not, at times, conform to conventional notation. Also, technical terms are *in italics* when they appear for the first time.

II. The Bayesian Paradigm

Classical, or *frequentist*, statistics is based on the basic principle that the parameters of a distribution are constants, and a statistical model is used, on which we base our analysis and inference. Usually, it is the normal model, and the most popular topics are *hypothesis testing, estimation* and *analysis of variance*. The meaning of the probability of an event is never clearly stated, but it could be a mixture of *Laplace symmetry principle* and *Von Mises limit proportion*. In the 20th century, other meanings of Probability have been suggested, in particular, subjective notions associated with betting. Works by Savage,and especially by De Finetti on exchangeable sequences, have provided a solid basis to the *subjective approach* to probability. There should be at least 3 kinds of Probability: physical, logical and subjective (Good, 1965), leading to 3 areas of Bayesian statistics: the *Empirical Bayes* (developed by Robbins), the *Logical Bayes* (Keynes, Carnap) and the *Subjectivist Bayes* (Ramsey, Savage, de Finetti). The interested reader can consult the thorough treatise on the foundations of different probability theories by Fine (1973), who concluded that *"of all the theories considered, subjective probability holds the best position with respect to the value of probability conclusions, however arrived at "* (page 240). Subjective probability concepts constitute the main engine that permitted the development of the Bayesian approach to statistics, which is based on Reverend Thomas Bayes (1702-1761) simple formula:

$$\Pr(A|B) = \Pr(A \cap B)/\Pr(B)$$

when considering single events A and B, and

$$\Pr(A_k|B) = \Pr(B|A_k) / \sum_{j=1}^{n} \Pr(A_j|B)$$

for the set of exhaustive events $\{A_j\}_{j=1}^{n}$. In the formal *Bayesian paradigm*, the parameter Θ under consideration is a random variable with a *prior distribution* $g(\theta)$, defined in $\Omega \subset \mathcal{R}^n$. In the statistical model $f(x|\theta)$, a value of Θ serves to determine the distribution, from which a sample $X = (X_1, \ldots, X_n)$ will be taken. The *likelihood func-*

tion $\prod_{i=1}^{n} f(x_i|\theta)$ can now be combined with the prior to give the *posterior distribution* of Θ, according to Bayes Theorem. This is the *conditional distribution* of Θ, given X, and we have:

$$g\left(\theta|X_1,\ldots,X_n\right) = g(\theta)\prod_{i=1}^{n} f\left(x_i|\theta\right) \bigg/ \int_{\Omega} g(\theta) \cdot \prod_{i=1}^{n} f\left(x_i|\theta\right) d\theta$$

$$(\text{II}.1)$$

Hence, we can see that the Posterior is proportional to the product Prior by Likelihood, denoted: *Posterior α Prior. Likelihood.*
Very often, a *sufficient statistic* of Θ, $\tau_n = k(X_1,\ldots,X_n)$ is considered, with density $h(\tau_n|\theta)$, and we have:

$$g\left(\theta|X_1,\ldots,X_n\right) = \frac{g(\theta)h\left(\tau_n|\theta\right)}{\int_{\Omega} g(\theta)h\left(\tau_n|\theta\right)} d\theta.$$

All information on the parameter Θ, i.e. the prior information as well as the sampling results, being contained in the posterior distribution, it is natural that any Bayesian inference should be based on this distribution.

III. The Prior Distribution

The prior is an indispensable component of the Bayesian paradigm and strictly speaking, it should be based on subjective probability only. However, in practice, its derivation can come from personal beliefs, historical data, or other means of information, or a combination of these sources. At the present time, although there is much debate on the nature and properties of the prior, there is still no concensus on how to derive it, not even general accepted guidelines to follow for that purpose. It can be fairly said that establishing the prior distribution is the *"weakest link"* in the Bayesian chain, a situation that, in many cases, has seriously prevented the application of Bayesian statistical methods. Concerning this problem, Fine (1973, page 240) wrote: *"The measurement problem in subjective probability is sizable and conceivably insurmountable".*

A. The Beta as a Prior: Its Elicitation

Elicitation of the prior, and in fact, the establishment of a probability distribution for an event, or a parameter, has been the subject of much study and debate. Since probability is taken here as a *degree of belief*, it has a subjective meaning and is closely associated with the *anchoring phenomenon* in human quantitative evaluations (Kahneman, Slovic and Tversky, 1981, page 14), with the consequence that different initial values suggested by the formulation of the problem may yield different estimates. Various biases inherent to human judgments, and evidenced by experiences in psychology, are also causes for concern in elicitation, and less subjective information, such as the histogram of relative frequencies (Raiffa and Schlaifer, 1961), can also serve as input to assist in making quantitative judgments. In general, this is a common area between statistics and psychology that still needs much attention, and closed cooperation between experts in these two disciplines has been called for, to alleviate serious difficulties that plague human assessment of probability. Hogarth (1975) should be consulted for a general view of the whole question. For the beta distribution, some *ground-hypotheses* for its choice as a prior have been proposed by Colombo and Constantini (1980), and a certain number of methods have been suggested to obtain its density, or *cumulative distribution function.*

i. Methods Based on Interviews

Some of the following methods are based on the *conjugacy* of the beta family w.r.t. *binomial sampling.* Hence, it is essential to understand this property, which is presented in later sections. Also, in practice, quantitative information is obtained from the *assessor* or *expert*, who can be different from the statistician in charge of obtaining the prior distribution.

a) *Equivalent Prior sample method (EPS):* It is a version of the well-known *Device of imaginary results* proposed by I. G. Good. Let α_0 and β_0 be the parameters of a *non-informative prior* that the assessor has chosen. According to this approach, he/she gives the size n_0 with r_0 successes (not necessarily integers) of an imaginary binomial sample that would reflect his/her belief of the number of successes out of a total number of trials. He/she then estimates the value of the mean, say π_0. Considering the posterior beta, we then have the posterior parameters: $\alpha_1 = \alpha_0 + r_0$ and $\beta_1 = \beta_0 + n_0 - r_0$ and $\pi_0 = (\alpha_0 + r_0)/(\alpha_0 + \beta_0 + n_0)$. From these equations, we have α_1 and β_1 the parameters of the assessor's beta prior, in function of the estimates she/he has provided:

$$\alpha_1 = \pi_0 \left(n_0 + \alpha_0 + \beta_0 \right),$$
$$\beta_1 = (1 - \pi_0) \left(n_0 + \alpha_0 + \beta_0 \right).$$

If $\alpha_0 = \beta_0 = 0$ and $0 < r_0 < n_0$, we have $\alpha_1 = \pi_0 n_0$ and $\beta_1 = (1 - \pi_0)n_0$. There is no guarantee, however, that this method will be free of cognitive assessment biases. Furthermore, there should also be some justification on the choice of the noninformative prior.

b) *Hypothetical Future Sample (HFS) method:* The assessor provides first the value of the (prior) mean of his beta, say π_0. He/she then provides the value of the mean if in a hypothetical future sample of size n, there are r successes. Equation (V.1) will then be used to find the values of ξ, and of $(\alpha + \beta)$, from which we obtain $\alpha = \pi_0(\alpha + \beta)$.

c) *de Finetti's exchangeable sequences method:* This approach is close to the elicitation approach based on the *predictive distribution*, but is based on long run concepts, that should, somehow, be grasped by the assessor (see Ferreri (1986)). There are several merits for using the predictive distribution, which deals with observables, and some authors have championed that cause (see, e.g., Winkler (1980), and the section of this chapter on the predictive).

d) *Using the mean absolute deviation:* This method is based on the idea that the *mean absolute deviation* about the mean of the beta, denoted δ_1, should be an information easier to obtain from the assessor, since it reflects an error about the mean, independent of the sign. From the assessed values of δ_1 and of the mean, we can deduce the values of the beta parameters (see Pham-Gia and Turkkan (1992a)).

e) *Aggregative approach*: Ferreri (1986) suggests using an aggregative approach, where two different assessors provide independent estimates of the two probabilities of success and failure, π and $1 - \pi$. A set of weights is used for this purpose. His method is related to the problem of obtaining the beta from several assessors, i.e. of reconciling several quantitative opinions, a topic on which a number of approaches have been proposed (Berger (1985), pages 272-286). Winkler (1968), for example, suggested a method by which exponents associated with different betas are directly combined, hence guaranteeing that the combination obtained is also a beta.

f) *The percentiles method* : Methods to determine directly the density, or the cumulative distribution of the prior beta are generally based on estimated percentiles, to be obtained from the assessor. Examples of questionnaires to be used in interviews for that purpose are given in Winkler (1967). For example, the assessor is asked to provide the numerical values of two percentiles of his prior beta, say $C_{.90}$ and $C_{.10}$. A computer program, or table (see Waller and Martz (1982)), will provide values of the parameters, by numerical search. A variation of this method asks for the values of the mean and the 5th (or 95th) percentile.

It is worth noticing that Kahneman et al. (1982, page 17) has reported that " *actual values of the assessed quantities are either smaller than $C_{.01}$ or larger than $C_{.99}$, reflecting an overconfidence of the expert in his judgment*". Corrective measures have been suggested. For example, Bunn(1975) suggested an hysteresis-based interview strategy, to anchor some of the biases, but recognized that other more important biases could still be present in the interviewing process.

For the beta, we have the following relations between α and β and the mean and variance, μ and σ^2, that can be used to determine α and β, in function of μ and σ^2.

$$\alpha = \left[\mu^2(1-\mu) - \mu\sigma^2\right]/\sigma^2,$$
$$\beta = (1-\mu)\left[\mu(1-\mu) - \sigma^2\right]/\sigma^2. \tag{III.1}$$

h) *The Pearson-Tukey approach:* In this approach to estimate densities in general, a fixed set of weights is used on the 3 percentiles $C_{.05}$, $C_{.50}$ and $C_{.95}$, provided by the assessor, to obtain the estimated mean:
$\hat{\mu} = 0.185(C_{.05} + C_{.95}) + 0.63C_{.50}$ and the estimated variance, in two steps:

$$\hat{\sigma}_0^2 = \left[(C_{.95} + C_{.05})/3.25\right]^2 \text{ and}$$
$$\hat{\sigma}^2 = (C_{.95} - C_{.05})/\left[3.29 - 0.1\left(C_{.95} + C_{.05} - 2C_{.50}\right)^2/\sigma_0^2\right].$$

From these values, using (III.1) we can obtain the values of α and β. In their investigation, Keefer and Bodily (1983) found that the above method is the most accurate among three-points estimation methods for a probability density. Zaino and d'Enrico (1989) claimed that their method, based on the 4th, 50th and 96th percentiles, provides an improvement over the above method.

j) *The PERT approach:* Finally, in classical Operations Research PERT Time (Program Evaluation and Review Technique), a simple procedure is used to elicit the general beta: the max, min and most likely (M) values of the distribution are obtained from the expert, and we take as approximate values,

$$\hat{\mu} = (max + 4M + min)/6 \text{ and } \hat{\sigma} = (max - min)/6.$$

This method has been highly criticized, however, because of its inaccuracy. Lau and Lau (1998) presented an improvement formula for the estimation of the standard deviation, based on the values of either 4 or 6 percentiles, provided by the assessor.

ii. Methods Based on Historical Data (Moments and Maximum Likelihood Methods and Their Variations)

These classical methods, well known in statistical estimation, strictly speaking, do not belong to Bayesian statistics, and neither do their variations/improvements. They are, at best, associated with *Parametric Empirical Bayes*. They also suppose that the past has a significant input into the future, and that the long range frequency interpretation of probability can be reconciled with its personal degree of belief interpretation. However, they can provide useful information to the assessor, on which he/she will build his/her prior, as do some other similar methods, such as the estimation from data of the α-th and $(1 - \alpha)$-th percentiles.

B. Non-Informative Priors

Since the prior introduces subjective information into the Bayesian process, it is thought that a prior distribution that brings in no information at all would provide a good starting reference point. Short of this complete absence of information, the least information possible brought in, would be acceptable. The general idea of a prior distribution in Bayesian Statistics, that would behave neutrally enough so that the *data can speak for itself* is a complex question that still has not found a satisfactory answer. Several methods have been suggested, the most accepted one is probably Jeffrey's, based on data invariance, which leads to a density proportional to the square root of Fisher's information

$$I(\theta|X) = -E_{X|\theta}\left(\partial^2 L/\partial\theta^2\right),$$

where $L(\Theta|X)$ is the log-likelihood function of the parameter. The basic argument used here is that if there is no information about θ, then there should be no information about a transformed ϕ of θ as well. More precisely, to define the non-informative prior of θ, we consider a transform ϕ of θ, such that the likelihood function of ϕ is *data-translated*, i.e. its form is completely determined, except for its location that depends only on X. We now assign a (locally) uniform distribution to ϕ, and by the reverse transformation, obtain the non-informative prior for θ. Considering Fisher's information in this transformation, we can show that the non-informative distribution for θ is proportional to $\sqrt{I(\theta|X)}$ (see Berger (1985, page 87)). For Bernoulli sampling, this transform of π is $\phi = arcsin\sqrt{\pi}$, which is well-known in variance stabilizing, and the non informative is $beta(1/2, 1/2)$. There are others, equally popular non-informative priors in the beta family, obtained with other arguments, such as the uniform, or $beta(1, 1)$ (based on the parameter itself, and also on the posterior mode), Villegas improper prior, or $beta(0, 0)$ (based a uniform prior for the log-odds transform) and Zellner's prior,

$$1.61856\theta^{\theta}(1 - \theta)^{1-\theta}.$$

For *Pascal sampling*, the associated non-informative prior $beta(0, 1/2)$ is improper.

C. The Least Informative Beta Prior

A number of authors have suggested that the least informative prior, instead of the non-informative prior, should be considered, in the case where some information is already available. That prior should then depend on the sample size, if this information is present. Using the expected *posterior variance* as a criterion it can be proven that, if the prior mean is fixed at μ_1 and the sample size is n, then the least informative beta prior is $beta(\mu_1\sqrt{n}, (1 - \mu_1)\sqrt{n})$. Since ignorance of the mean would imply $\mu_1 = 1/2$, rather than any other value, the least informative prior is now beta $(\sqrt{n}/2, \sqrt{n}/2)$, which, in turn, gives Villegas, Jeffrey and uniform prior, for $n = 0, 1$ and 4 (Pham-Gia, 1994). Dyer and Chiou (1984), on the other hand, used a computer method to show that the best family of distributions to use as priors in binomial sampling is, precisely, the beta family, and also derived the least informative betas under different hypotheses on the information available.

D. The Reference Prior

The *reference prior* is based on the amount of *entropy information* measure on the parameter of interest, that an experiment is expected to provide, and on the missing information about the parameter, as a function of the prior $p(\theta)$. The reference prior is then the one that maximizes the missing information functional. It is obtained usually via a limiting process, by applying Bayes theorem to the posterior (Bernardo and Smith, 1994, page 306). Under some regularity conditions, the reference prior can be characterized in terms of the parametric model, and it agrees with Jeffrey's prior, both being proportional to the square root of Fisher's information. This leads to $beta(1/2, 1/2)$ for Bernoulli sampling, and $beta(0, 1/2)$ for Pascal sampling. The Reference prior can be shown to be independent of the sample size, invariant under one-to-one transformation, and is compatible with sufficient statistics (Bernardo and Smith, 1994, page 309).

IV. Sampling Phase

The *sampling phase* is associated with the likelihood function of Classical statistics. If $X_1 \ldots, X_n$ is a sample from a distribution, with density $f(x|\theta)$ indexed by θ, then the likelihood function is $\Pi_{j=1}^n f(x_j|\theta)$. The *Likelihood principle* (Press, 1989, page 56) states that only information contained in the sample already obtained should be considered, not the one contained in potential samples. The Bayesian approach is hence in conformity with the *Likelihood principle*, unlike several important concepts in frequentist statistics. The *Likelihood Principle* itself is the result of 2 other principles, the *Weak conditionality principle* and the *Weak sufficiency principle*. Naturally, associated with the beta prior can be any sampling model. We will only discuss Bernoulli (more precisely, *binomial*) and Pascal samplings here: Let the population under consideration consist of elements having a specified character C (or in state 1), and forming the theoretical population proportion π, and those not having it (or in state 2). Thus, there is a dichotomous state, e.g. *pass* or *fail*, or *good* or *bad*. Let N be the size of the population.

Bernoulli sampling: We assign to the first state the value $X = 1$ and to the other the value $X = 0$. In a sample of fixed size n, $\{X_1, X_2, \ldots, X_n\}$, taken from the above population, where there are k elements in state 1 with $0 \le k \le N$, let $Y = \sum_{i=1}^{n} X_i$. Then $Y \sim H(N, n; k)$, the *hypergeometric distribution*, with density

$$\Pr(Y = j) = \frac{\binom{k}{j}\binom{N-k}{n-j}}{\binom{N}{n}}, 0 \le j \le n, \quad 0 \le j \le k.$$

Its mean is $\mu = n\frac{k}{N}$ and its variance

$$\sigma^2 = n\frac{k}{N}\left(1 - \frac{k}{N}\right)\left(\frac{N-n}{N-1}\right).$$

If N is very large, or the probability $\pi = k/N$ remains constant during sampling, we have: $X \sim Ber(\pi)$, the Bernoulli distribution, and $Y \sim Bin(n, \pi)$, the *binomial distribution*, with density

$$\Pr(Y = j) = \binom{n}{j}\pi^j(1 - \pi)^{n-j}, \quad 0 \le j \le n,$$

with mean $\mu = n\pi$ and variance $\sigma^2 = n\pi(1 - \pi)$.

Pascal sampling: Let π be the constant probability of *success*, for which $x = 1$, and s be a positive fixed integer. We are interested in the first value of n, in sampling from a large universe, when we have the sth *success*. Hence n becomes random here, and we have: $n \sim Pas(\pi, s)$, the Pascal distribution (also called Negative-binomial), with density

$$\Pr(n = j) = \binom{j-1}{s-1}(1 - \pi)^{j-s}\pi^s, \quad j = s, s+1, \ldots.$$

It has, as mean, $\mu = \frac{s}{\pi}$ and, as variance, $\sigma^2 = \frac{s(1 - \pi)}{\pi^2}$. For $s = 1$, it is called the *Geometric distribution*. The relation between $X \sim Bin(n, \pi)$ and $W \sim Pas(\pi, s)$ is $\Pr(W \le n) = \Pr(X \le s)$. We can see that both these sampling schemes have mathematical expressions similar to that of the beta. When there are noises, or dependency, in the sampling phase, the likelihood of an element to be classified as *"success"* or *"failure"* will be affected. Winkler (1985) has studied the effects of these noises on the effective sample size and Pham-Gia and Turkkan (1992d) have generalized these results, and studied them in a Bayesian context as well.

Conjugacy: In the general case, when likelihood functions and prior distributions bear no mathematical similarities, the integral in (II.1) has to be computed numerically, and no general mathematical expression can be given to the posterior distribution. In some cases, we have the posterior in closed form, thanks to the above-mentioned similarity, which gives rise to the proportionality between the likelihood and the prior. The posterior is then obtained by merely updating the parameters of the prior with sampling results. The prior is then called *natural-conjugate to the likelihood function*, or more simply *conjugate to sampling*. For the exponential family of distributions, of the form $f(x|\theta) = h(\theta)g(x)exp\{\psi(\theta)t(x)\}$, we can see that the likelihood is $l(\theta|X) = h(\theta)^n exp\left\{\sum t(x_i)\psi(\theta)\right\}$ and hence, the conjugate prior has the form:

$$h(\theta)^v \exp\left(\tau\psi(\theta)\right).$$

A variant of the natural conjugate prior family is the *g*-prior distribution family. Since it is mainly used in regression, it has, so far, little relation with the beta distribution (Press, 1989, page 54).

The notion of conjugacy is important because it allows the convenient handling of mathematical computations. Also, since most densities can be adequately approximated by a beta in practical situations, approximate posteriors can then be handily obtained.

1. Prior Beta type1 (see(I.1)): When sampling is either Bernoulli or Pascal, we have conjugacy. For $\pi \sim beta(\alpha, \beta)$, the posterior distribution is

    $$beta(\alpha + r, \beta + n - r)$$

    if Bernoulli sampling gives r favorable results over n trials with n fixed, and the posterior is

    $$beta(\alpha + s, \beta + n - s)$$

    if in Pascal sampling, it takes n trials to obtain s favorable outcomes, with s fixed. More generally, it is $beta(\alpha + ks, \sum_{i=1}^{k} n_i - ks)$ for a sample $\{n_1, n_2, \ldots, n_k\}$ of trials, where we obtain the sth favorable outcome at each n_i.

2. For $ibetap(\alpha, \beta, 1)$ as prior (see (I.5)), if sampling is Bernoulli in $1/\pi$, there is conjugacy and the posterior is $ibetap(\alpha + r, \beta + n - r, 1)$ (Raiffa and Schlaifer, 1961, pages 264 and 267).

3. Prior Beta type2 (see (I.3)): There is no known conjugacy for this distribution with any sampling scheme, but Ghosh and Parsian (1981) used it as a prior in the *Bayes minimax estimation* of multiple Poisson parameters, because of some of its special asymptotic properties.

V. The Posterior Distribution

1. Other sampling schemes frequently encountered in the literature include *Gamma Sampling, Poisson sampling, Exponential* and *Uniform sampling*. For Normal sampling we usually distinguish between the cases of known mean, or known precision, or both parameters unknown. When the beta is taken as prior for these sampling schemes, numerical methods have to be used to derive the posterior and predictive distributions. This fact could be a hindrance for the establishment of further properties of the posterior, or for variables depending on it. In the last two decades, however, advances in computer technology have allowed the application of *computer intensive methods* in the determination of the posterior. Techniques such as the *Gibbs sampler*, the *Metropolis-Hastings algorithm* and the *Monte Carlo Markov Chains* simulation approach, certainly help derive the posterior distribution, under very extreme conditions. Still, the analytic form of the latter, if available, would be very useful.

2. We have seen that it is logical that the posterior has a key role in any analysis and inference. It is immediate that if one of the two factors dominates in this product *Prior × Likeli-*

hood, the posterior will inherit more features from that factor. Dominance here can be thought of as having more peakedness in the region of interest. Hence, for a small sample, the prior will dominate while as n increases, the likelihood function takes over and at the limit, the posterior will be close to a *normalized likelihood function*. By virtue of the Central limit Theorem, that limit is also close to a normal distribution (see Bernardo and Smith (1994, Section 5.3)).

3. Let π have a $beta(\alpha, \beta)$ prior, with posterior $beta(\alpha + r, \beta + n - r)$. This distribution has mean

$$\mu = \frac{\alpha + r}{\alpha + \beta + n},$$

variance

$$\sigma^2 = \frac{(\alpha + r)(\beta + n - r)}{[(\alpha + \beta + n)^2(\alpha + \beta + n + 1)]},$$

and mean absolute deviation

$$\delta_1 = \frac{2(\alpha + r)^{\alpha+r}(\beta + n - r)^{\beta+n-r}}{[B(\alpha + r, \beta + n - r)(\alpha + \beta + n)^{\alpha+\beta+n+1}]}.$$

The form of this posterior beta depends on the value of r, i.e. on the sampling result. If there is agreement between the two sources, sampling and prior, the posterior will be tighter than the prior (there will be less dispersion) while if the 2 information sources are conflicting, the posterior will be more spread out. Since r/n is the *likelihood estimate* of π, it is very informative to identify the contribution of each source in the expressions of the mean, the variance and the mean absolute deviation of the posterior distribution, also called *posterior mean, posterior variance* and *posterior mean absolute deviation*. For the posterior mean, we have:

$$\mu_{post} = \xi \left(\frac{r}{n}\right) + (1 - \xi) \left(\frac{\alpha}{\alpha + \beta}\right), \tag{V.1}$$

where $\xi = \frac{n}{(\alpha+\beta+n)}$. We have a similar decomposition of the posterior mode.

Hence the posterior mean is the weighted average of the likeli-

hood estimate and the prior mean, with the weight being the
ratio of the real sample size to the total sample size. Here,
$\alpha + \beta$ is considered as the size of a (fictive) sample, equivalent
to the prior distribution. Again, we can see that the source of
information that dominates in the expression of μ_{post} depends
on the relative values of n and $\alpha + \beta$. A similar discussion can
be made concerning Pascal sampling (see Raiffa and Schlaifer
(1961)). The above form also shows that the posterior mean
of the beta is a linear function of the sufficient statistic r. This
result is typical of some families of conjugate priors, and results in this direction, in a more general context, have been
obtained by some authors. Diaconis and Ylvisaker (1979), for
example, showed that if τ is a prior distribution, and for each
$n = 1, 2, \ldots$, there exist a_n and b_n such that

$$\int_0^1 p^{k+1}(1-p)^{n-k} d\tau(p)$$

$$= (a_n k + b_n) \int_0^1 p^{k+1}(1-p)^{n-k} d\tau(p),$$

then τ is a beta distribution whose parameters are related
to a_n and b_n by the relations $a_n = \alpha/[1 + \alpha(n-1)]$ and $b_n = \beta/[1 + \alpha(n-1)]$.

4. For the posterior variance, a similar decomposition is possible, although not as clearcut. But in the representation of the
   posterior variance as a function of r and n, with $0 \le r \le n$, as
   given in Pham-Gia (1995a), it is shown, that, when the maximum likelihood estimate and the prior mean agree, then the
   posterior variance decreases continuously, reflecting the fact
   that coherent contributing factors help reduce the uncertainty.
   For other values of r, the posterior variance differs from the
   prior variance, and goes through a maximum value which is
   even larger than the prior variance when the two contributing
   sources are at their most conflicting points. A similar conclusion applies for the posterior mean absolute deviation.

5. In Statistics, the concept of information contained in a distribution is a very important one, and information can be mea-

sured by the reciprocal of the variance since, intuitively, the smaller the variance, the more certain we are about the variation range of the random variable. Hence, two prior betas can be measured in terms of the information they respectively contain, and also, relatively to each other. Similarly, when posterior distributions are considered for information purpose, the expected valued of their variances, $E_X(Var_{post}^{(n,x)})$, should be considered. Such a study concerning the beta, together with some related questions, was carried out by Pham-Gia (1994).

6. The highest posterior density region: In relation with the posterior distribution of a parameter, the region with $(1 - \gamma)$ probability is called the $(1 - \gamma)100\%$ *credible region*. This really means that the probability that the parameter lies within that region is $(1 - \gamma)$, unlike the $(1 - \gamma)100\%$ confidence region in Classical statistics. Furthermore, if any point inside the region has a higher probability than any point outside it, we have a *highest posterior density*, or *hpd*, region, which can consist of a single interval, or a set of disjoint intervals. Only for symmetrical densities that this interval is the same as the equal-tailed ($\gamma/2$ area on each side) interval. Otherwise, we have to solve the set of equations:

$$\int_{\theta_a}^{\theta_b} f(\theta|x_1, \ldots, x_n) \, d\theta = 1 - \alpha,$$

and

$$f(\theta_a|x_1, \ldots, x_n) \, d\theta = f(\theta_b|x_1, \ldots, x_n) \, d\theta,$$

with $\alpha = \gamma$ or $1 - \gamma$, depending on the presence of a mode, or an antimode in the beta density. An adjustment is needed in other cases. Turkkan and Pham-Gia (1993) provided an algorithm, based on another approach, to compute the hpd region for general distributions, and an extension of it to the bivariate case is given in Turkkan and Pham-Gia (1997).

7. Reference Posterior: The reference posterior is obtained from the reference prior. Hence, we have, respectively, for Bernoulli

sampling,

$$beta\left(\frac{1}{2}+r,\frac{1}{2}+n-r\right),$$

and for Pascal sampling,

$$beta\left(s,\frac{1}{2}+n-s\right),$$

where r and n observed values.

8. An extended form of the beta of type2, the generalized F-distribution (see (I.4)), is the *predictive distribution* for a sampling model from an exponential distribution with a Gamma prior, as follows: Let the Gamma variable Y, denoted $Y \sim Ga(m,\theta)$, have a density of the form:

$$p(y|\theta) = \frac{\theta^m y^{m-1}\exp(-\theta y)}{\Gamma(m)}, \quad y > 0.$$

Y is, in this context, the sum of m replicates of an exponential random variable, with mean $1/\theta$. Let $x_{(n)}$ be the sum of the observations in a sample of size n, taken from the same exponential distribution, and let $Ga(g,h)$ be the prior density of θ.

Then setting $G = g + n$ and $H = h + x_{(n)}$, we have the *marginal distribution* of $x_{(n)}$, $X_{(n)} \sim G3F(n, g; 1/h)$, the predictive distribution $Y \sim G3F(m, g; 1/h)$, and the predictive posterior $Y|x_{(n)} \sim G3F(m, G; 1/H)$ (Amaral-Turkman and Dunsmore, 1985). These results come from the simple fact that the gamma-gamma distribution is in fact a generalized F distribution.

9. For normal models, let $X \sim N(W, R)$, with W being the mean and R the precision. Let W be known and R have $Ga(\gamma, \epsilon)$ as prior. If $(X_1, \ldots, X_n)$ is a random sample, then

$$\sum_{i=1}^{n}(X_i - W)^2 \sim G3F\left(n/2, \gamma; 1/(2\epsilon)\right).$$

When both W and R are unknown, and have as prior the compound density

$$p(w, r) = N\left(w|\mu_0, n_0 r\right) Ga(r|\gamma, \epsilon),$$

then

$$ns^2 = \sum_{i=1}^{n} \left(X_i - \overline{X}\right)^2 \sim G3F\left((n-1)/2, \gamma; 1/(2\epsilon)\right).$$

Other similar results and their applications can be found in Pham-Gia (2003).

VI.   The Predictive Distribution

A.   The Beta-Binomial and Beta-Pascal Distributions

The predictive distribution of X is also the continuous mixture of the sampling distribution with the prior, or, for some authors, with the posterior. Here, we consider the mixing of the binomial probability of each value of the sample (of fixed size n), with the prior distribution $beta(\alpha, \beta)$ of the parameter π. In other terms, if $(X|\pi) \sim Bin(n, \pi)$ and $\pi \sim beta(\alpha, \beta)$, then $X \sim Bbin(\alpha, \beta; n)$, called the *beta-binomial*, or *Polya*, distribution, with density:

$$\Pr(R = r) = \binom{n}{r} \frac{B(\alpha + r, \beta + n - r)}{B(\alpha, \beta)}, \quad 0 \leq r \leq n.$$

It has, as mean

$$\mu_{pred} = n\alpha/(\alpha + \beta),$$

and, as variance

$$\sigma^2_{pred} = n\alpha\beta(\alpha + \beta + n)/\left[(\alpha + \beta)^2(\alpha + \beta + 1)\right]. \tag{VI.1}$$

Hence, we have the following relations: $\mu_{pred} = n\mu_{prior}$ and $\sigma^2_{pred} = n(\alpha + \beta + n)\sigma^2_{prior}$, which reflect the relative importance of the prior information and the sampling size. If the posterior beta is used in the mixing operation, α and β are replaced by $\alpha + r$ and $\beta + n - r$ respectively.

The predictive distribution, which is the marginal distribution of R, is used to compute the expected value w.r.t. data, in a sample of size n, of any random variable whose value is function of the posterior parameters.

If sampling is $Pascal(\pi, s)$, the number n of trials required to achieve the sth success has a $Beta - Pascal$ as predictive distribution, denoted $n \sim Bpas(\alpha, \beta; s)$, of density:

$$\Pr(n = j) = \binom{j-1}{s-1} \frac{B(\alpha + s, \beta + j - s)}{B(\alpha, \beta)}, \quad j = s, s+1, \ldots.$$

It has, as mean,

$$\mu_{pred} = s(\alpha + \beta - 1)/(\alpha - 1),$$

and, as variance,

$$\sigma^2_{pred} = s\beta(s + \alpha - 1)(\alpha + \beta - 1)/\left[(\alpha - 1)^2(\alpha - 2)\right].$$

The relation between the two cumulative distribution functions is: $F_{Bpas}(n|\alpha, \beta; s) = F_{Bbin}(s|\alpha, \beta; n)$ and both distributions belong to the *General Hypergeometric* family of discrete distributions.

For a *single observable*, we also have, respectively, the *Conjugate Posterior Predictive distribution* and the *Reference Posterior Predictive distribution*

a)

$$Bbin(\alpha + r, \beta + n - r; 1), \quad \text{and}$$

$$Bbin\left(\frac{1}{2} + r, \frac{1}{2} + n - r; 1\right), \quad \text{for Bernoulli sampling, and}$$

b)

$$Bpas(\alpha + s, \beta + n; s), \quad \text{and}$$

$$Bpas\left(s, \frac{1}{2} + n; s\right), \quad \text{for Pascal sampling.}$$

B.   Uses of the Predictive Distribution

One important use of the predictive distribution is for the elicitation of the prior beta since observables are usually easier to deal with than parameters. This has been the argument of proponents of the use of predictive distributions, which are based on de Finetti

Theorem, to elicit the prior distribution, with the case of the beta included. "*The major advantage of asking about predictive distributions rather than asking directly about prior distributions is that potentially observable statistics are often easier to understand and relate more directly to an expert's knowledge than unobservable parameters* " stated Winkler (1980), who proposed using consecutive values of $n = 1, 2$, to assess the probability of k successes in n trials. This corresponds to the values $\alpha/(\alpha + \beta)$ for $n = k = 1$, and $\beta(\beta + 1)/[(\alpha + \beta)(\alpha + \beta + 1)]$, $2\alpha\beta/[(\alpha + \beta)(\alpha + \beta + 1)]$ and $\alpha(\alpha + 1)/[(\alpha + \beta)(\alpha + \beta + 1)]$ for $n = 2$ and $k = 0$, 1 and 2, respectively. The parameters α and β can then be obtained from these expressions. He further recommended that: "*It seems best to ask about relatively simple statistics from small samples*".

Moreover, information obtained from the expert on the predictive distribution can also provide some flexibility in terms of model building since we can contemplate different combinations of likelihood functions and prior distributions, such that the statistics given by the corresponding predictive distribution would best fit the expert's given values. As an example, Winkler (1980) gave the choice between a *beta(13,13)* prior with Bernoulli sampling, and a Markov model in its steady-state, with a degenerate matrix-beta prior distribution.

Chaloner and Duncan (1983) also used the beta-binomial, but were essentially concerned with the mode. Their method is called *predictive modal* and uses an interactive computer program. Adjustments are made by asking the expert whether a $50p.c.$ predictive interval is too long or too short.

In Poisson sampling the natural conjugate prior is the exponential distribution, which is generally used. However, if for some reason the parameter can be considered as a beta variable of type1 or type2, then the predictive distribution will be the discrete *Beta-Poisson* distribution of type1, or type2, whose density can be readily derived. The *beta-binomial* distribution has an interesting application in *Bayesian Quality Sampling*, as presented in Section XI, beside its use to compute the expected value of random quantities whose values depend on the posterior distribution, as already mentioned. Geisser

(1993) can be consulted for more results on prediction in general.

VII. Preposterior Analysis

The posterior is well defined once the sample results are known. However, prior to taking the sample (and settle with its results) the posterior, and its parameters and characteristics, can be consider as random, depending on sampling results, i.e on data. In the case of a beta prior, with Bernoulli sampling of fixed size n, we have:

$$E_R\left(\mu_{post}^{(n,r)}\right) = \mu_{prior}$$

$$E_R\left(Var_{post}^{(n,r)}\right) = \frac{\alpha+\beta}{\alpha+\beta+n}Var_{prior}$$

$$Var\left(\mu_{post}^{(n,r)}\right) = \frac{n}{\alpha+\beta+n}Var_{prior}, \text{ and}$$

$$Var_R\left(Var_{post}^{(n,r)}\right) = \frac{A}{B} - \left[E\left(Var_{post}^{n,r}\right)\right]^2,$$

with

$$A = \left\{m_4 + 2(\alpha-n-\beta)m_3 + \left[\alpha^2 + (n+\beta)(\beta+n-4\alpha)\right]m_2\right\}$$
$$+ \left\{2\alpha(\beta+n)(\beta+n-\alpha)m_1 + \alpha^2(\beta+n)^2\right\}$$

and

$$B = (\alpha+\beta+n)^4(\alpha+\beta+n+1)^2,$$

where $m_i = E_R[(Var_{post}^{(n,r)})^i]$, $i = 1,\ldots,4$. These results can be directly established (Pham-Gia and Turkkan, 1992c), or they are applications to the beta of more general results on conjugate families (see Dickey (1982)). They can be used in several contexts, especially when various scenarios have to be considered in the planning a *Bayesian experiment*, and, in particular, in *Bayesian Decision Theory*, when a quadratic loss function is used. Raiffa and Schlaifer (1961) should be consulted for more details. Furthermore, it can be established, in *Bayesian sequential analysis*, that, for n fixed, if r_0

has been observed, then the expected value of $Var_{post}^{(n+1,r)}$ is

$$E_R\left(Var_{post}^{(n+1,r)}\right) = \frac{\alpha + \beta + n}{\alpha + \beta + n + 1}Var_{post}^{(n,r_0)}. \tag{VII.1}$$

VIII. Bayesian Decision Theory

A. Statistical Decision Theory

Statistical Decision Theory is a well-developed subfield of statistics and a *decision-theoretic* approach can be taken in most topics of classical statistics. Ferguson (1967) is a concrete example. We consider a *parameter space* Ω (also called *state of nature*) and a *decision space* D. For any value $\theta \in \Omega$, which determines a probability distribution P, and any decision $d \in D$, let $\gamma(d, \theta)$ be the "*consequence*" of choosing d when the parameter has value θ. The decision d is chosen on the basis of an observation $x \in X$, subsequent to an experience e. The subjective notion of *utility* is usually used to deal with that consequence. The real-valued function $U(\theta, d)$ is called an *utility function* if, for two distributions P_1 and P_2, P_1 is not preferred to P_2 if and only if $E(U|P_1) \leq E(U|P_2)$. Hence, between two distributions, we would prefer the distribution for which the expected utility of the consequence, or $E(U|P, d)$, is the larger one. If P is fixed we then choose the decision d^*, called Bayes decision, such that $E(U|P, d^*)$ would be maximum. *Bayesian Decision Theory* is the Bayesian approach to *Statistical Decision Theory*, and introduces a prior distribution on θ. The basic decision problem can be presented as a decision graph, where there are two decision nodes and two random nodes, and is worked on backward to maximize the utility function (see Lindley (1972)), as follows:

$$\max_{e} \int_X \left[\max_{d} \int_\Theta U(d, \theta, e, x)p(\theta|x, e)p(x|e)d\theta\right] dx,$$

where U is the utility function.

Alternately, the *loss function* $L(\theta, d)$ is the penalty of making decision d when the value of the parameter is θ. The expected loss is called the risk $E(L|P, d)$, and we should minimize that risk for a given value of P. This duality between utility and loss functions is very convenient and permits looking at any decision problem from two points of view. A comprehensive comparative table is provided by Raiffa and Schlaifer (1961), and for the sample size problem, by Pham-Gia (1997).

B. Admissibility

This notion plays an important role since it selects the best decision, according to an effectiveness criterion. Loosely speaking, a decision is admissible only if there is no decision strictly more effective than itself. *Bayes decisions* can be proved to be admissible when the prior is proper, although admissibility is often not accepted as part of the Bayesian approach since it violates the *Likelihood principle*. However, the converse, that every admissible rule is the Bayes rule for some prior distribution, is generally not true.

C. Bayesian Decision Criteria

When the decision itself is the choice of the value of the parameter, i.e the Decision space and the Parameter space are identical, we have *Bayesian Estimation Theory*. The loss function approach seems to be more prominent in recent years, possibly due to lesser requirement in subjective personal input than the use of utility functions. Although loss functions can be of any form, complex ones naturally leads to mathematical intractability and the solution has to be worked out numerically. An estimation, denoted a here, of the parameter θ is made, based on our knowledge of that parameter, represented by the prior distribution $F(\theta)$, on a loss function $L(\theta, a)$ and on a density $p(x, \theta)$, of which θ is the parameter. Among the most convenient loss functions used, let's note the *quadratic loss function* $L(\theta, a) = K_q(\theta - a)^2$ and the *linear loss function* $L(\theta, a) = K_u(\theta - a)$ for $\theta \geq a$ and $L(\theta, a) = K_l(a - \theta)$ for $\theta \leq a$, where K_q, K_u and K_l are positive constants. A particular case of the linear loss function is the *absolute value loss function* $L(\theta, a) = K|\theta - a|$, which is intimately related to the *absolute distance* and the L_1-*approach* in statistics (Pham-Gia, 1995b), and has the clear advantage of being more robust than the quadratic one. Naturally, the best estimate of θ is the one that minimizes the average value of the loss function over data, denoted $R(\theta) = E_X[L(\theta, a)]$, called the *risk function*. With that value of the estimate, denoted a^*, the corresponding value of the expected minimum risk (w.r.t. the prior), $E_\theta(R(\theta))$, called the *Bayes risk*. In the absence of any information X, it is the *Expected Value of Perfect Information* (denoted *EVPI*), since this is average value attributed to the information required, at the start, to obtain perfect knowledge.

In the presence of observations, we change the order of integration in the derivation of the *Bayes risk*. A random sample of observations $X = (X_1, \ldots, X_n)$ gives rise to the posterior distribution of θ, denoted $g(\theta|X)$, and a revised estimation of θ using the same loss function. Again, the optimal estimation a^{**} of θ is the one that minimizes the average value of the loss function over the posterior distribution, or posterior risk $E_{\theta|X}[L(\theta, a^{**})]$. The expected value of this posterior risk, w.r.t. the marginal distribution of X, is the *Bayes risk*, denoted by $\rho(n)$, and depends only on n. However, prior to taking a sample, in preposterior analysis, we are interested in knowing the average worth of a sample of size n. The reduction in risk, represented by the difference between the prior risk and the above average posterior risk, or $EVPI - \rho(n)$, is that average worth. It is also the maximum value attributed to a sample of size n, in order to acquire more information, and reduce the uncertainty present in the prior distribution. This difference is called the *Expected Value of Sample Information*, or $EVSI(n)$.

If we consider the real cost of sampling, $C(X_1, \ldots, X_n)$ for a sample of size n, the *Expected Net Gain in Sampling* (denoted $ENGS(n)$) is just the difference between $EVSI(n)$ and the expected cost $E(C(x_1, \ldots, X_n))$. As long as $ENGS(n)$ is positive, it is advantageous to sample. $ENGS(n)$ is usually a concave function presenting a maximum at n_0, which is the optimal sample size, at which point we would gain the most out of sampling (see Raiffa and Schlaifer (1961) for an in-depth treatment of those topics).

D.   Quadratic Loss Function

For the case of a quadratic loss function of the form $L(\theta, a) = K_q(a - \theta)^2$, $K_q > 0$, the above quantities are well-known, and are related very naturally to the means and variances of different distributions. For convenience, and without loss of generality, we will suppose $K_q = 1$. We then have, with the notation from Subsection C:

$a^* = $ Mean of prior distribution (denoted μ_{prior}),and
$EVPI$: Variance of prior distribution (denoted Var_{prior}).
The posterior distribution of θ being $f(\theta|X)$, we have:
$a^{**} = $ Mean of posterior distribution (denoted $\mu_{post}^{(n,x)}$)
Posterior risk $=$ Variance of posterior distribution
$$(\text{denoted } Var_{post}^{(n,x)})$$
Bayes risk (denoted $\rho(n)$) $=$ Expectation w.r.t. X of the
 Posterior risk $= E_X(Var_{post}^{(n,x)}))$
$EVSI(n) = EVPI - \rho(n) = Var(\mu_{post}^{(n,x)})$, with the last part of this
double equality easy to establish, and
 $ENGS(n) = EVSI(n) - E(C(X_1, \ldots, X_n))$.

For the beta as prior and binomial sampling, formula (of Section VII) above apply to the above quantities (see Pham-Gia and Turkkan (1992c)). In the Bayes sequential approach to estimation with the same loss function, very similar considerations apply in the determination of the optimal stopping rule (see De Groot (1970)).

E.   Absolute Value Loss Function

Let us consider a random variable X with distribution $F(x)$ and finite mean μ. Its median $Md(X)$ is defined as the value such that $\int_{-\infty}^{Md(x)} dF(t) = 0.5$, or $F^{-1}(Md(X)) = 0.5$, where F^{-1} is the quantile function, but $Md(X)$ is not necessarily unique. The *mean absolute deviation* of X about its mean is

$$\delta_1(X) = \int_{-\infty}^{\infty} |x - \mu| dF(x)$$

(see Stuart and Ord (1987)), and about its median, is

$$\delta_2(X) = \int_{-\infty}^{\infty} |x - Md(X)| dF(x)$$

$$= \int_0^{\frac{1}{2}} \left[F^{-1}(1 - \alpha) - F^{-1}(\alpha) \right] d\alpha.$$

Using Liapunov's inequality, we always have: $\delta_2 \leq \delta_1 \leq \sigma$, where σ is the standard deviation. For the normal $N(\mu, \sigma^2)$,

$$\delta_1(X) = \delta_2(X) = \sqrt{\frac{2}{\pi}} \sigma = 0.7978\sigma.$$

Let's now consider a distribution $f(x|\theta)$ with its parameter θ having prior distribution $F(\theta)$ and loss function $L(\theta, a) = |\theta - a|$. We have: $a^* = $ Median of prior distribution (denoted Md_{prior}), and $EVPI = \delta_{2,prior}(\theta)$: Mean absolute deviation of prior distribution. With the posterior distribution $f(\theta|X)$, we have: $a^{**} = $ Median of posterior distribution (denoted $Md_{post}^{(n,x)}$). Posterior risk = Mean absolute deviation of posterior distribution $\delta_{2,post}^{(n,x)}(\theta)$, and Bayes risk = $\rho(n) = E_X(\delta_{2,post}^{(n,x)}(\theta))$. Due to the difficulty in obtaining the value of the median in closed form, δ_2 itself is not often available in closed form, but δ_1 frequently is, and a table of δ_1 for common distributions is given in Pham-Gia and Turkkan (1994b). For a general survey of up-to-date results available on δ_1 and δ_2, and their sample estimations, see Pham-Gia and Tranloc (2001).

Finally, for the *zero-one loss function*, we have: $L(\theta, a) = 0$ if $|\theta - a| \leq \varepsilon$ and $L(\theta, a) = 1$ if $|\theta - a| > \varepsilon$, where ε is a constant. Then $a^* = $ *mode* of the prior, and $Mo_{prior} = (\alpha - 1)/(\alpha + \beta - 2)$, for the unimodal beta type1. The posterior risk is then $\Pr(|\theta - Mo_{post}^{(n,r)}| > \varepsilon)$, which can be computed numerically, as well as the corresponding Bayes risk. Hence, all the three measures of central tendency can arise as Bayes estimators.

IX.   Sequential Bayes, Empirical Bayes

A.   Sequential Bayes

The *sequential approach* in Classical statistics can be applied in Bayesian statistics and the *Stopping rule principle*, which follows from the Likelihood principle, is a logical consequence of the Bayesian approach. The beta again plays an important role when dichotomous variables are considered. Basically, here, we take one observation at a time, and continuously compare the expected cost of taking the next observation (or its Bayes risk) with the cost of making an immediate decision. We only proceed if the former is less than the latter. More precisely, let (τ, δ) be a *sequential sampling* plan, with τ being the *stopping rule* and δ the *decision rule*. In the case of the proportion π, let $beta(\alpha, \beta)$ be its prior, and let r be the number of "successes" out of n observations. Under a quadratic loss function of the form $(\theta - a)^2$, the truncation value of N can be determined first, so that the sample size will not exceed N. Let c be the cost of one observation and let's suppose $C(X_1, \ldots, X_n) = nc$ and $Var_{post}^{(n,r_0)}$ be the posterior risk when we observed r_0 successes in n observations. By (VII.1) we have

$$E_R(Var_{post}^{(n+1,r)} | Var_{post}^{(n,r_0)}) = \frac{\alpha + \beta + n}{\alpha + \beta + n + 1} Var_{post}^{(n,r_0)}$$

and hence, a decision rule can be obtained by setting

$$\frac{\alpha + \beta + n}{\alpha + \beta + n + 1} Var_{post}^{(n,r_0)} + c \leq Var_{post}^{(n,r_0)}.$$

Berger (1985, page 442) and Pham-Gia (1998) has studied this case and Ferguson (1967, pages 319-324) has computed the distribution of $X | Var_{post}^{(n,r_0)}$, and obtained the decision associated with each value of the new observation x_{n+1}, when $\alpha = \beta = 1$. He also considered the loss function $K(\theta - a)^2 / [\theta(1 - \theta)]$ which provides an interesting application. Berger (1985, Chapter 7) should be consulted for more applications, and de Groot (1970), for more general discussions on this topic. Govindarajulu (1987) is also an excellent source of reference. Concerning the *mixture proportion* π of 2 normal distributions, of the form:$\pi f_1 + (1 - \pi) f_2$, with

$$f_i(x) = \exp\left\{-(x - \mu_i)^2 / 2\sigma_i^2\right\} / \left(\sigma_i \sqrt{2\pi}\right), \quad i = 1, 2,$$

Titterington et al. (1985) presented a sequential sampling scheme based on a prior beta distribution for π, by which new observations are classified into one of the 2 populations, depending on the value of a chosen criterion. The posterior is, hence, constantly updated, but has to be approximated by another beta every time, to avoid computational difficulties. Other criteria were also considered.

B.   Empirical Bayes

Empirical Bayes is concerned with the case where the prior is not known explicitly, only as possibly belonging to a family of distributions, and has to be obtained from data or by another mean. Due to Robbins (1955), it originally dealt with the *Compound decision problem*, and data could be either current or past. It is further divided into *Parametric Empirical Bayes*, where the family of distributions is known but not its *hyperparameters*, and *non-parametric Empirical Bayes*, where no other information is available. In this case, it constitutes a class of decision theoretic procedures that use past or current data to bypass the necessity of identifying the unknown prior, which has a *frequency interpretation*, however. The use of the beta distribution in *Parametric Empirical Bayes*, if needed, reduces to estimating its parameters using data (Myhre and Rennie, 1988; Maritz and Lwin, 1989, pages 88-102) and will proceed subsequently as in previous chapters (see Waller and Martz (1982, Chapter 13)). To illustrate the use of EB methods, Casella (1985)used the beta to model the distribution of true purchasing intentions. Although Empirical Bayes has certain similarities with Bayesian methods in general, and has relations with important topics in Statistics, such as the *Stein estimator*, it has been frequently argued that Empirical Bayes methods should not be considered as Bayesian.

X. Mixtures of Betas

A. Discrete Mixture Distributions

Mixture distributions play an important role in many statistical applications. Let's consider a *discrete mixture* (or finite mixture) of several densities of the form: $g(\pi) = \sum_{i=1}^{m} \omega_i f_i(\pi)$ with $\sum_{i=1}^{m} \omega_i = 1$. The density $g(\pi)$ can be taken as the prior distribution of the parameter π, from which the posterior will be derived. If each f_i is a beta of type1, $beta(\alpha_i, \beta_i)$, $i = 1, \ldots, m$, we have a finite mixture of betas, which can present one, or several modes. For Bernoulli sampling, the posterior density $g(\pi|n, r)$ can be obtained very simply by using the conjugate property of each individual beta, with, however, changes in the posterior weights. We have

$$g(\pi|n, r) = \sum_{i=1}^{m} \omega_i^* beta\left(\pi|\alpha_i + r, \beta_i + n - r\right)$$

with

$$\omega_i^* = A_i \Big/ \sum_{j=1}^{m} A_j$$

where

$$A_i = \omega_i \frac{Beta\left(\alpha_i + r, \beta_i + n - r\right)}{Beta\left(\alpha_i, \beta_i\right)}, i = 1, \ldots, m.$$

In general, we can prove that

$$A_i = \omega_i predf_i \Big/ \sum_{j=1}^{m} predf_i$$

for the posterior mixture $g(\pi|X)$, where $predf_i$ refers to the predictive distribution associated with f_i alone as prior.

For the predictive distribution of X, however, the associated weights remain the same as ω_i, i.e. the predictive distribution of π in the above case, is

$$\psi(x|n) = \sum_{i=1}^{m} \omega_i Bbin_i \left(x|\alpha_i, \beta_i; n\right).$$

When considering only two densities, the posterior weights in the mixture can also be determined from the Bayes factor in favor of f_1, which is the ratio of the posterior odds to the prior odds. We have: $BF = \frac{w^*}{1-w^*} / \frac{w}{1-w}$. Since we also have

$$BF = \int f(x|\theta)f_1(\theta)d\theta / \int f(x|\theta)f_2(\theta)d\theta,$$

by equaling these two expressions we have the same expression of w^*, as obtained previously.

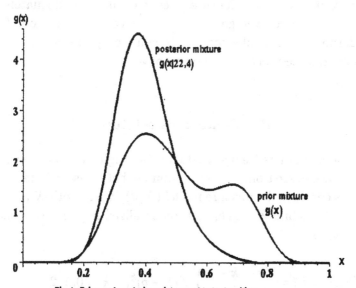

Fig.1: Prior and posterior mixtures of beta densities

Figure 1 shows the mixture of 3 betas,

$$g(x) = 0.45 beta(x; 12, 20) + 0.25 beta(x; 17, 17)$$
$$+ 0.30 beta(x; 22, 10)$$

and the posterior mixture, for $n = 22$ and $r = 4$, i.e. $g(x|22, 4)$. Similarly, with Pascal sampling, where s is fixed, and n is observed, $n > s$, and with the above mixture prior, we have, as posterior,

$$g(\pi|s, n) = \sum_{i=1}^{m} \omega_i^* beta_i (\pi|\alpha_i + s, \beta_i + n - s), n \geq s,$$

with $\omega_i^* = B_i / \sum_{j=1}^{m} B_j$, where

$$B_i = \omega_i \frac{Beta (\alpha_i + s, \beta_i + n - s)}{Beta (\alpha_i, \beta_i)}, i = 1, \ldots, m.$$

Mixtures of beta distributions could be very useful in many respects. First, since they can display one or several modes, they can be used to approximate complex densities. Second, they can adequately represent the weighted average of several sources of information. Third, with Bernoulli or Pascal sampling, they are easy to use, and provide solutions in closed forms, as shown above.

B.   Continuous Mixtures

The beta type1 is frequently taken as weight in *continuous mixtures*, as we can see from the beta-binomial, beta-Pascal, beta-normal, etc. Let's consider the mixture model (X, θ), where both X and θ are random. We define $X \sim Nb_1(r, p)$ (generalized negative binomial) if its density is

$$\Pr(X = k) = \binom{r + k - 1}{k} p^r (1 - p)^k, \quad k = 0, 1, 2, \ldots$$

and $Y \sim GH_{IV}(A, B, C)$ (generalized hypergeometric) if

$$\Pr(Y = k) = \frac{\binom{A}{k}\binom{B}{C-k}}{\binom{A+B}{C}}, k = 0, 1, \ldots.$$

There are some interesting results on identifiability related to the beta of both types, as follows (Papageorgiou and Wesolowski, 1997): Let $\mu_{X|\theta}$ denote the conditional mean of X, given θ. If $\mu_{X|\theta}$ is $Nb_1(r, 1 - \theta)$ and $E(\theta|X) = \frac{X+a}{X+a+b+r}$, $a, b, r > 0$, then the prior distribution of θ is $beta(a, b)$ and $X \sim GH_{IV}(-a, a + b - 1, -r)$. A similar result holds for the beta type2, and hence, under the above hypotheses, the posterior mean can identify the prior beta distribution of θ. In the same spirit, Gupta and Wesolowski (2001) characterized completely the family of models with linear posterior means.

XI. Bayesian Applications of the Beta in Other Fields

Several fields of applied sciences have successfully applied Bayesian techniques. Among them, let us mention *Bayesian Reliability, Bayesia Quality Sampling Plans* and *Pattern recognition*, where the beta distribution plays a certain role.

A. Reliability

This field was basically concerned with the life distribution of a component, and of a system, through a determined configuration, and the probability that it functions at a certain time t_0. In the last fifty years, it has grown into a very dynamic field, which includes now other subdomains such as *Availability, Preventive Maintenance, Optimal Design* and *Reliability Growth*. It has maintained close relations with other fields of Statistics, since several concepts used there share the same statistical basis and structures. In relation to the beta, its basic use there, is still as a prior for the reliability of a component (or system), or as its approximate distribution, obtained from binomial data. Myhre and Rennie (1988), for example,used Empirical Bayes methods for the second purpose. Whenever the structure is too complex, a beta density can be used to represent the reliability of a each substructure, and the system reliability can be evaluated from the configuration and the substructure betas, leading to the system "*induced prior beta*", which can then be combined with the system "*native prior beta*", using an accepted weight. Combination can be carried out here, according to Winkler's proposed approach (1968), in order to obtain another approximate beta, now considered as system prior. Binomial sampling, to be performed at system level, leads to the corresponding posterior beta for system reliability (Waller, Martz and Fickas, 1988).

Components are usually supposed independent and various models of dependence have been proposed. For example, considering Common cause failures, Hokstad (1988) proposed a *Random Probability Shock* (RPS) model, where independent failures occur, in addition to group failures, and there are various degrees of dependence in the outcomes of shocks. The number of components that fail has a binomial distribution with probability π, which, itself, has a beta prior distribution.

The Bayesian approach to Reliability (Waller and Martz, 1982) has given rise to a multitude of results and continues to attract the interest of researchers.

B.   Pattern Recognition

One of the basic questions in *Statistical Discriminant Analysis* and *Pattern Recognition* is the establishment of a criterion to classify a new observation as belonging to one among two, or several, given distributions. The Bayes error is defined as the minimum error in classification. For the two-distribution case in R, it is given by $\int_R min\{\pi f_1(\theta), (1 - \pi)f_2(\theta)\}d\theta$, where π is the prior probability assigned to f_1. In the Bayesian approach to estimating π, a beta distribution can be used as its prior distribution, as in Pham-Gia,

Turkkan and Bekker (2002), or as in Titterington et al. (1985). Duda, Hart and Stork (2001) can be consulted for more advanced questions related to *Pattern Recognition*.

C. Bayesian Quality Sampling Plans

Sometimes, in industrial production, *between batches* quality variation is greater than *within batch* variation, and random fluctuations in sampling come from the variability inherent to the process, as well as from the random selection of one item within the batch. Whereas in classical sampling plans we suppose that the process quality parameter is a fixed constant, and there is only one source of variation, in *Bayesian sampling plans*, there are variations from batch to batch, and within batches. Hence, the process quality, reflected by the proportion of defective items, is a random variable, having, in general, a beta distribution, taken as the prior distribution. Here, the prior beta is usually obtained from a combination of historical data and subjective judgment.

Decision on either to accept or to reject a batch is based on the number of defectives in a sample taken from that batch and, on either statistical approach: the beta posterior distribution of the process quality within the sample, or the predictive *beta-binomial* (or *Polya*) distribution of batch items. The parallels in conventional sampling plans are, respectively, the hypergeometric distribution (when we consider a batch with a finite number of elements and, within a sample taken from that batch, the probability of detecting a defective item changes after the removal of the previous defective item), and the binomial distribution (where that probability does not change).

If the posterior beta is used, it can be discretized in order to find the probability of obtaining k "successes". Hence, for p and n chosen, and with c as the acceptance number, we can take

$$\Pr(Accept.)$$
$$= 1 - \sum_{j=c+1}^{n} \frac{\Gamma(\alpha + \beta + n)}{\Gamma(\alpha + j)\Gamma(\beta + n - j)} p^{\alpha+j-1}(1 - p)^{\beta+n-j-1}.$$

When using the beta-binomial, we have a nice interpretation of the variance of the proportion of defectives as (approximately) the sum of the process variance (of the prior beta) and the sampling binomial variance (using the prior mean as the probability of "success"), as expected: Taking $p = x/n$, from (VI.1) we have,

$$\sigma^2(p) = \frac{\alpha\beta}{n(\alpha+\beta)^2} \frac{(\alpha+\beta+n)}{(\alpha+\beta+1)}. \qquad \text{(XI.1)}$$

With $p^* = \mu_{prior} = \alpha/(\alpha+\beta)$ taken as the probability of "success", we then have:

$$\sigma^2(p) = Var_{prior} + \frac{p^*(1-p^*)}{n} \frac{(\alpha+\beta)}{(\alpha+\beta+1)}.$$

Once the prior distribution is adopted, *Operation Characteristic* (OC) curves for *Bayesian acceptance sampling*, for either approach, can be drawn, from which the *Accepted Quality Level* (AQL) and the *Rejectable Quality Level* (RQL) can be derived. Both families of curves depend on n, the sample size, and c, the acceptance number. If the posterior beta is used, the curves also depend on α and β and the batch size N, and is function of the batch real proportion of defectives. If the beta-binomial is used, they depend on the ratio of the Polya variance to the binomial variance of p, $R = \frac{\alpha+\beta+n}{\alpha+\beta+1}$, obtained from (XI.1), and is function of the process proportion of defectives.

Bayesian sampling plans can be combined with economic factors to arrive at cost optimal Bayesian plans, and classical tools from *Bayesian Decision theory* can be used here. The interested reader should consult Calvin (1984) for more details, and he/she is referred to Hald (1981) for a more advanced treatment of this topic.

D. Bayes Operator

The Bayes paradigm hence depends on two components: the prior and the sampling model. The prior and the sampling scheme could belong to mathematically similar, or completely different, families of parametric distributions, i.e. there could be conjugacy or not. However, in all cases, we have a prior distribution mapped into a posterior one, and, under some restrictive hypotheses on the sampling phase, we can define an operator from the space of prior densities to the one of posterior densities. Cuevas and Sanz (1988), for example, have considered such an approach. Although this approach could be highly theoretical, based, for example, on *Fréchet derivatives*, the L^1-metric can be used to measure *distances* between different densities, and between the priors and posteriors. Also, the use of conjugacy can also help make the study simpler. Under some hypotheses, the very important topic of *Bayesian Robustness*, can be tackled with this approach, and the beta distribution is particularly useful for such a quantitative study of the Bayes operator (see Pham-Gia, Turkkan and Bekker (2002)).

XII. Sum, Difference, Product and Quotient of Independent Betas

Let $\pi_1 \sim beta(\alpha_1, \beta_1)$ and $\pi_2 \sim beta(\alpha_2, \beta_2)$ be 2 independent beta variables of type1. The expressions of the densities of $W_1 = \pi_1 + \pi_2$, defined in $[0, 2]$, $W_2 = \pi_1 - \pi_2$, defined in $[-1, 1]$, $W_3 = \pi_1\pi_2$, defined in $[0, 1]$ and $W_4 = \pi_1/\pi_2$, defined in $[0, \infty]$ are now available in closed form. Similarly, expressions for the densities of these four variables when the more general form of the beta-prime, the G3F (see (I.4)), is considered, are also available. Pham-Gia (2000) and Pham-Gia and Turkkan (1993, 1994a and 2002a) should be consulted for details, since the expressions of the densities of these variables are too cumbersome to be reproduced here. A computer code, developed for hypergeometric series by our research group, permits, however, the efficient handling of these derived densities.

: If π_1 and π_2 are the parameters of two separate independent Bernoulli sampling schemes then, with (n_1, r_1) and (n_2, r_2), we can obtain the expressions of the posterior distributions of W_1, W_2, W_3 and W_4. There are several potential applications of these expressions. For example, in Reliability, $\Pr(W_2 \geq 0)$ provides the reliability of a *stress-strength* system, when both stress and strength are beta distributed. Subsequent to some engineering improvement process, and the gathering of the associated binomial test data, the posterior distribution of W_2 can be computed, and a reevaluation of the system reliability can be undertaken. Other applications include the derivation of the distributions of complex test statistics, as well as those of quadratic forms, and the determination of sample sizes in the Bayesian study of the difference of two independent parameters (Pham-Gia and Turkkan, 2003). Extensions of the above operations are being carried out, by other authors, to multivariate beta, and matrix-variate beta distributions. Applications of these results in complex Physics, Computer, Engineering and Social Science problems can be anticipated.

XIII.   Distributions Derived from the Beta

Several distributions directly derived from the betas of both types, are encountered in the literature. Some are obtained by a change of variable, or by multiplication of the beta density with a mathematical expression. Others are obtained by continuous mixing, using the beta as mixing weight. Their uses in Bayesian statistics are, for the present time, quite limited, but below are some, with more potential to be used there, thanks to the nature of their mathematical expressions.

1. The Inverted Beta of the first kind: Y is defined on $(1/b, 1/a)$ by

$$f(y; a, b; \alpha, \beta) = \frac{(1 - ay)^{\alpha-1}(by - 1)^{\beta-1}}{B(\alpha, \beta)(b - a)^{\alpha+\beta-1}y^{\alpha+\beta}},$$

with $0 \le a < b$ and $\alpha, \beta > 0$. It is a generalization of the *ibetap* (see (I.5)), which is its particular case when $a = 0$.

2. The Non-central Beta of first type, defined on $(0, 1)$

$$f(t; \alpha, \beta, \lambda) = beta(t; \alpha, \beta)\Phi(\lambda t, \alpha + \beta, \alpha)/\exp(\lambda),$$

with $\alpha, \beta, \lambda > 0$ and the Non-central Beta of second type, defined on $(0, \infty)$ by:

$$f(t; \alpha, \beta, \lambda) = \frac{\exp\left(-\lambda^2/2\right) x^{\alpha-1}}{B(\alpha, \beta)(1+x)^{\alpha+\beta}} \Phi\left(\alpha + \beta; \alpha; \frac{\lambda^2}{2} \frac{x}{1+x}\right)$$

with $\alpha, \beta, \lambda > 0$, where Φ is *Kummer confluent hypergeometric function*.

3. The $Hyperbeta((a), (b); \alpha, \beta; \lambda)$, defined on the unit interval, with density

$$f(y; (a), (b); \alpha, \beta; \lambda)$$
$$= beta(y; \alpha, \beta) \frac{{}_{p+1}F_{q+1}\left(([(a), \alpha + \beta]); ([(b), \alpha]); \lambda y\right)}{{}_pF_q\left((a); (b); \lambda\right)},$$

where the generalized hypergeometric function ${}_pF_q$ is defined by:

$$_pF_q((a); (b); z) = \sum_{k=0}^{\infty} \frac{((a), k) z^k}{((b), k) k!},$$

where $(a) = (a_1, \ldots, a_p)$ and $(b) = (b_1, \ldots, b_q)$. For $(c) = (c_1, \ldots, c_m)$,

$$((c), k) = \prod_{i=1}^{m} (c_i, k).$$

We define:

$$(a, k) = a(a+1) \cdots (a + k - 1) = \Gamma(a + k)/\Gamma(a),$$

the ascending factorial, or Pochhammer coefficient, with $(a, 0) = 1$, by definition, and $([(a), c]) = (a_1, \ldots, a_p, c)$. Gauss hypergeometric function ${}_2F_1$, is a particular case. Hence, $Y \sim Hyperbeta((a), (b); \alpha, \beta; \lambda)$ is obtained by setting $Y|X = x \sim beta(y; \alpha + x, \beta)$, with $X \sim Hyp((a), (b), \lambda)$, i.e. X has as discrete density

$$p_\lambda(x) = \frac{1}{{}_pF_q((a); (b); \lambda)} \frac{((a), x)}{((b), x)} \frac{\lambda^x}{x!}$$

with $x = 0, 1, 2, \ldots$.

4. Fisher's Z-distribution, defined on $(-\infty, \infty)$:

$$f(z; \nu_1, \nu_2) = \frac{2(\nu_1/\nu_2)^{\nu_1/2}}{B(\nu_1/2, \nu_2/2)} \frac{\exp(\nu_1 z)}{\{1 + (\nu_1/\nu_2)\exp(2x)\}^{(\nu_1+\nu_2)/2}}$$

with $\nu_1, \nu_2 > 0$.

5. The folded Student distribution defined in $(0, \infty)$:

$$f(t; \nu) = \frac{2}{\sqrt{\pi\nu}} \frac{\Gamma\left(\frac{\nu+1}{2}\right)}{\Gamma\left(\frac{\nu}{2}\right)} \frac{1}{\left(1 + \frac{t^2}{\nu}\right)^{(\nu+1)/2}},$$

with $\nu > 0$.

XIV. Generalizations of the Beta

The beta has been generalized in different ways. A partial list of these generalizations is given by Pham-Gia and Turkkan (1998). In this section, we present the most frequently encountered forms of the generalized beta, again, under 4 successive parts: prior, sampling, posterior and predictive, as in previous sections.

A. The Generalized Beta as a Prior

i. Generalizations in One Variable

We will only consider generalizations which are conjugate to Binomial or Pascal sampling here, i.e. whose posterior w.r.t. these samplings can be obtained in closed form. This is the case of most derived distributions in the previous section, which are obtained by simple multiplication of the standard beta with another function, with the normalizing constant usually obtained by using hypergeometric functions.

1. The more general form of the beta is defined on any interval $[c, d]$ of R and its density is given by (I.2). This general

beta distribution in 4 parameters is well utilized in *Operations research* and its properties are studied by Whitby (1972). It contains the *Beta warning time* distribution as a particular case when $c = 0$, *distended beta* distribution when $\beta = 1$, and *Mielke beta-kappa* after re-parameterization. (Johnson, Kotz and Balakrishnan, 1995, volume 2, page 350). The density of a linear combination of random variables having independent general beta distributions is obtained by Pham-Gia and Turkkan (1998), and the densities of their product and quotient by Pham-Gia and Turkkan (2002b).

2. The Generalized beta in 3 parameters $G3B(\alpha, \beta, \lambda)$, defined on $(0, 1)$, as introduced by Chen and Novick (1984), has its density under the convenient form:

$$f(y; \alpha, \beta, \lambda)$$
$$= \frac{beta(\alpha, \beta)}{{}_2F_1(\alpha; \alpha + \beta, \alpha + \beta; 1 - \lambda)} \frac{1}{[1 - (1 - \lambda)y]^{\alpha + \beta}} \quad \text{(XIV.1)}$$

with $0 \leq y \leq 1$ and α, β, $\lambda > 0$ (the normalizing constant here, equals λ^{α}). It is sometimes called the Euler distribution (Thompson and Haynes, 1980). We can see that $G3B$ is a special case of the general family of $G(2n + 2)B$ distributions in one variable, studied by Pham-Gia and Duong (1989), which is conjugate to Bernoulli sampling, and has as density in $(0, 1)$:

$$f(y; \alpha, \beta; \gamma_1, \ldots, \gamma_n; \delta_1, \ldots, \delta_n)$$
$$= \frac{beta(\alpha, \beta) \prod_{j=1}^{n} (1 - \delta_j x)^{\gamma_j - 1}}{F_D^{(n)}(\alpha; 1 - \gamma_1, \ldots, 1 - \gamma_n; \alpha + \beta; \delta_1, \ldots, \delta_n)},$$

where the Lauricella D-function in n variables is

$$F_D^{(n)}(x_1, \ldots, x_n; b; d_1, \ldots, d_n; c)$$

$$= \sum_{m_n=0}^{\infty} \cdots \sum_{m_1=0}^{\infty} \frac{(b, m_1 + \cdots + m_n)}{(c, m_1 + \cdots + m_n)}$$

$$\times (d_1, m_1) \cdots (d_n, m_n) \frac{x_1^{m_1}}{m_1!} \cdots \frac{x_1^{m_n}}{m_n!},$$

and reduces to Appell's first function F_1 for $n = 2$ and to Gauss hypergeometric function $_2F_1$ for $n = 1$.

ii. Generalizations in Several Variables

1. The generalized beta in m variables is often known as the *standard Dirichlet distribution*, defined on a simplex. However, it is not the only form known as Generalized beta.

    Let $\pi = (\pi_1, \ldots, \pi_m)$ be a vector of m variables, and let $\Theta = (\theta_0, \theta_1, \ldots \theta_m)$ be a vector of $(m+1)$ parameters $\theta_i > 0$, $i = 0, 1, \ldots, m$. Then $\pi \sim Dir(\theta_0, \theta_1, \ldots \theta_m)$ has its density defined in the simplex $\sum_{i=1}^{m} \pi_i \leq 1$, with $0 \leq \pi_j \leq 1$, $1 \leq j \leq m$, by:

    $$f(\pi_1, \ldots, \pi_m, \theta_0, \ldots, \theta_m)$$

    $$= \frac{\left(\prod_{j=1}^{m} \pi_j^{\theta_j - 1} \right) \cdot \left(1 - \sum_{j=1}^{m} \pi_j \right)^{\theta_0 - 1}}{B(\theta_0, \theta_1, \ldots, \theta_m)},$$

    where

    $$B(\theta_0, \theta_1, \ldots, \theta_m) = \frac{\prod_{j=0}^{m} \Gamma(\theta_j)}{\Gamma\left(\sum_{j=0}^{m} \theta_j \right)}$$

    is the generalized beta function of m+1 variables. Then each of its marginal variables π_j has a standard beta distribution, $beta(\theta_j, (\sum_{i=0}^{m} \theta_i) - \theta_j)$, but they are not independent of each other. Elicitation of the prior Dirichlet, from one or several assessors, has been a topic of much attention, but remains a challenge in real-life applications. The interested reader should consult examples given in Bunn (1978) and Press (1989). Chaloner and Duncan (1987) gave an approach based on the associated predictive distribution, known as the Dirichlet-Multinomial. A list of important applications of this distribution is given in Johnson et al. (2000).

2. The Dirichlet Distribution of second kind, or Inverted Dirichlet: This is the generalization to m dimensions of the beta prime. It can be obtained from the above Dirichlet by considering

$$w_r = \frac{\pi_r}{\left[1 - \sum_{j=1}^{m} \pi_j\right]}, r = 1, 2, \ldots, m.$$

It can also be obtained from $m + 1$ independent Gamma variables $Y_0, Y_1, \ldots, Y_m$, with common scale parameter. It is then the joint distribution of $X_1, X_2, \ldots, X_m$, where $X_i = Y_i/Y_0$. Hence, $W = (w_1, \ldots, w_r)$ has an Inverted Dirichlet distribution, i.e. $W \sim IDir(\theta_0, \theta_1, \ldots, \theta_m)$, if its density is:

$$f(w_1, \ldots, w_m) = \frac{\Gamma\left(\sum_{i=0}^{m} \theta_i\right) \prod_{i=1}^{m} w_i^{\theta_i - 1}}{\prod_{i=0}^{m} \Gamma(\theta_i) \left(1 + \sum_{i=1}^{m} w_i\right)^{\sum_{i=0}^{m} \theta_i}},$$

with $0 < w_i < \infty$, and $\theta_i > 0$, for $i = 0, 1, \ldots, m$. This is a particular case of the more general Liouville-Dirichlet distribution. In the linear model $y = X\beta + e$, if an appropriate prior is assigned to (β, σ^2), then it can be proven that two sample quantities, namely b and W, the estimators of β and of the residual sum of squares respectively, have the Inverted Dirichlet distribution (Tiao and Guttman, 1965).

Generalizations to Matrices

The most general extension of the standard beta is, at present, the trix variate beta distribution. Several types of matrix variates can considered.

A) Symmetric positive random matrices

    1. A single random matrix as variable: Considering random matrices as variables, we define a matrix-variate Beta distribution of type1 as the distribution of the form:

$$f(X) = \frac{|X|^{\alpha - \frac{p+1}{2}} |1 - X|^{\beta - \frac{p+1}{2}}}{B_p(\alpha, \beta)},$$

$0 < X < I$, where $\alpha > (p-1)/2$, $\beta > (p-1)/2$, X is a symmetric positive random matrix, with $I - X > 0$ and $|A| = det(A)$. Also,

$$B_p(\alpha, \beta) = \Gamma_p(\alpha)\Gamma_p(\beta)/\Gamma_p(\alpha + \beta),$$

with

$$\Gamma_p(\alpha) = \pi^{p(p-1)/4}\Gamma(\alpha)\Gamma(\alpha - 1) \cdots \Gamma\left(\alpha - \frac{p-1}{2}\right).$$

Similarly, the matrix-variate Beta distribution of type2 has the form:

$$f(X) = \frac{|X|^{\alpha - \frac{p+1}{2}} |1 + X|^{-(\alpha+\beta)}}{B_p(\alpha, \beta)},$$

with X being a positive random matrix with $\alpha + \beta > (p-1)/2$.

    2. A vector with matrix components as variable: Let $Y = (Y_1, \ldots, Y_g)$ be a vector of g symmetric matrices Y_j, with $j = 1, 2, \ldots, g$, defined as follows: Let $S_{(0)}$, $S_{(1)}$, $\ldots$, $S_{(g)}$ be independent m by m matrices, with $S_{(j)}$ having the Wishart distribution $W_m(\nu_j S; I)$, and let

$$Y_j = \left(\sum_{i=0}^{g} S_{(i)}\right)^{-1/2} S_{(j)} \left(\sum_{i=0}^{g} S_{(i)}\right)^{-1/2}.$$

Then $Y = (Y_1, \ldots, Y_g)$ has as density:

$$f_{Y_1,\ldots,Y_g}(y_1,\ldots,y_g)$$

$$= \frac{\prod_{j=1}^{g} K(\nu_j; I) |y_j|^{(\nu_j-m-1)/2}}{K(\nu; I)} \left| I - \sum_{j=1}^{g} y_j \right|^{(\nu_0-m-1)/2} \quad \text{(XIV.2)}$$

where $y_1, \ldots, y_g$ and $I - \sum_{j=1}^{g} y_j$ are positive definite, with

$$\nu = \sum_{j=0}^{g} \nu_j, \text{ and } K(\nu_j; I) \text{ be such that}$$

$$\prod_{j=0}^{g} K(\nu_j; I) / K(\nu; I) = \Gamma_m(\nu/2) \left/ \prod_{j=0}^{g} \Gamma_m(\nu_j/2) \right. .$$

This form is due to Olkin and Rubin (see Johnson and Kotz (1972, page 235)) and can be considered a generalization to matrices of the Dirichlet distribution. Similarly, the joint distribution of the g matrices $\mathbf{W}=(W_1,\ldots,W_g)$, with $W_j = S_{(0)}^{-1/2} S_{(j)} S_{(0)}^{-1/2}$ is a generalization of the beta type 2. It has as density:

$$f_{W_1,\ldots,W_g}(w_1,\ldots,w_g)$$

$$= \frac{\prod_{j=0}^{g} \Gamma_m(\nu_j/2) |w_j|^{(\nu_j-m-1)/2}}{\Gamma_m(\nu/2)} \left| I + \sum_{j=1}^{g} w_j \right|^{-\frac{\nu-m-1}{2}} .$$

B) A general $K \times N$ stochastic matrix as variable: Martin (1967) considered the case of a $K \times N$ random stochastic matrix P, having a matrix beta distribution with parameter M. Its density is:

$$f_{M,\beta}^{K,N}(P|M) = \prod_{i,j=1}^{N} \left\{ \prod_{k=1}^{K_i} B_N\left(m_i^k\right) \left(p_{i,j}^k\right)^{m_{i,j}^k - 1} \right\} \quad \text{(XIV.3)}$$

with $p \in S_{K,N}$. Here, M is a $K \times N$ matrix with $m_{ij}^k > 0$, $k = 1, \ldots, K_i$, and $i, j = 1, \ldots, N$. Also, m_i^k are the rows of M, and are N-dimensional vectors, while

$$B_N(m_i^k) = \Gamma \left(\sum_{i=1}^N m_i^k \right) \Big/ \prod_{i=1}^N \Gamma \left(m_i^k \right).$$

We have $K = \sum_{i=1}^N K_i$ as the total number of rows for both P and M. The above density is, in fact, the joint distribution of $K(N-1)$ random variables p_{ij}^k since the remaining K elements of P are determined by $\sum_{i=1}^N p_{ij}^k = 1$. Furthermore, it can be considered as the product of K Dirichlet densities $Dir(m_i^k)$. The above matrix beta distribution can be taken as a prior for the *transition matrix* P of a *Markov chain*.

Thanks to the recent advances in personal computing technology, that have made rapid complex computations very accessible, at low cost, matrix-variate distributions is a field in expansion, with several applications in Multivariate analysis and in neighboring fields. We will not discuss further generalizations of the beta, or the Dirichlet, since they outside the scope of this chapter, but refer the interested reader to Mathai (1997), Gupta and Nagar (2000), Dickey (1983) and the very interesting original work of Martin (1967).

B.   Sampling Model

Depending on the parameter under consideration, an appropriate ımpling scheme is adopted.

1. Binomial and Pascal samplings apply to the parameter π having the general beta, as given by (I.2), as prior.
2. We use the beta type1, $beta(\alpha, \beta)$, as a prior for ρ^2, the coefficient of determination of $N_p(\mu, \Sigma)$. Let R^2 be an observed

value of ρ^2, from a random sample of size n taken from that population, with $R^2 = S_{11}^{-1} S_{12}' S_{22}^{-1} S_{12}$, where S_{ij} are from the sample covariance matrix

$$S = \begin{bmatrix} S_{11} & S_{12}' \\ S_{12} & S_{22} \end{bmatrix}$$

and S_{11} is a scalar(Marchand, 2001).

3. An extension of Bernoulli sampling is *multinomial sampling*, associated with the *multinomial distribution*: Let $\pi = (\pi_0, \pi_1, \ldots, \pi_m)$ be a vector of probabilities, such that $\sum_{j=0}^{m} \pi_j = 1$. Let n be a fixed positive integer and $x = (x_0, x_1, \ldots, x_m))$ such that $\sum_{j=0}^{m} x_j = n$. We have:

$$\Pr(X_0 = x_0, X_1 = x_1, \ldots, X_m = x_m)$$

$$= \frac{n!}{\prod\limits_{i=0}^{m} x_j!} \prod_{i=0}^{m} \pi_i^{x_i}. \qquad (\text{XIV}.4)$$

Because of their sums equal 1 and n respectively, there are in fact, only m variables π_i and x_i.

4. Considering (XIV.2) as the prior of a vector of matrix variates, a sampling scheme based on the multinomial distribution for matrices can be adopted.

5. For a general matrix-variate prior with expression (XIV.3), the two sampling rules considered as: *consecutive sampling of size n* and *ν- step sampling rule of size n*. Loosely speaking, the first rule, for n fixed, consists of n consecutive observations of the states of a Markov chain, with alternatives under a sequence of policies which is selected in advance of sampling. Let $x = (x_0, x_1, \ldots, x_n)$ be a sample of n transitions observed under this sampling rule, where x_0 is the initial state. Let f_{ij}^k be the number of *transitions* in the sample from state i to state j under the kth alternative in state i ($k = 1, \ldots, K_i$; $i, j =$

$1, \ldots, N$). We define the transition count of the sample as the $K \times N$ matrix $F = \left[f_{ij}^k \right]$. For $k = 1$, F is a $N \times N$ matrix, and follows the discrete multivariate distribution called the *Whittle distribution* (Martin, 1967, page 120) with density:

$$f_W^{(N)}(F|u, n, P) = F_{\nu u}^* \frac{\displaystyle\prod_{i=1}^{N} f_i!}{\displaystyle\prod_{i,j=1}^{N} f_{ij}!} p_{ij}^{f_{ij}}, \qquad (XIV.5)$$

$u = 1, 2, \ldots, N$, $n = 1, 2, \ldots$, where $F_{\nu u}^*$ is the (v, u)th cofactor of the $N \times N$ matrix $F^* = \left[f_{ij}^* \right]$ defined by: $f_{ij}^* = \delta_{ij} - (f_{ij}/f_i)$ if $f_i > 0$, and $= 0$ otherwise.

C.   Posterior Distribution

The corresponding posterior distributions are as follows:

1. When the general beta, defined on an interval (c, d), with $0 < c < d < 1$ (see (I.2)), is used as prior in Bernoulli sampling, the posterior distribution can still be expressed in closed form. However, its expression is more complex and an argument, based on *Appell's first hypergeometric function* in 2 variables (see Pham-Gia and Turkkan (1992b)), and Picard's representation of Appell's function in two variables as an integral of a single variable, gives the following result: Let π have a $gbeta(\pi; \alpha, \beta; c, d)$ prior, and let sampling give r successes out of n observations. Then the posterior distribution of π is:

$$\phi(\pi; \alpha, \beta; c, d | r, n)$$
$$= gbeta(\pi; \alpha, \beta; c, d) \frac{(\pi/c)^r \left\{ (1 - \pi)/(1 - c) \right\}^{n-r}}{P_0(r; \eta, \xi)} \quad (XIV.6)$$

where

$$P_0(r; \eta, \xi) = \sum_{k=0}^{n-r} \sum_{m=0}^{r} \frac{(\alpha, m + k)(-r, m)(r - n, k)}{(\alpha + \beta, m + k)} \frac{\eta^m}{m!} \frac{\xi^k}{k!},$$

$\eta = -(d-c)/c$ and $\xi = (d-c)/(1-c)$.

2. Using the $G3B(\alpha, \beta; \lambda)$ as a prior (see (XIV.1)), the posterior also comes out simply, under Bernoulli sampling. For the $G(2n+2)B$ family, we also have a closed form for the posterior.

3. With a $beta(\alpha, \beta)$ prior for ρ^2, and R^2 obtained from a sample of size n from $N_p(\mu, \Sigma)$, the posterior distribution of ρ^2 is

$$Hyperbeta\left((\alpha, \beta+t); (t,t,\alpha); (\alpha+\beta+t,\nu); R^2\right) \text{(XIV.7)}$$

as shown by Marchand (2001), where $t = (n-1)/2$, $u = (n-p)/2$ and $\nu = (p-1)/2$. The distribution Hyperbeta is already encountered in Section XIII.

4. With Dirichlet prior and multinomial sampling (equation (XIV.4 the posterior is also Dirichlet, i.e.

$$\pi|X \sim Dir\left(\theta_0 + n - \sum_{i=1}^{m} x_i, \theta_1 + x_1, \ldots, \theta_m + x_m\right)$$

$$\text{(XIV.8)}$$

Figure 2 shows, in three dimensions, the prior $Dir(x, y; 7, 3, 8)$, where $(\theta_0, \theta_1, \theta_2) = (7, 3, 8)$, and the posterior $Dir(x, y; 10, 8, 20)$, which results from sampling data: (3,5,12). We can see that the posterior is more concentrated about its mean than the prior.

5. Subsequent to a multinomial matrix sampling scheme with a matrix Dirichlet prior, as given by (XIV.2), the posterior is a matrix distribution of Dirichlet type.

6. Let P have as prior the matrix beta distribution with parameter M', as given by (XIV.3), and let a sample with transition count F be observed under the *consecutive sampling rule*, with *non informative stopping*. Then the posterior distribution of P is also a matrix variate beta of the same type, with parameter $M'' = M' + F$. In other words, the family of matrix beta distributions, as defined by (XIV.3), is conjugate to *consecutive sampling* rule.

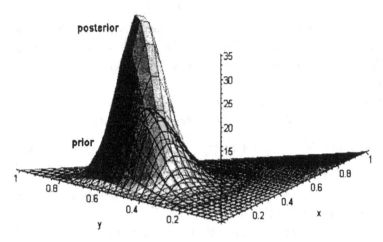

Fig.2: prior and posterior Dirichlet distributions

D. Predictive Distribution

The associated predictive distributions are:

1. The predictive distribution, when the prior is the four- parameter beta and sampling is Bernoulli, can be shown to be

$$\psi(r; \alpha, \beta, d; n) = \binom{n}{r} c^r (1-c)^{n-r} P_0(r; \eta, \xi)$$

with $r = 0, 1, \cdots, n$, where $P_0(r; \eta, \xi)$ is given in (XIV.6).

2. Under the same hypotheses as (XIV.7) above, it can be established that the marginal distribution of R^2 is *Hyperbeta* $((\nu, u); (t, \beta); \alpha + \beta + t; 1)$.

3. Subsequent to (XIV.8) above, the Predictive distribution is the discrete Dirichlet-Multinomial $DirMul(\theta_0 + n - \sum_{i=1}^{m} x_i, \theta_1 + x_1, \ldots, \theta_m + x_m; n)$. In general, the Dirichlet-Multinomial,

$DirMul\,(\alpha_0,\,\alpha_1,\,\ldots,\,\alpha_m;n)$ has density:

$$\Pr\left(X_0 = x_0, X_1 = x_1, \ldots, X_m = x_m; n\right)$$

$$= \frac{n!}{\left(\sum_{j=0}^{m} \alpha_j, n\right)} \cdot \frac{\left(\alpha_0, n - \sum_{j=1}^{m} x_j\right)}{\left(n - \sum_{j=1}^{m} x_j\right)!} \prod_{j=1}^{m} \left(\frac{(\alpha_j, x_j)}{x_j!}\right),$$

where $0 \le x_j \le n$, with $\sum_{i=0}^{m} x_i = n$, and (α, x) is the Pochhammer notation already encountered.

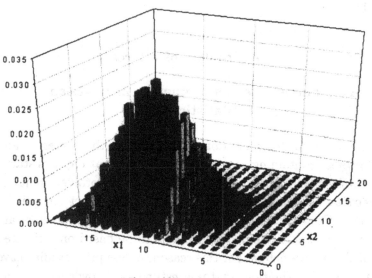

Fig.3: Dirichlet-Multinomial distribution

Figure 3 gives the graph of $DirMul(10, 8, 20; 20)$, which results from the prior $Dir(7, 3, 8)$ and sampling data $(3, 5, 12)$, already considered.

4. The Dirichlet-Multinomial for matrices is the corresponding predictive distribution when matrix variates are considered, with a matrix Dirichlet prior and Multinomial matrix sampling.

5. In association with (XIV.5), the Beta-Whittle distribution is the predictive distribution of the transition count F considered in previous sections. We have

$$f_{\beta W}^{(N)}(F|u,n,M) = F_{\nu u}^* \frac{\displaystyle\prod_{i=1}^{N} f_i.Beta\,(f_i.,m_i.)}{\displaystyle\prod_{i,j=1}^{N} f_{ij}Beta\,(f_{ij},m_{ij})} \qquad (XIV.9)$$

for $F \in \phi_N(u,n)$, with v being the unique solutions of the equations: $f_i. - f._i = \delta_{iu} - \delta_{i\nu}, i = 1,\ldots,N$ (Martin, 1967, page 151).

XV.   A Problem of Interest

A.   Sample Size Determination in Bayesian Statistics

The sample size determination problem is at the center of any applied project containing an experimental part that requires statistical analysis. Basically, in classical statistics, this problem requires that precision conditions be explicitly set, and a power analysis be undertaken. However, since the exact values of the parameters are unknown, the sample size, in general, has to be based on estimated values of these parameters. For this reason, unless prior studies have provided some information for these estimates, knowledge the key influencing factors within the application domain, is often required to make subjective estimations. For theoretical statistics, the exact sampling distributions of various statistics, which are dependent of the size of the random sample $X = (X_1 \ldots, X_n)$, is also a complex problem, and asymptotic considerations have to be used in many studies, when the exact distributions do not come in closed form.

In Bayesian statistics, when considering hpd intervals, for parameter estimation purpose, a condition imposed on that interval leads to the required sample size. Since the sample results are not known before sampling is undertaken, the criteria used are based on average (or expected) values. We have the *Expected length* criterion, the *Expected coverage* criterion, the *Worst of cases* criterion (Joseph, Wolfson and du Berger, 1995). Computations of sample sizes usually rely heavily on simulation methods, unless hypergeometric functions are used (Pham-Gia and Turkkan, 2003). The posterior variance can also be used as a criterion, especially since the posterior converges to a normal distribution as n increases. Decision theoretic criteria include the *Bayes risk*, the *Expected Value of Sample Information*, the *Expected Net Gain from Sampling* already discussed. These criteria come under very simple forms with the quadratic loss function. Again, the beta distribution with binomial sampling provide excellent tools, and illustrations, for these concepts, but normal and other common distributions are increasingly used in recent works. The results on these distributions have attracted the attention of researchers in other areas, who also use the Bayesian approach, and recent works on this topic, in quite different contexts, include Lee and Zelen (2000) on *clinical trials samples*, and Wang and Gelfand (2002) on *model identification*.

When hypothesis testing is considered, the *Bayes factor*, which is the ratio of the *posterior odds*, on H_0 against H_a, over the *prior odds*, serves to determine the sample sizes. DasGupta and Vidakovic (1997) gave some sample size computations concerning *Bayesian Analysis of variance*, and Kass and Raftery (1995) present a thorough discussion on the Bayes factor.

XVI. Conclusion

The beta distribution, and its generalizations, have provided very useful and interesting examples for theory, as well as for applications, in Bayesian analysis. Being simple but versatile, the beta lends itself to various operations. This chapter presents the main roles that the beta, and its generalizations, have played in Bayesian Statistics, but it is, by no means, exhaustive. Indeed, in view of the volume of new results appearing in the Bayesian statistics literature in recent years, quite frequently involving the beta distribution, an update of this article will certainly be required in the near future.

References

1. Amaral-Turkman, M.A.; Dunsmore, I.R. Measures of information in the predictive distribution, in *Bayesian Statistics 2* (editor J.M. Bernardo) **1985**, pp. 603-612, North-Holland: Amsterdam.

2. Berger, J. *Statistical Decision Theory and Bayesian Analysis*; Springer Verlag: New York, 1985.

3. Bernardo, J.M.; Smith, A.F.M. *Bayesian Theory*; John Wiley and Sons: New York, 1994.

4. Bunn, D.W. Anchoring bias in the assessment of subjective probability, Operations Research **1975**, *26*, 449-454.

5. Bunn, D.W. Estimation of a Dirichlet prior, Omega **1978**, 6, 371-373.

6. Calvin, T.W. *How and When to Perform Bayesian Acceptance Sampling* (volume 7); American Society for Quality Control: Milwaukee, 1984.

7. Casella, G. An introduction to empirical Bayes data analysis, The American Statistician **1985**, *39*, 83-87.

8. Chaloner, K.; Duncan, G.T. Assessment of a Beta prior distribution: PM elicitation, The Statistician **1983**, *32*, 174-180.

9. Chaloner, K.; Duncan, G.T. Some properties of the Dirichletmultinomial distribution and its use in prior elicitation, Communications in Statics–Theory and Methods **1987**, *16*, 511-523.

10. Chen, J.C.; Novick, M.R. Bayesian analysis for binomial models with generalized beta prior distributions, Journal of Educational Statistics **1984**, *9*, 163-175.

11. Colombo, A.G.; Constantini, D. Ground-hypotheses for beta distribution as Bayesian prior, IEEE Transactions on Reliability **1980**, *1*, 17-21.

12. Cuevas, A.; Sanz, P. On differentiability properties of Bayes operators, in *Bayesian Statistics 3* (editors J.M. Bernardo, M.H. DeGroot, D.V. Lindley and A.F.M. Smith) **1988**, pp. 569-570,

Oxford University Press: London.

13. DasGupta, A.; Vidakovic, B. Sample sizes problem in ANOVA Bayesian point of view, Journal of Statistical Planning and Inference, **1997**, *65*, 335-347.

14. De Groot, M. *Optimal Statistical Decisions*; Mc Graw-Hill: New York, 1970.

15. Diaconis, P.; Ylvisaker, D., *Conjugate priors for exponential families*, Annals of Statistics **1979**, *7*, 269-281.

16. Dickey, J.M. Conjugate family of distributions, in *Encyclopedia of Statistics*, volume 2 (editors N.L. Johnson and S. Kotz), **1982**, 135-145.

17. Dickey, J.M. Multiple hypergeometric functions: probabilistic interpretations and statistical uses, Journal of the American Statistical Association **1983**, *78*, 628-637.

18. Duda, R.O.; Hart, P.E.; Stork, D.G. *Pattern Classification* (second edition); 2002.

19. Ferguson, T. *Mathematical statistics: a Decision Theoretic Approach*; Academic Press: New York, 1967.

20. Ferreri, C. Elicitation procedures of a beta prior distribution, Statistica, **1986**, *46*, 277-290.

21. Fine, T. *Theories of Probability*; Academic Press: New York, 1973.

22. Geisser, S. *Predictive Inference: An Introduction*; Chapman and Hall: London, 1993.

23. Ghosh M.; Parsian, A. *Bayes minimax estimation of multiple Poisson parameters*; Journal of Multivariate Analysis **1981**, 280-288.

24. Good, I.G. *The Estimation of Probabilities: an Essay on Modern Bayesian Methods*; MIT Press: Cambridge, MA, 1965.

25. Govindarajulu, Z. *Sequential Statistical Analysis*; American Science Press: Columbus, Ohio, 1987.

26. Gupta, A.K.; Nagar, D.K. *Matrix Variate Distributions*; Chapman and Hall/CRC: Boca Raton, 2000.

27. Gupta, A.K.; Wesolowski, J. Regressional identifiability and identification for beta mixtures, Statistics and Decisions **2001**, *19*, 71-82.

28. Hald, A. *Statistical Theory of Sampling Inspection by Attribute*; Academic Press: London, 1981.

29. Hogarth, R.M. Cognitive processes and the assessment of subjective probabilities, Journal of the American Statistical Association **1975**, *70*, 271-289.

30. Hokstad, P. A shock model for common-cause failure, Reliab. Engin. and System Safety **1988**, *23*, 127-145.

31. Johnson N.; Kotz, S. *Distributions in Statistics: Continuous Multivariate Distributions*; John Wiley and Sons: New York, 1972.

32. Johnson, N.; Kotz, S.; Balakrishnan, N. *Continuous Univariate Distributions* (volume 2, second edition); John Wiley and Sons: New York, 1995.

33. Johnson N.; Kotz, S.; Balakrishnan, N. *Continuous Multivariate Distributions: Models and Applications* (volume 1); John Wiley and Sons: New York, 2000.

34. Kahneman, D.; Slovic, P.; Tversky, A. *Judgement under Uncertainty, Heuristics and Biases*; Cambridge University Press: London, 1982.

35. Kass, R.E.; Raftery, A.E. Bayes factors, Journal of the American Statistical Association, **1995**, *90*, 773-795.

36. Keefer, D.L.; Bodily, S.E. Three point approximation for continuous random variables, Management Science **1983**, *29*, 595-609.

37. Lau, H.S.; Lau, A.H.L. An improved pert-type formula for standard deviation, Inst. Industr. Engin. Transactions **1998**, *30*, 273-275.

38. Lee, S.J.; Zelen, M. Clinical trials and sample size considerations: another perspective, Statistical Science **2000**, *15*, 95-110.

39. Lindley, D. *Bayesian Statistics, A Review*; Society for Industrial and Applied Mathematics: Philadelphia, 1972.

40. Marchand, E. Point estimation of the coefficient of determination, Statistics and Decisions **2001**, *19*, 137-154.

41. Maritz, J.S.; Lwin, T. *Empirical Bayes Methods*; Chapman and Hall: London, 1989.

42. Martin, J.J. *Bayesian Decision Problems and Markov Chains*; John Wiley and sons: New York, 1967.

43. Mathai, A.N. *Jacobians of Matrix Transformations and Functions of Matrix Argument*; World Scientific: Singapore, 1997.

44. Myhre, J.M.; Rennie, M.W. Confidence bounds for reliability of coherent systems based on binomially distributed component data, in *Reliability and Quality Control* (editor A.P. Basu) **1986**, pp. 265-279, Elsevier Science Publishers B.V.

45. Papageorgiou, H.; Wesolowski, J. Posterior mean identifies the prior distribution in NB and related models, Statistics and Probability Letters **1997**, *36*, 127-134.

46. Pham-Gia, T. Value of the beta prior information, Communications in Statistics–Theory and Methods **1994**, 2175-2195.

47. Pham-Gia, T. Sample size determination in Bayesian statistics- a commentary, The Statistician **1995a**, *44*, 163-166.

48. Pham-Gia, T. Some applications of the Lorenz curve in Bayesian analysis, American Journal of Mathematical and Management Sciences **1995b**, *15*, 1-34.

49. Pham-Gia, T. On Bayesian analysis, Bayesian decision theory and the sample size problem, *The Statistician* **1997**, *46*, 139-144.

50. Pham-Gia, T. Distribution of the stopping time in Bayesian sequential analysis, Australian and New Zealand Journal of Statistics **1998**, *40*, 221-227.

51. Pham-Gia, T. Distributions of the ratios of independent beta variables, Communications in Statistics–Theory and Methods **2000**, *29*, 2693-2715.

52. Pham-Gia, T. Decision criteria and sample sizes, submitted for publication, **2003**.

53. Pham-Gia, T.; Duong, Q. P. The generalized beta and *F*-distributions in statistical modelling, Mathematical and Computer Modelling **1989**, *12*, 1613-1625.

54. Pham-Gia, T.; Tranloc, H. The mean and median absolute deviations, Mathematical and Computer Modelling **2001**, *34*, 921-936.

55. Pham-Gia, T.; Turkkan, N. Using the mean absolute deviation to determine the beta prior distribution, Statistics and Probability Letters **1992a**, 373-381.

56. Pham-Gia, T.; Turkkan, N. Bayes binomial sampling with the general beta prior distribution, IEEE Transactions on Reliability **1992b**, *41*, 2001-2018.

57. Pham-Gia, T.; Turkkan,N. Sample size determination in Bayesian statistics, The Statistician **1992c**, *41*, 389-397.

58. Pham-Gia, T.; Turkkan, N. Information loss for a noisy dichotomous sampling process, Communications in Statistics–Theory and Methods **1992d**, *21*, 2001-2018.

59. Pham-Gia, T.; Turkkan,N. Bayesian analysis of the difference of two proportions, Communications in Statistics–Theory and Methods **1993**, *26*, 1755-1771.

60. Pham-Gia, T.; Turkkan, N. Reliability of a standby system with beta component lifelengths, IEEE Transactions on Reliability **1994a**, 71-75.

61. Pham-Gia, T.; Turkkan, N. The Lorenz and T4-curves: a unified approach, IEEE Transactions on Reliability **1994b**, 76-84.

62. Pham-Gia, T.; Turkkan, N. Distribution of the linear combination of two general beta variables and applications, Communications in Statistics–Theory and Methods **1998**, *27*, 1851-1869.

63. Pham-Gia, T.; Turkkan, N. Operations on the generalized *F*-variables and applications, Statistics **2002a**, *36*, 195-209.

64. Pham-Gia, T.; Turkkan, N. The product and ratio of independent general beta variables, Statistical Papers **2002b**, *43*, 537-550.

65. Pham-Gia, T.; Turkkan, N. Determination of the exact sample sizes in the Bayesian study of the difference of two proportions, The Statistician **2003**, 1-20.

66. Pham-Gia, T.; Turkkan, N.; Bekker, A. *Bayesian analysis in the L_1-norm of the mixing proportion*, submitted to Scandinavian Journal of Statistics.

67. Press, S.J. *Bayesian Statistics, Principles, Models and Applications*; John Wiley and Sons: New York, 1989.

68. Raiffa, H.; Schlaifer, R. *Applied Statistical Decision Theory*; Harvard University Press: Cambridge, MA, 1961.

69. Robbins, H. An empirical Bayes approach to statistics, in *Proceedings of the Third Berkeley Symposium on Mathematical*

Statistics and Probability **1955**, *1*, 157-164.

70. Stuart, A.; Ord, J.K. *Kendall's Advanced Theory of Statistics* (volume 1); Oxford University Press: London, 1987.

71. Thompson, W.E.; Haynes, R.D. On the reliability, availability, and Bayes confidence intervals for multicomponent systems, Naval Research Logistics Quarterly **1980**, *27*, 345-358.

72. Tiao, G.G.; Guttman, I. *The inverted Dirichlet distribution with applications*, Journal of the American Statistical Association **1965**, *60*, 793-805.

73. Titterington, D.M.; Smith, A.F.M.; Makov, U.E. *Statistical Analysis of Mixture Distributions*; John Wiley and Sons: Chichester, 1985.

74. Turkkan, N.; Pham-Gia, T. Computation of the highest posterior density region in Bayesian analysis, Journal of Statistical Computation and Simulation **1993**, *44*, 243-250.

75. Turkkan, N.; Pham-Gia, T. Highest posterior density and minimum volume confidence region: the bivariate case, Journal of the Royal Statistical Society, C **1997**, *46*, 131-140.

76. Waller R.; Martz H. *Bayesian Reliability Analysis*; John Wiley and Sons: New York, 1982.

77. Waller, R.; Martz H.; Fickas, E.T. Bayesian reliability analysis of series systems of binomial subsystems and components, Technometrics **1988**; *30*, 143-154.

78. Wang, F.; Gelfand. A.E. A simulation-based approach to Bayesian sample size determination for performance under a given model and for separating models, Statistical Science **2002**, *17*, 193-208.

79. Whitby, O. Estimation of parameters in the generalized beta distribution, *PhD Thesis*, Stanford University, 1972.

80. Winkler, R.L. The assessment of prior distribution in Bayesian analyses, Journal of the American Statistical Association **1967**, *62*, 776-800.

81. Winkler, R.L. The concensus of subjective probabilities distributions, Management Science **1968**, *15*, B61-B75.

82. Winkler, R.L. Prior information, predictive distributions and Bayesian model building, in *Bayesian Analysis in Economet-*

rics and Statistics (editor A. Zellner) **1980**, pp. 95-109, North-Holland: New York.

83. Winkler, R.L. Information loss in noisy and dependent processes, in *Bayesian Statistics 2* (editors J.M. Bernardo, M.H. DeGroot, D.V. Lindley and A.F.M. Smith) **1985**, pp. 559-570, North-Holland: New York.

84. Zaino, N.A.; d'Enrico, J. Optimal discrete approximations for continuous outcomes with applications in decision and risk analysis, Journal of the Operations Research Society **1989**, *40*, 379-388.

Applications of the Beta Distribution

Sudhir R. Paul

Department of Mathematics and Statistics
University of Windsor
Windsor, Ontario
Canada N9B 3P4

I. Introduction

The beta distribution is important in many real life data analyzes. For example, data in the form of proportions arise in Toxicology (Weil, 1970; Williams, 1975) and other similar fields (Crowder, 1978; Otake and Prentice, 1984). These proportions often exhibit extra variation than can be explained by a simple binomial distribution. In other applications, such as in fecundability studies, the number of cycles required to achieve pregnancy would be distributed as a geometric distribution with parameter p. Here also in real life data situations the actual variation of the data exceeds that of the geometric distribution. The idea behind it is that the binomial parameter p or the parameter p of the geometric distribution does not remain constant in the course of collecting the data. In this situation it is useful to assume that the parameter p varies from observation to observation. One can assume one of many continuous distributions for p in the parameter space $0 < p < 1$. But the most convenient and most sensible distribution for p is the beta distribution, because it is the natural conjugate prior distribution in the Bayesian sense. It also produces a convenient mixed distribution leading to practical interpretation of the parameters of the mixed distribution. In what follows we study two distributions: the beta-binomial and the beta-geometric. These two distributions are useful in practice, particularly in situations described above, and arise as a beta mixture of a binomial or a geometric distribution.

Suppose that a continuous random variable X has a beta distribution with parameters α and β, where $0 < \alpha < 1$ and $0 < \beta < 1$. Then the probability density function of X has the form

$$f(x|\alpha,\beta) = \frac{x^{\alpha-1}(1-x)^{\beta-1}}{B(\alpha,\beta)}, \qquad 0 < x < 1, \tag{I.1}$$

where $B(\alpha,\beta) = \frac{\Gamma(\alpha+\beta)}{\Gamma(\alpha)\Gamma(\beta)}$ is the beta function and where $\Gamma(a)$ is the gamma function:

$$\Gamma(a) = \int_0^\infty x^{a-1}e^{-x}dx.$$

The mean and variance of the beta random variable X are

$$\mu = \frac{\alpha}{\alpha+\beta}$$

and

$$\sigma^2 = \frac{\alpha\beta}{(\alpha+\beta)^2(\alpha+\beta+1)},$$

respectively.

A.   The Beta-binomial Distribution

We assume that $Y|p \sim$ binomial (n,p). That is,

$$\Pr(Y = y|p) = \binom{n}{y} p^y (1-p)^{n-y}.$$

Note that the binomial parameter p may not remain constant in the course of collecting the data. So, we assume that the binomial probability p is a random variable distributed as a beta distribution with parameters α and β having probability density function (I.1).

Then, the marginal distribution of Y is

$$
\begin{aligned}
\Pr(Y = y | n) &= \int_0^1 \Pr(Y = y | p) f(p | \alpha, \beta) dx \\
&= \frac{1}{B(\alpha, \beta)} \binom{n}{y} \int_0^1 p^{y+\alpha-1} (1-p)^{n-y+\beta-1} dx \\
&= \binom{n}{y} \frac{B(\alpha + y, n + \beta - y)}{B(\alpha, \beta)}.
\end{aligned}
$$

This distribution is known as the beta-binomial distribution. The beta-binomial distribution in the form given above was initially given by Skellam (1948). Asymptotic forms of the distribution are obtained under the following sets of conditions (see Paul and Plackett (1978)).

(i) Fix α, write $\beta = n\gamma$, where γ is constant, and let $n \to \infty$. The limit is a negative binomial distribution

$$
\Pr(Y = y) = \frac{\Gamma(\alpha + y)\gamma^\alpha}{n!\Gamma(\alpha)(1 + \gamma)^{y+\alpha}}.
$$

This result can be interpreted as a beta-mixture of binomials tending to a gamma-mixture of Poissons.

(ii) Fix α and β, and let $n \to \infty$. The asymptotic distribution of $Z = Y/n$ is that of a mixing beta

$$
f(z | \alpha, \beta) = \frac{z^{\alpha-1}(1 - z)^{\beta-1}}{B(\alpha, \beta)}.
$$

(iii) Fix n, write $\alpha = t\gamma_1$, $\beta = t\gamma_2$ and let $n \to \infty$. The asymptotic distribution of Y is binomial with index n and probability $\gamma_1/(\gamma_1 + \gamma_2)$.

(iv) Write $\alpha = t\gamma_1$, $\beta = t\gamma_2$ and let $n \to \infty$. The asymptotic distribution of

$$
X = \{Y - n\gamma_1/(\gamma_1 + \gamma_2)\} \Big/ \sqrt{n\gamma_1/(\gamma_1 + \gamma_2)^2}
$$

is normal with mean zero and variance $(1 + \gamma_1 + \gamma_2)/(\gamma_1 + \gamma_2)$.

Now, reparameterize α and β as $\pi = \alpha/(\alpha + \beta)$ and $\theta = 1/(\alpha + \beta)$. Then, the mean and the variance of Y can be written as $n\pi$ and $n\pi(1 - \pi)\frac{1+n\theta}{1+\theta}$ respectively. The parameter θ is called the over-dispersion parameter. Since $\alpha > 0$, $\beta > 0$, it is evident that $0 < p < 1$ and $\theta > 0$. In the limit as $\theta \to 0$ the beta-binomial distribution with index n parameters π and θ becomes a binomial distribution with index n and parameter p. Thus, the parameter θ of the beta-binomial distribution can assume only positive values, hence the name over-dispersion. Note that the random variable Y is the sum of n binary random variables. If the n binary random variables are independent then the distribution of Y is binomial with index n and probability parameter p. On the other hand, if the binary variables are correlated with intraclass correlation $\phi = \theta/(1 + \theta)$, then, the distribution of Y is beta-binomial with index n and intraclass correlation ϕ. This, then, constraints ϕ so that $\phi > 0$.

Prentice (1986) extended the beta-binomial distribution so that the parameter θ or ϕ can assume positive as well negative values. Note that the distribution of Y can be written in a computationally convenient form

$$\Pr(Y = y|n) = \binom{n}{y} \prod_{r=0}^{y-1} \{\pi + r\theta\} \prod_{r=0}^{n-y-1} \{(1 - \pi) + r\theta\} \Big/ \prod_{r=0}^{n-1} \{1 + r\theta\}$$

or

$$\Pr(Y = y|n) = \binom{n}{y} \prod_{r=0}^{y-1} \{(1 - \phi)\pi + r\phi\}$$
$$\times \prod_{r=0}^{n-y-1} \{(1 - \phi)(1 - \pi) + r\phi\} \Big/ \prod_{r=0}^{n-1} \{1 + \phi(r - 1)\}.$$

The mean and the variance of Y can now be written as $n\pi$ and $n\pi(1 - \pi)\{1 + (n - 1)\phi\}$ respectively. Further, it can be seen that $0 \le \pi_i \le 1$ and $(\frac{-1}{n-1}) < \phi < 1$ (see Prentice (1986) for more details). The parameter ϕ is called the intraclass correlation parameter between the n binary responses. In the context of clustered correlated data the parameter ϕ is interpreted as a measure of similarity of responses within a cluster or within a litter in the context of Toxicology data. Thus, it can take positive as well as negative value. In this sense the parameter ϕ is called the dispersion parameter indicating that it can represent either over-dispersion when the binary responses are positively correlated or under-dispersion when these responses are negatively correlated.

Data that arise in practice are often of the form (y_i, n_i), $i = 1, \ldots, k$. We assume that given n_i, Y_i is distributed as an extended beta-binomial distribution. Many methods of estimation are available in the literature. For a most recent article see Paul, Saha and Balasoorya (2003) where 26 different estimators are compared in terms of bias and efficiency. Here we discuss estimation by the method of maximum likelihood. The likelihood function for the data is given as

$$L = \prod_{i=1}^{k} \binom{n_i}{y_i} \prod_{r=0}^{y_i-1} \{(1-\phi)\pi + r\phi\}$$

$$\times \prod_{r=0}^{n_i-y_i-1} \{(1-\phi)(1-\pi) + r\phi\} \Big/ \prod_{r=0}^{n_i-1} \{1 + \phi(r-1)\}$$

and the corresponding log-likelihood, apart from a constant, can be written as

$$l = \sum_{i=1}^{k} [\sum_{r=1}^{y_i} \log\{(1-\phi)\pi + (r-1)\phi\}$$

$$+ \sum_{r=1}^{n_i-y_i} \log\{(1-\phi)(1-\pi) + (r-1)\phi\}$$

$$- \sum_{r=1}^{n_i} \log\{1 - \phi + (r-1)\phi\}].$$

The maximum likelihood estimates $\hat{\pi}$ and $\hat{\phi}$ of the parameters π and ϕ are obtained by solving the maximum likelihood estimating equations $\frac{\partial l}{\partial \pi} = 0$ and $\frac{\partial l}{\partial \phi} = 0$ simultaneously, that is, by solving

$$\sum_{i=1}^{k} \{\sum_{r=0}^{y_i-1} \frac{1-\phi}{(1-\phi)\pi_i + r\phi} - \sum_{r=0}^{n_i-y_i-1} \frac{1-\phi}{(1-\phi)(1-\pi_i) + r\phi}\} = 0$$

(I.2)

and

$$\sum_{i=1}^{k}\left\{\sum_{r=1}^{y_i-1}\frac{-\pi_i+r}{(1-\phi)\pi_i+r\phi}+\sum_{r=0}^{n_i-y_i-1}\frac{-(1-\pi_i)+r}{(1-\phi)(1-\pi_i)+r\phi}\right.$$

$$\left.-\sum_{r=0}^{n_i-1}\frac{r-1}{(1-\phi)+r\phi}\right\}=0,\qquad(I.3)$$

simultaneously subject to the constraints $0<\pi<1$ and $\max(\frac{-1}{n_i-1})<\phi<1$. Obviously, no closed form solution exists. So these equations are to be solved using a numerical procedure such as the Newton-Raphson method or a numerical subroutine, such as the IMSL subroutine ZBRENT or NEQNF. The estimation procedure can be extended when covariates are involved. See Paul et al. (2003) for more details.

Example 1: The data taken from Williams (1975) refer to an experiment where a group of 16 pregnant female rats were treated with a particular chemical. For each rat the number n of pups alive at 4 days and the number y of pups that survived the 21 day lactation period were recorded. The data, given here as fractions y/n, are: 12/12,11/11,10/10,9/9,10/11,9/10,9/10,8/9,8/9,4/5,7/9,4/7,5/10, 3/6,3/10,0/7.

We used IMSL subroutine NEQNF to solve the equations (I.2) and (I.3) for these data and obtained $\hat{\pi}=0.740$ and $\hat{\phi}=0.317$.

In Toxicology and similar fields, often, it is of interest to test whether there is over-dispersion in the data. For this a likelihood ratio test can be applied. However, this statistic often shows conservative or liberal behavior. That is, is applications, the likelihood ratio statistic may spuriously show over-dispersion for data that are not over-dispersed and vice-versa. The score test statistic (Rao, 1947), on the other hand, has good properties, such as it often produces a simple statistic and it holds nominal level well.

Tarone (1979) developed such a statistic. Suppose we wish to test $H_0:\theta=0$ against $H_A:\theta>0$. Now, let l be the log-likelihood. Further, define

$$\psi=\left.\frac{\partial l}{\partial\theta}\right|_{\theta=0},$$

$$I_{\pi\pi} = E\left(\left.\frac{\partial^2 l}{\partial \pi^2}\right|_{\theta=0}\right),$$

$$I_{\pi\phi} = E\left(\left.\frac{\partial^2 l}{\partial \pi \partial \phi}\right|_{\theta=0}\right)$$

and

$$I_{\theta\theta} = E\left(\left.\frac{\partial^2 l}{\partial \theta^2}\right|_{\theta=0}\right).$$

Then, a score test statistic for testing $H_0 : \theta = 0$ against $H_A : \theta > 0$ is given by $Z = \psi/\sqrt{(I_{\theta\theta} - I_{\pi\phi}^2/I_{\pi\pi})}$. If the nuisance parameter π is replaced by its maximum likelihood estimate under the null hypothesis, then, asymptotically, as $n \to \infty$, the distribution of Z is standard normal.

Suppose we obtain data similar to what was described earlier, that is, (y_i, n_i), $i = 1, \ldots, k$. Let $\hat{p} = \sum_{i=1}^{k} y_i / \sum_{i=1}^{k} n_i$ and $S = \sum_{i=1}^{k} (y_1 - n_i\hat{p})^2/\hat{p}\hat{q}$, where, $\hat{q} = 1 - \hat{p}$. Then, it can be shown that the score test statistic for testing $H_0 : \theta = 0$, against $H_A : \theta > 0$ is

$$Z = \left(S - \sum_{i=1}^{k} n_i\right) \bigg/ \sqrt{2\sum_{i=1}^{k} n_i(n_i - 1)}.$$

Since the test is one-sided we reject H_0 in favor of H_A, at $100(1 - \alpha)\%$ level of significance, if $Z > z_\alpha$, where, z_t is the $100(1-t)\%$ point of the standard normal distribution. Since the parameter ϕ and hence the parameter θ can also take negative values within the constraint given earlier a two-sided test can similarly be obtained.

Example 2: For the data in example 1 test whether a binomial model is adequate. For these data we obtain Z=7.93, indicating presence of highly significant over-dispersion.

Confidence intervals for the parameter π and the parameter ϕ can be constructed by using the estimated variances of $\hat{\pi}$ and $\hat{\phi}$. These variances can be obtained by inverting the observed or the expected Fisher information matrix. Paul and Islam (1978) obtained the elements of the expected Fisher information matrix. For the type of data described above the expected Fisher information matrix has elements I_{11}, I_{12}, I_{22}, which are given in what follows.

$$I_{11} = E\left(\frac{-\partial^2 l}{\partial \pi^2}\right)$$

$$= (1-\phi)^2 \sum_{i=1}^{k}\left[\sum_{r=1}^{n_i}\frac{P\left(Y_i \geq r\right)}{\{(1-\phi)\pi + (r-1)\phi\}^2}\right.$$

$$\left. + \sum_{r=1}^{n_i}\frac{P\left(Y_i \leq n_i - r\right)}{\{(1-\phi)(1-\pi) + (r-1)\phi\}^2}\right],$$

$$I_{12} = E\left(\frac{-\partial^2 l}{\partial \pi \partial \phi}\right)$$

$$= \frac{\phi-1)}{\phi}\sum_{i=1}^{k}\left[\sum_{r=1}^{n_i}\frac{\pi P\left(Y_i \geq r\right)}{\{(1-\phi)\pi + (r-1)\phi\}^2}\right.$$

$$\left. - \sum_{r=1}^{m_i}\frac{(1-\pi)P\left(Y_i \leq n_i - r\right)}{\{(1-\pi)(1-\phi) + (r-1)\phi\}^2}\right]$$

and

$$I_{22} = E\left(\frac{-\partial^2 l}{\partial \phi^2}\right)$$

$$= \frac{1}{\phi^2}\sum_{i=1}^{k}\left[\sum_{r=1}^{n_i}\frac{\pi^2 P\left(Y_i \geq r\right)}{\{(1-\phi)\pi + (r-1)\phi\}^2}\right.$$

$$+ \sum_{r=1}^{n_i}\frac{(1-\pi)^2 P\left(Y_i \leq n_i - r\right)}{\{(1-\phi)(1-\pi) + (r-1)\phi\}^2}$$

$$\left. - \sum_{r=1}^{n_i}\frac{1}{\{1-\phi + (r-1)\phi\}^2}\right].$$

Thus, the estimated variance of $\hat{\pi}$ and $\hat{\phi}$ are $Var(\hat{\pi}) = \hat{I}_{22}/(\hat{I}_{11}\hat{I}_{22} - \hat{I}_{12}^2)$ and $Var(\hat{\phi}) = \hat{I}_{11}/(\hat{I}_{11}\hat{I}_{22} - \hat{I}_{12}^2)$, respectively, where $\hat{I}_{11}$, $\hat{I}_{12}$ and $\hat{I}_{22}$ are estimates of I_{11}, I_{12} and I_{22}, respectively, obtained by replacing the parameters by their maximum likelihood estimates.

Example 3. For the data in example 1 we obtain $\hat{I}_{11} = 233.073$, $\hat{I}_{12} = 40.459$ and $\hat{I}_{22} = 88.593$. From these it can be seen that the standard errors of π and ϕ are .06826 and .11072, respectively.

B. The Beta-Geometric Distribution

We assume that $Y|p \sim$ geometric distribution. Let $q = 1 - p$. Then, the probability function of Y is

$$\Pr(Y = y|q) = q^{y-1}p.$$

In this case also the parameter p of the geometric distribution may not be assumed to remain constant throughout the course of collecting the data. For example, in human reproduction the random variable Y may be the number of menstrual cycles required for conception in which the parameter p may be interpreted as the pre-cycle conception probability or a measure of fecundability (see Weinberg and Gladen (1986)). We assume that the parameter p is fixed for a given couple, but across couples it varies according to some unspecified underlying distribution which we assume to be beta with probability density function given by (I.1). Then, the marginal distribution of Y is

$$\Pr(Y = y) = \int_0^1 \Pr(Y = y|p) f(p|\alpha, \beta) dx$$

$$= \frac{1}{B(\alpha, \beta)} \int_0^1 p^\alpha (1 - p)^{y+\beta-2} dx$$

$$= \frac{B(\alpha + 1, y + \beta - 1)}{B(\alpha, \beta)}.$$

This distribution is known as the beta-geometric distribution. In the human reproduction literature $\Pr(Y = y)$ is the probability that conception occurs at y for a randomly selected couple. As in the case of beta-binomial distribution, the beta-geometric distribution also can be written in terms of the parameter $\pi = \alpha/(\alpha + \beta)$ and $\theta = 1/(\alpha + \beta)$, where p is interpreted as the mean parameter and θ as the shape parameter (see Weinberg and Gladen (1986) for more details), which is given in what follows.

$$\Pr(Y = y|n) = \frac{\pi \prod_{r=0}^{y-2} \{(1 - \pi) + r\theta\}}{\prod_{r=0}^{y-1} \{1 + r\theta\}}.$$

The distribution has mean

$$\frac{1-\theta}{\pi-\theta}$$

and variance

$$\frac{\pi(1-\pi)(1-\theta)}{(\pi-\theta)^2(\pi-2\theta)}.$$

Obviously, $\theta = 0$ corresponds to the geometric distribution with mean $\frac{1}{p}$ and variance $\frac{1-p}{p^2}$.

Suppose data are available on n individuals as y_i, $i = 1,\ldots,n$. Then, the likelihood function for the data is given as

$$L = \pi^n \prod_{i=1}^n \frac{\prod_{r=1}^{y_i-1} \{1-\pi+(r-1)\theta\}}{\prod_{r=1}^{y_i} \{1+(r-1)\theta\}}$$

and the corresponding log-likelihood, apart from a constant, can be written as

$$l = n\log(\pi) + \sum_{i=1}^n \left[\sum_{r=1}^{y_i-1} \log\{1-\pi+(r-1)\theta\} - \sum_{r=1}^{y_i} \log\{1+(r-1)\theta\} \right].$$

The maximum likelihood estimates of the parameters π and θ are obtained by solving the maximum likelihood estimating equations $\frac{\partial l}{\partial \pi} = 0$ and $\frac{\partial l}{\partial \theta} = 0$ simultaneously. That is, by solving

$$\frac{n}{\pi} - \sum_{i=1}^n \left\{ \sum_{r=1}^{y_i-1} \frac{1}{1-\pi+(r-1)\theta} \right\} = 0, \qquad (I.4)$$

and

$$\sum_{i=1}^n \left[\sum_{r=1}^{y_i-1} \frac{r-1}{1-\pi+(r-1)\theta} - \sum_{r=1}^{y_i} \frac{r-1}{1+(r-1)\theta} \right] = 0, \qquad (I.5)$$

simultaneously subject to the constraints $0 < p < 1$ and $\theta > 0$. In this case also no closed form solution exists. So these equations are to be solved using a numerical procedure such as the Newton-Raphson method or a numerical subroutine, such as the IMSL subroutine ZBRENT or NEQNF.

Paul (2003) developed the score and the likelihood ratio tests of the goodness of fit of the geometric distribution against the beta-geometric distribution. Using the parameterization this is equivalent to testing the null hypothesis $H_0 : \theta = 0$ against the alternative $H_A : \theta > 0$. In what follows we give the score test.

Now, define

$$S = \frac{1}{1-p} \sum_{i=1}^{n} \sum_{r=1}^{y_i-1} (r-1) - \sum_{i=1}^{n} \sum_{r=1}^{y_i} (r-1) = 0,$$

$$I_{11}(p) = \frac{n}{p^2(1-p)},$$

$$I_{12}(p) = -\frac{n}{p^2}$$

and

$$I_{22}(p) = \frac{n\left\{2 - 5p + p^2(4-p) - (p-1)(p-2)(1-p)^2\right\}}{p^3(1-p)^2}.$$

It can be seen that $Var(S) = I_{22}(p) - (I_{12}(p))^2/I_{11}(p) = n/p^2$. Then, the score test statistic for testing $H_0 : \theta = 0$ against $H_A : \theta > 0$ is given by $Z = S/\sqrt{(n/p^2)}$. If we replace p by $\hat{p}$, where $\hat{p}$ is the maximum likelihood estimate of the parameter p of the geometric distribution, in Z, then, under the null hypothesis $H_0 : \theta = 0$, the statistic Z will have an asymptotic standard normal distribution. We reject the null hypothesis $H_0 : \theta = 0$ in favor of $H_A : \theta > 0$ at $100(1 - \alpha)\%$ level of significance if $Z > z_\alpha$, where, z_t is the $100(1 - t)\%$ point of the standard normal distribution.

Paul (2003) also obtained the elements of the expected Fisher information matrix which are given in what follows.

$$I_{11} = E\left(\frac{-\partial^2 l}{\partial \pi^2}\right)$$

$$= \frac{n}{\pi^2} + n\sum_{r=2}^{\infty} \frac{\Pr(Y \geq r)}{\{(1 - \pi + (r-2)\theta\}^2},$$

$$I_{12} = E\left(\frac{-\partial^2 l}{\partial \pi \partial \phi}\right)$$

$$= -n\sum_{r=3}^{\infty} \frac{(r-2)\Pr(Y \geq r)}{\{1 - \pi + (r-2)\theta\}^2}$$

and

$$I_{22} = E\left(\frac{-\partial^2 l}{\partial \phi^2}\right)$$

$$= n\left[\sum_{r=3}^{\infty} \frac{(r-2)^2 \Pr(Y \geq r)}{\{1 - \pi + (r-2)\theta\}^2} - \sum_{r=2}^{\infty} \frac{(r-1)^2 \Pr(Y \geq r)}{\{1 + (r-2)\theta\}^2}\right].$$

Calculation of the above terms does not pose any problem if ∞ in the upper limit of the summation is replaced by a sufficiently large number, say, 5000. Thus, the estimated variance of $\hat{\pi}$ and $\hat{\theta}$ are $Var(\hat{\pi}) = \hat{I}_{22}/(\hat{I}_{11}\hat{I}_{22} - \hat{I}_{12}^2)$ and $Var(\hat{\theta}) = \hat{I}_{11}/(\hat{I}_{11}\hat{I}_{22} - \hat{I}_{12}^2)$, respectively, where $\hat{I}_{11}$, $\hat{I}_{12}$ and $\hat{I}_{22}$ are estimates of I_{11}, I_{12} and I_{22}, respectively, obtained by replacing the parameters by their maximum likelihood estimates.

Example 4: The data given in the Table 1 from Weinberg and Gladden (1986) refer to times taken by couples who were attempting to conceive, until pregnancy results. The data were obtained retrospectively, starting from a pregnancy in each case. Weinberg and Gladen (1986) analyzed fecundity data for a total of 586 women, contributing a total of 1844 cycles. See Weinberg and Gladen (1986) for more details regarding the data. For these data we have combined data for 12 or more cycles. The value of the score test statistic is $Z = 3.42$ indicating a very strong evidence in favor of the beta-geometric distribution. The maximum likelihood estimates of the parameters π and θ obtained by solving equations (I.4) and (I.5) are $\hat{\pi} = 0.36596$ and $\hat{\theta} = 0.0745$ and the standard errors of the estimates

$\hat{\pi}$ and $\hat{\theta}$ are .0162 and .0204, respectively.

Table 1. Data from Weinberg and Gladen (1986) on the number of menstrual cycles to pregnancy

Cycles	Number of Women
1	227
2	123
3	72
4	42
5	21
6	31
7	11
8	14
9	6
10	4
11	7
12	28

Acknowledgments

This research was partially supported by the Natural Sciences and Engineering Research Council of Canada. The work of this chapter was completed while the author was visiting The National University of Singapore.

References

1. Crowder, M.J. Beta-binomial ANOVA for proportions, Applied Statistics **1978**, *27*, 34-37.
2. International Mathematical and Statistical Libraries (IMSL) Manual, 1994.
3. Otake, M.; Prentice, R.L. The analysis of chromosomally aberrant cells based on beta-binomial distribution, Radiation Research **1984**, *98*, 456-470.
4. Paul, S.R. Testing the Goodness of fit of the Geometric Distri-

bution, Unpublished manuscript **2003**.

5. Paul, S.R.; Islam, A.S. Joint estimation of the mean and dispersion parameters in the analysis of proportions: a comparison of efficiency and bias, Canadian Journal of Statistics **1998**, *26*, 83-94.

6. Paul, S.R.; Plackett, R.L. Inference sensitivity for Poisson mixtures, Biometrika **1978**, *65*, 591-602.

7. Paul, S.R.; Saha, K.K.; Balasoorya, U. An empirical investigation of different operating characteristics of several estimators of the intraclass correlation in the analysis of binary data, Journal of Statistical Computation and Simulation, accepted for publication.

8. Prentice, R.L. Binary regression using an extended beta-binomial distribution, with discussion of correlation induced by covariate measurement errors, Journal of the American Statistical Association **1986**, *81*, 321-327.

9. Rao, C.R. Large sample tests of statistical hypotheses concerning several parameters with applications to problems of estimation, Proceedings of the Cambridge Philosophical Society **1947**, *44*, 50-57.

10. Skellam, J.G. A probability distribution derived from the binomial distribution by regarding the probability of success as a variable between the sets of trials, Journal of the Royal Statistical Society **1948**, *10*, 257-261.

11. Tarone, R.E. Testing the goodness of fit of the binomial distribution, Biometrika **1979**, *66*, 585-590.

12. Weil, C.S. Selection of valid number of sampling units and a consideration of their combination in toxicological studies involving reproduction, teratogenesis or carcinogenesis reproduction, teratogenesis, Food and Cosmetic Toxicology **1970**, *8*, 177-182.

13. Weinberg, P.; Gladen, B.C. The beta-geometric distribution applied to comparative fecundability studies, Biometrics **1986**, *42*, 547-560.

14. Williams, D.A. Analysis of binary responses from toxicological experiments involving reproduction and teratogenicity, Biometrics **1975**, *31*, 949-952.

Beta Distribution in Bioassay

R. T. Smythe

Department of Statistics
Oregon State University
Corvallis, Oregon 97331

I. Introduction

One of the many applications of the beta distribution in statistics is in the context of bioassay (biological assays). The most common use of the beta in bioassay is in modeling dispersion of a Bernoulli parameter p. A typical application is in quantal bioassay, where "success" may constitute detection of a tumor of a certain type in a certain organ. In more general settings, such as a multinomial response vector, the multivariate generalization of the beta distribution, the Dirichlet, is often used as a model for the response vector; in this case the beta will of course appear as a model for the marginal distributions.

It is well known that the beta (Dirichlet) is the conjugate prior of the binomial (multinomial). Because of this, most appearances of the beta distribution in bioassay are in Bayesian or Empirical Bayes approaches, where the beta (Dirichlet) serves as a prior distribution for a binomial (multinomial) parameter of interest p. In the Bayesian context, the use of the Dirichlet in bioassay goes back

at least to Ramsey (1972). The best-known use of the beta distribution in an empirical Bayes setting is probably Tarone's (1982)
model incorporating historical controls into a trend test in quantal
bioassay. Our review will concentrate on Bayesian approaches and
on the considerable body of work spawned by Tarone's (1982) paper
aimed primarily at carcinogen bioassay for rodents. For background
on the conduct and analysis of bioassays, the numerous resources include the books of Finney (1964) and Govindarajulu (2001), and the
collections edited by Krewski and Franklin (1991) and Milman and
Weisburger (1985). We first set out some notation and assumptions
for a "typical" bioassay. A substance will be administered (or tested)
at $K + 1$ dose levels $0 = d_0 < d_1 < \cdots < d_K$. (It will be convenient
to represent the possible dose range as $[0, d_{K+1}]$.) There are n_i subjects at dose level $d_i, i = 0, \ldots, K$. We assume that the probability of
response at dose d is represented by $P(d)$. In general, the goal of the
bioassay is to make inferences about the unknown response function
$P(d)$. We take the dose levels, the number of doses, and the number
of subjects at each dose as given. Clearly the efficiency and inter-
pretability of the bioassay depends critically on these choices; for
some purposes, notably carcinogenicity testing on rodents, there are
detailed protocols governing these choices. The responses of subjects
at different dose levels are assumed independent. In most contexts
that we consider, the responses of different subjects at the same dose
level are assumed independent. (This would not be the case, for example, in mutagenicity testing, where "litter effects" are presumed
to be present.)

In quantal bioassay, a two-step analysis is often used. First, a trend
test is made to determine if indeed the response is dose-dependent.
If the answer appears affirmative (i.e. the hypothesis of no trend
is rejected), then one attempts to estimate the dose-response curve
$P(d)$. In Empirical Bayes approaches, the use of the beta distribution
is concentrated mainly in the first of these steps.

II. Bayesian Bioassay

Ramsey's Approach. Ramsey (1972) considers a quantal response bioassay where there may be only one quantal response available at each dose level. He assumes that $P(d)$ is increasing with dose and places a Dirichlet prior distribution on the dose increments $P(d_i) - P(d_{i-1})$, with $P(0) = 0$, $P(d_{K+1}) = 1$. Ramsey constructs this distribution as follows:

Let $\alpha_i, i = 1, \ldots, K+1$ be non-negative constants summing to 1. Let $P(d_1)$ have a beta distribution with parameters $M\alpha_1$ and Ma_1. For $i > 1$, the conditional distribution of $P(d_i)$ given $P(d_1)$, $\ldots$, $P(d_{i-1})$ is a "translated beta distribution" over the interval $(P(d_{i-1}, 1)$, that is,

$$P(d_i)|P(d_{i-1}) = P(d_{i-1}) + (1 - P(d_{i-1})Y,$$

where Y has a beta distribution on $(0, 1)$ with parameters $M\alpha_i$ and Ma_i, and $M, \alpha_i, a_i \geq 0$. Then the marginal distribution for $P(d_i)$ for any i is the beta distribution with parameters MA_i and $M(1 - A_i)$, where $A_i = \sum_{j=1}^{i} \alpha_j$. The parameter A_i is the mean of this distribution and the mode of the density. If one's prior "best guess" for $P(d)$ is $P^*(d)$, then taking

$$\alpha_1 = P^*(d_1), \quad \alpha_i = P^*(d_i) - P^*(d_{i-1}) \ for \ i = 2, \ldots, K,$$

gives a prior distribution with $P^*(d)$ as both the prior modal and mean function. Selection of a prior distribution thus requires specification of a prior modal function $P^*(d)$ and a parameter M, representing in this context the degree of smoothing.

Using the joint mode of the posterior density to summarize the posterior distribution, Ramsey maximizes the joint posterior at $(d_1, \ldots, d_K)$. Because the joint posterior densities are convex and unimodal, this leads to a unique solution $\hat{P}(d)$ when $0 < M < \infty$. Solution of a set of nonlinear equations then provides the values $\hat{P}(d_i)$, and an interpolation formula gives the values of $P(d)$ for d not an observational dose. (In the limiting case $M = 0$, the mode of the posterior is the *isotonic regression* estimator introduced by Ayer et al. (1955).)

Ramsey presents a number of examples to illustrate the method and, in particular, the role of the parameter M. In one of these, he takes the standard cumulative normal for the prior mode and assumes that the actual dose-response curve is a location-shifted standard normal c.d.f. Four possible experimental designs with 6 subjects are considered for estimating the $ED50$, i.e. the dose $\hat{d}$ giving $P(\hat{d}) = .50$. He presents evidence that designs with one observation per dose are superior in this context, an interesting conclusion in view of the fact that standard probit and logit transforms cannot be used in such designs (cf. Finney (1964)). It also appears from Ramsey's examples that the smoothing accomplished by the parameter M in the Bayesian approach improves on isotonic regression, even when an originally poor guess is used to parameterize the prior distribution.

Extending the Bayesian approach: computing posterior distributions. The use of the Dirichlet prior was further developed by Ferguson (1973), who formally introduced the Dirichlet process prior for the dose-response function $P(d)$. Using Ramsey's terminology and notation, given a "prior modal function" P^* and a parameter M (> 0), the induced prior on $P(d)$ will then be a beta distribution with parameters $MP^*(d)$ and $M(1 - P^*(d))$. The induced prior on $\mathbf{p} \equiv (p_1, p_2, \ldots, p_K)$, where $p_i = P(d_i)$, is then an *ordered Dirichlet*

$$\pi_D(\mathbf{p}) = \frac{\Gamma\left(\sum \gamma_i\right)}{\Pi\,\Gamma\left(\gamma_i\right)} p_1^{\gamma_1 - 1}\,(p_2 - p_1)^{\gamma_2 - 1} \cdots$$
$$\times (p_K - p_{K-1})^{\gamma_K - 1}(1 - p_K)^{\gamma_{K+1} - 1},$$

where the sum and product are both over the range $i = 1, \ldots, K + 1$ and

$$\gamma_i = M\{P^*(d_i) - P^*(d_{i-1})\}, \quad i = 1, \ldots, K + 1.$$

Here $P^*(0) = 0$, $P^*(d_{K+1}) = 1$, so that $\sum \gamma_i = M$.

Further extending the Bayesian approach, Gelfand and Kuo (1991) consider this induced prior on $\mathbf{p}$ as well as a product-beta prior, taking the form

$$\pi_B(\mathbf{p}) = c_K(\alpha, \delta) \, \Pi_{i=1}^K \, p_i^{\alpha_i - 1}(1 - p_i)^{\delta_i - 1},$$

where $\alpha = (\alpha_1, \ldots, \alpha_K)$, $\delta = (\delta_1, \ldots, \delta_K)$ and c_K is a normalizing constant. Given that x_i subjects respond at dose d_i, Gelfand and Kuo derive the posterior distribution of $P(d)$ under these classes of prior specifications, and extend the quantal assay approach to ordered polytomous response arising from stochastically ordered dose-response curves.

Let $\mathbf{X}$ denote the response vector $(X_1, \ldots, X_K)$. Computational difficulties in calculating the marginal posterior distributions of $p_i|\mathbf{X}$ and $P(d)|\mathbf{X}$ for specified d had impeded the development of fully Bayesian solutions until the advent of Markovian sampling based approaches (cf. Gelfand and Smith (1990), for example). Gelfand and Kuo use a Gibbs sampler to obtain posterior density estimates under π_D and π_B. The prior π_D is not conjugate for the product binomial likelihood, but the introduction of a set of unobserved multinomial variables simplifies the required sampling and leads to an estimate for the marginal posterior density of p_i. This gives an estimate for the posterior means $E_D(p_i|\mathbf{X})$, and once these have been computed, an interpolation produces the values of $E_D(P(d)|\mathbf{X})$. Gelfand and Kuo apply their method to an an example with 150 subjects at 5 different stimulus levels, obtaining interval estimates for $P(d_i)$ using the empirical distribution from the 1000 replications with the Gibbs sampler.

Sampling using the conjugate prior π_B is somewhat easier because the complete conditional distribution of $p_i|\mathbf{X}, p_j$ $(j \neq i)$ is a translated beta distribution on (p_{i-1}, p_{i+1}) with parameters $\alpha_i + X_i$, $\delta_i + n_i - X_i$. Again a posterior density estimate for p_i is obtained, but $E_B(p_i|\mathbf{X})$ in this case is best obtained directly from the parallel replications in the sampling. A posterior density estimate is obtained as before at, say, $d = d^*$, by including $P(d^*)$ as an additional model parameter, but no simple interpolation formula holds in this case.

Gelfand and Kuo also develop a generalization from the quantal response setting to polytomous response for two different "nested" outcomes. This requires a fairly natural extension of the priors π_D

and π_B. Estimation of the (now) two dose-response curves employs Gibbs sampling much as before. Gelfand and Kuo note that, with either prior, missing data can be readily accommodated in this procedure: should X_i be missing, it can be included as an additional parameter in the model.

Capturing features of the dose-response curve. Further development of nonparametric Bayesian bioassay was carried out by Ramgopal, Laud and Smith (1993), building on an observation by Shaked and Singpurwalla (1990) that the Dirichlet specification denoted above by π_D fails to capture features of the dose-response curve $P(d)$ that are often known to be present in particular applications. They consider three cases: $P(d)$ convex, $P(d)$ concave, and $P(d)$ ogive (changing at some point from convex to concave), and develop priors for p by considering the implications of these three cases for the nonnegative "slope parameters"

$$z_i = \frac{P(d_i) - P(d_{i-1})}{d_i - d_{i-1}}, \quad i = 1, \ldots, K+1.$$

In the case of convexity of $P(d)$, Ramgopal *et al* assign a Dirichlet prior to $\mathbf{u} = (u_1, \ldots, u_{K+1})$, where

$$u_i = (d_{K+1} - d_{i-1})(z_i - z_{i-1}), \quad i = 1, \ldots, K+1,$$

and $z_0 \equiv 0$. Then $u_i | u_1, \ldots, u_{i-1}$ has a translated beta distribution on $(0, 1 - \sum_{j=1}^{i-1} u_j)$ and some algebra reveals that $p_i | p_1, \ldots, p_{i-1}$ also has a translated beta distribution. For the concave prior, the definition of $\mathbf{u}$ is changed to

$$u_i = d_i(z_i - z_{i-1})$$

and the same Dirichlet prior is used for this $\mathbf{u}$. In this case $u_i | u_{i+1}$, $\ldots, u_{K+1}$ has a translated beta distribution on $(0, 1 - \sum_{j=i+1}^{K+1} u_j)$ and $p_i | p_{i+1}, \ldots, p_k$ again has a translated beta distribution. The ogive case combines features of the convex and concave cases, again yielding conditional translated beta distributions for the p_i, but now with $i*$, the change-point, as an added conditioning variable.

The convex and concave cases may be viewed as special cases of the ogive case, with $i* = K + 1$ or $i* = 0$, respectively. To compute posterior distributions for the ogive case, Ramgopal et al. use a Monte Carlo simulation analysis, combining aspects of Gibbs sampling and sampling-importance-resampling (cf. Rubin (1988)). First a Gibbs sampling approach is used for simulating from the joint posterior of p and $i*$, and then sampling-importance-resampling is used to sample efficiently from the resulting one-dimensional translated distributions. Re-analysis of the example used by Gelfand and Kuo reveals a definite ogive form for the dose-response curve, but considerable uncertainty regarding the location of the change-point.

Adaptive estimation. The parameter M in the Dirichlet prior has an interpretation as the strength of the prior belief. (In the limiting case $M \to \infty$, the posterior distribution is concentrated on the prior.) The choice of M is thus a matter of some significance in a Bayesian bioassay. Wesley (1976) (see also Govindarajulu (2001), p. 180) chooses a value of M based on the data, assuming equal sample sizes n at each dose, and equally spaced doses a distance d^* apart. Treating each dose as an independent binomial experiment, the conjugate prior for a dose level d is taken to be a beta distribution with parameters $M\alpha(d)$, $M(1 - \alpha(d))$. The Bayes estimate of p, the probability of response at dose d_i, is then

$$\tilde{p}_i = [M\alpha(d_i) + x]/(M + n),$$

where x is the number of responses in the n trials. With $\alpha_i = \alpha(d_i)$, Wesley's Bayes-binomial estimator for the mean response is then

$$\tilde{\mu} = d_{K+1} + d^*/2 - d^* \sum_{i=1}^{K} \tilde{p}_i.$$

This is similar in form to the nonparametric Spearman-Karber estimate of the mean response, given by

$$\hat{\mu} = d_{K+1} + d^*/2 - d^* \sum_{ij=1}^{K} \hat{p}_i,$$

where $\hat{p}_i = x_i/n$, and in fact we have

$$\tilde{\mu} = \hat{\mu} + \frac{d^* M}{M + n} \sum (\hat{p}_i - \alpha_i).$$

The two estimators will be close if the observed proportion of responses at each dose is close to the prior.

The desired value of M is then found by minimizing (as a function of M) the mean-square error of the Bayes estimate $\tilde{\mu}$. If μ is estimated by $\hat{\mu}$ and p_i by $\hat{p}_i = x_i/n$ in this process, the minimizing value of M is found to be

$$\tilde{M} = \frac{\displaystyle\sum_{i=1}^{K} \left(\hat{p}_i - \hat{p}_i^2\right)}{\left[\displaystyle\sum_{i=1}^{K} (\hat{p}_i - \alpha_i)\right]^2}.$$

The adaptive estimator of the probability of response at d_j is then taken to be

$$\tilde{p}_j(\tilde{M}) = \frac{\tilde{M}\alpha_j}{\tilde{M} + n} + \frac{n}{\tilde{M} + n}\frac{x_j}{n}$$

and the adaptive estimate of the mean is

$$\tilde{\mu}(\tilde{M}) = d_{K+1} + d^*/2 - d^* \sum \tilde{p}_j(\tilde{M}).$$

III.   Empirical Bayes Approaches: Bioassay for Carcinogens

Over the past several decades, the U.S. National Toxicology Program (NTP) has carried out a large-scale program of rodent bioassays, in which suspected carcinogens are administered at several dose levels, generally through feeding (National Toxicology Program (1984)). Similar tests have also been carried out under other auspices. Most of these are lifetime bioassays with the response of interest being the detection of a tumor of a specified type in a given organ (time-to-tumor data is also used, but this will receive only brief mention in our account). This screening of chemicals for their toxicological properties is an important component of environmental health protection. The statistical analysis of these studies typically requires a test to determine if increasing doses of the chemical produce increasing responses. If a trend is observed, an appropriate dose-response model is fitted, which may be used to extrapolate risks to lower measures of exposure more representative of human levels.

Protocols for these experiments require a control (unexposed) group and usually two or more positive levels of exposure, often with equal numbers of subjects at each level. Tests for trend are usually carried out using Fisher's exact test or a Cochran-Armitage test (Cochran, 1954; Armitage, 1955). With notation as before, let $\hat{p}_i = x_i/n_i$, $x. = \sum_{i=0}^{K} x_i$, $n. = \sum_{i=0}^{K} n_i$, $\hat{p} = x./n.$, $\hat{q} = 1 - \hat{p}$. The test statistic for the Cochran Armitage (CA) test is

$$X = \frac{\sum_{i=1}^{K} x_i d_i - \hat{p} \sum_{i=1}^{K} n_i d_i}{\hat{p}\hat{q}\left\{\sum_{i=1}^{K} n_i d_i^2 - \dfrac{\left(\sum_{i=1}^{K} n_i d_i\right)^2}{n.}\right\}},$$

which is asymptotically standard normal under the null hypothesis of no differences in the probability of developing a tumor among the K+1 groups. Tarone and Gart (1980) showed that this test is asymptotically locally optimal against alternatives that can be expressed as a smooth, increasing function of dose.

a. *Historical controls in trend tests*

Tarone's approach. For a particular rodent bioassay, there may be other studies involving the same rodent strain and sex that also contain information about the spontaneous rate of occurrence of the lesion of interest. These "historical controls" may be useful in evaluating results in the exposed groups of the current experiment. Historical control data have long been used in a qualitative way as a supplement to statistical analyses, especially in cases of rare tumors or in interpreting marginally significant results relative to the concurrent controls (Haseman et al., 1984).

The first formal statistical procedure for incorporating historical control data in testing for carcinogenic effects in rodent bioassay was proposed by Tarone (1982). He assumed that the probability p of a tumor occurring in the controls varied across studies in accordance with a beta distribution. This leads to a beta-binomial distribution for the number of animals y_j with tumors in the jth historical control group, out of a total of m_j in the jth group. The mean of this beta-binomial is $m_j\theta$, where θ is the mean of the beta "prior", and the intrastudy correlation is denoted by $0 < \rho < 1$. The intrastudy correlation provides a measure of dispersion in the historical controls, with small values of ρ corresponding to low dispersion.

Assuming that the dose-response function $P(d)$ is given by $H(a + bd)$ for some smooth distribution function H, the trend test amounts to a test of the hypothesis $H_0 : b = 0$. Let $\alpha := \theta/\rho$ and $\beta := (1 - \theta)/\rho$ be the usual parameters of the beta distribution for p. Under the beta-binomial model and assuming a logistic dose-response, the score statistic analogue of the numerator of the CA statistic is

$$T_{HC} = \sum x_i d_i - \tilde{p} \sum n_i d_i,$$

where $\tilde{p} = (x. + \alpha)/(n. + \alpha + \beta)$. Tarone(1982) normalized this statistic by replacing pq in the denominator of the CA statistic with $\tilde{p}\tilde{q}$ and $n.$ with $n. + \alpha + \beta$, where $\tilde{q} = 1 - \tilde{p}$ and α and β are estimated from the historical control series by maximum likelihood. The resulting test statistic can be shown to be asymptotically normal. Tarone applied this procedure to two experiments using rats, one on lung tumor rates with 70 historical controls and another on endometrial stromal polyp, also with 70 historical controls. In the first example, the historical controls are fairly homogeneous ($\hat{\rho} = .002$), and a marginally significant current experiment (P-value $= .044$) becomes highly significant ($P < 10^{-5}$). In the polyp experiment, the historical rates are highly variable ($\hat{\rho} = .062$), and a P-value of .038 is actually raised to .072 by inclusion of the historical data. Tarone notes that "An unstated assumption in the above development has been that the historical control rates used in the analysis come from experiments which are similar to the current experiment in factors known to affect the magnitude of tumor rates." This is an important caveat regarding the use of these methods (cf. Haseman et al. (1984), Gart et al. (1979)).

The variance of the statistic T_{HC} given above is

$$V(T_{HC}) = \frac{\alpha\beta}{(\alpha+\beta)(\alpha+\beta+1)}\left\{\sum n_i d_i^2 - \left(\sum n_i d_i\right)^2/(n+\alpha+\beta)\right\}.$$

An estimator $\hat{V}(T_{HC})$ is obtained by replacing α and β by their maximum likelihood estimates obtained from the historical control series. Instead of using the observed information to normalize as Tarone did, Yanagawa and Hoel (1985) consider the statistic

$$S_{HC} = T_{HC}/[\hat{V}(T_{HC}]^{1/2}.$$

The asymptotic distribution of S_{HC} is a mixture of normal distributions; Krewski et al. (1985) show that it is generally well approximated by a standard normal.

Modifications and extensions: conditioning on x_0. A potential difficulty in the use of historical controls is that the current control study group may appear to be incompatible with the beta prior estimated from the historical control series. In these cases (and possibly more generally) it may be advisable to condition on the value of x_0, the observed tumor incidence in the concurrent controls. The fact that x_0 is an ancillary statistic for the test of H_0 strengthens the case for conditioning (Hoel (1983); Yanagawa and Hoel (1985).) It was noted by Hoel and Yanagawa (1986) that for very small values of θ, the mean of the beta prior, the test statistics T_{HC} may be quite skewed. In such cases an exact conditional test, given x_0, gave P-values considerably less than those given by the Cochran-Armitage test (without historical controls) but greater than those given by using the standard normal approximation to S_{HC}. If t_0 is the observed value of T_{HC}, exact P-values for the conditional test are given by

$$\Sigma'\Pi_{i=1}^{K}\binom{n_i}{x_i}\frac{\Gamma(x.+\alpha)\Gamma(n.-\beta-x.)\Gamma(n_0+\alpha+\beta)}{\Gamma(x_0+\alpha)\Gamma(n_0-\beta-x_0)\Gamma(n.+\alpha+\beta)},$$

where the summation is over all $(x_1, \ldots, x_K)$ satisfying $T_{HC} \geq t_0$ for fixed x_0. Yanagawa, Hoel and Brooks (1989) showed that if the estimate of $\alpha + \beta$ is large compared with n, a slight change in the estimated mean $\hat{\theta}$ of the beta prior can alter the resulting P-values considerably, particularly in the case when x_0/n_0 is quite different from $\hat{\theta}$. These authors proposed a conservative method based on the conditional test that evaluates the P-values in the region constructed by confidence intervals for θ and $\alpha + \beta$.

Hoel and Yanagawa (1986) derive the conditional mean $E(T_{HC}|x_0)$ $= 0$ and the conditional variance

$$V(T_{HC}|x_0) = \rho_0 \tilde{p}_0 (1 - \tilde{p}_0)\{\sum n_i d_i^2 - (\sum n_i d_i)^2/(n. + \alpha + \beta)\},$$

where

$$\rho_0 = (n_0 + \alpha + \beta) / (n_0 + \alpha + \beta + 1)$$

and

$$\tilde{p}_0 = (x_0 + \alpha) / (n_0 + \alpha + \beta) = E(p|x_0).$$

Krewski et al. (1991) derive the limit of $T_{HC}/[V(T_{HC})]^{1/2}$ as n (and x_0) become large, and investigate the finite sample properties of the resulting approximation. The average power of this procedure is shown to be comparable to the power of the unconditional test.

An alternative approach to conditioning (Krewski et al. (1987)) is to select a "tolerance interval" for x_0 based on the beta-binomial distribution resulting from the prior. If x_0 falls in a $100\gamma\%$ interval, where $0 < \gamma < 1$, the test statistic S_{HC} would be used; if x_0 falls outside this interval, the historical control information would be regarded with suspicion and the CA statistic would be used, without historical controls. The optimal choice of γ will depend on the difference between the historical and concurrent control distributions.

Accounting for error in historical control rates. An implicit assumption in the procedures described using historical controls is that the historical control series is sufficiently large that the sampling errors in $\hat{\theta}$ and $\hat{p}$ (equivalently, $\hat{\alpha}$ and $\hat{\beta}$) are negligible. It seems evident, however, that this sampling error should be taken into account in computing $V(T_{HC})$. In addition, simulation evidence

(Tamura and Young, 1986) suggested that there may be considerable bias and variability in the maximum likelihood estimates of α and β, resulting in inflated false positive rates. Although later work of Prentice et al. (1992) indicated that bias can be minimized with appropriate numerical procedures, further simulations by Krewski et al. (1991) confirmed that Type I error rates tend to be inflated with Tarone's method. Furthermore, the likelihood for α and β is quite flat, so the maximum likelihood estimates are rather unstable; slight alterations in the series of historical controls may produce substantial change in the P-values of the test statistics (Smythe et al., 1987). Use of the (θ, ρ) parameterization results in stable estimates of the mean θ but not of the dispersion parameter ρ.

Krewski et al. (1991) used bootstrapping to account for sampling errors, and found that taking variation in $\hat{\alpha}$ and $\hat{\beta}$ into account results in higher P-values in most cases for the observed test statistic (although often lower than P-values without using historical controls). Tamura and Young (1987) proposed the use of a stabilized moment estimator for $\alpha + \beta$, instead of the maximum likelihood estimator. Their simulations indicated that this estimator is less sensitive to small perturbations of the data than the maximum likelihood estimate, as well as being more efficient for small sample sizes. Simulations by Fung et al. (1996) showed that the resulting test statistic controls Type I inflation successfully and has reasonably good power.

Prentice *et al* (1992) developed several new tests for trend in proportions using historical controls. Two of these use beta "priors", but in different ways. In one of these tests, the current experimental responses are assumed not to be overdispersed, but to have binomial distributions with the concurrent control response rate equal to the historical control mean. Thus the historical controls are considered to be independent observations from a beta distribution with mean θ, the same value as the concurrent control rate. Under this assumption, the overall likelihood function

$$L = L_1 L_2$$

is the product of the binomial likelihood L_1 for the current experiment (with a logistic dose-response model) and L_2, a product of beta-binomial likelihoods for the historical control studies. This leads to a test statistic similar to T_{HC}, with $\bar{p}$ replaced by the maximum likelihood estimate of θ. Not surprisingly, the resulting test has high power when the current experiment exhibits little or no overdispersion and the historical control rates are compatible with concurrent controls, but has highly inflated Type I error rate in the presence of significant overdispersion (Fung et al., 1996).

The second method allows for overdispersion in the current experiment by assuming that the current study control response probability θ, like the response probabilities from the historical controls, arises as an independent variate from a beta distribution. The likelihood is again written as $L_1 L_2$, with L_2 as before but with L_1 now reflecting the overdispersion from the beta sampling. The combining of the current and historical likelihoods incorporates error in the beta parameter estimates into the variance of the test statistic. Simulation indicates that the resulting statistic has minimal Type I error inflation and generally good power (Fung et al., 1996).

Priors with heavier tails. A distribution with heavier tails than the beta might be considered to model variability in the historical controls. Such a distribution would be expected to be somewhat less sensitive to perturbations in the historical control series used. Smythe et al. (1987) used a mixture of two beta distributions as a prior distribution for p, using a Γ-minimax criterion to select the mixing proportion. The resulting test statistics correspond roughly to a weighted average (with data-dependent weights) of "with historical controls" and "without historical controls". Their analysis suggests that P-values resulting from using the more robust priors to form the likelihood are less sensitive than beta priors to small perturbations in the historical controls.

Tiwari and Zalkikar (1999) carried this idea further, positing that $p|\eta$ has a beta distribution with parameters $M\eta$ and $M(1-\eta)$, where η follows a beta distribution with parameters b_1, b_2. Although they derive a test statistic conditionally on x_0, they replace x_0 by x. and n_0 by n. in showing asymptotic normality of the statistic. Taking $b_1 = 1$, they consider both moment estimates and maximum likelihood estimates of M and b_2 and conclude that bias of the estimates is not significant, except for estimating b_2 in the case of rare tumors, where some Type I error inflation is observed. For moderate tumors, Type I error rates are generally satisfactory. Their results suggest that their test is more robust under small perturbations of the historical controls than previous tests, although apparently at some price in power of the test.

Time-adjusted trend tests. In order to take into account possible toxic effects of the test chemical, Ibrahim and Ryan (1996) used historical controls in time-adjusted trend tests for carcinogenicity. They assume a multinomial distribution to model the number of animals dying with tumors in each discretized time interval, and put a Dirichlet prior on this multinomial. Then the conditional probability of a control animal dying with a tumor in a given time interval, given that it is alive at the start of the interval, is given by a beta distribution. Because these are a posteriori independent, the resulting Dirichlet-multinomial likelihood reduces to a product of beta-binomial likelihoods. A score test for trend is derived which generalizes that given by Tarone (1982). The method is applied to an example with rats exposed to bromide in feed for two years; an effect not significant with a log-rank test is detected by their analysis using historical controls.

b. Historical controls in modeling dose-response relationships

Thus far we have described the use of historical controls, based on a beta prior, in testing for trend in a quantal bioassay. In cases in which a dose-response relationship exists, similar methods may be employed to model the dose-response curve and to estimate quantiles of the curve such as the *ED10*. Smythe et al. (1986) use a beta prior on the control response and an uninformative prior on the remaining parameters of the dose-response curve. Estimating the beta parameters from the historical control series and regarding them as known, the tumor incidences in the dose groups will have beta-binomial distributions, and the parameters of the dose-response curve may be estimated using this historical control information.

For a two-parameter logistic dose-response, comparison of the empirical Bayes estimates with the maximum likelihood estimates without historical control information shows appreciable gains in efficiency using historical controls, in the case where the prior is centered on the response rate in the concurrent control group. For a three-parameter independent background logistic model, the historical control series provides little information on parameters other than the control response rate, and efficiency gains in the quantile estimates were smaller. It should be noted that here, as in Tarone's use of historical controls in testing for trend, error in the estimation of the beta parameters has not been accounted for. In addition, the problem of possible incompatibility between the historical and concurrent controls is again present.

IV. Discussion and Summary

The beta distribution (and its generalization, the Dirichlet) provide a model for variability of a success parameter (or parameters) in some widely used types of bioassay. Bayesian and Empirical Bayes approaches have been described involving the use of beta (Dirichlet) priors and mixtures of betas. In the Bayes context, sampling approaches developed within the last fifteen years have made calculation of posterior distributions feasible. The principal application using Empirical Bayes approaches has been the incorporation of historical control data into tests for trend in quantal bioassay for carcinogens. A number of issues associated with this are discussed, including methods for estimating parameters, conditioning on ancillary statistics, and robustness of estimates. The research described has made important contributions to the use of Bayesian and Empirical Bayes methods in bioassay.

References

1. Armitage, P. Tests for linear trend in proportions and frequencies, Biometrics **1955**, *11*, 375-386.

2. Ayer, M.; Brunk, H.D.; Ewing, G.M.; Reid, W.T.; Silverman, E. An empirical distribution function for sampling with incomplete information, Annals of Mathematical Statistics **1955**, *26*, 641-647.

3. Cochran, W.G. Some methods of strengthening the common χ^2 tests, Biometrics **1954**, *10*, 417-451.

4. Ferguson, T.S. A Bayesian analysis of some nonparametric problems, Annals of Statistics **1973**, *1*, 209-230.

5. Finney, D.J. *Statistical Method in Biological Assay* (second edition); Hafner: New York, 1964.

6. Fung, K.Y.; Krewski, D.; Smythe, R.T. A comparison of tests for trend with historical controls in carcinogen bioassay, Canadian Journal of Statistics **1996**, *24*, 431-454.

7. Gart, J.J.; Chu, K.C.; Tarone, R.E. Statistical issues in interpretation of chronic bioassay tests for carcinogenicity, Journal of National Cancer Institute **1979**, *62*, 957-974.

8. Gelfand, A.E.; Kuo, L. Nonparametric Bayesian bioassay including ordered polytomous response, Biometrika **1991**, *78*, 657-666.

9. Gelfand, A.E.; Smith, A.F.M. Sampling based approaches to calculating marginal densities, Journal of the American Statistical Association **1990**, *85*, 398-409.

10. Govindarajulu, Z. *Statistical Techniques in Bioassay* (second edition); Karger: Basel, Freiburg, 2001.

11. Haseman, J.K.; Huff, J.; Boorman, G.A. Use of historical control data in carcinogenicity studies in rodents, Toxicologic Pathology **1984**, *12*, 126-135.

12. Hoel, D.G. Conditional two-sample tests with historical controls, in *Contributions to Statistics: Essays in Honor of Norman L. Johnson* (editor P.K. Sen) **1983**, pp. 229-236, North-Holland: Amsterdam.

13. Hoel, D.G.; Yanagawa, T. Incorporating historical controls in

testing for trend in proportions, Journal of the American Statistical Association **1986**, *81*, 1095-1099.

14. Ibrahim, J.G.; Ryan, L.M. Use of historical controls in time-adjusted trend tests for carcinogenicity, Biometrics **1996**, *52*, 1478-1485.

15. Krewski, D.; Franklin, C. (editors) *Statistics in Toxicology*, Gordon and Breach: New York, 1991.

16. Krewski, D.; Smythe, R.T.; Burnett, R.T. The use of historical control information in testing for trend in quantal response carcinogenicity data, in *Proceedings of the Symposium on Long-Term Animal Carcinogenicity Studies: A Statistical Perspective* **1985**, pp. 56-62, American Statistical Association: Washington, DC.

17. Krewski, D.; Smythe, R.T.; Colin, D. Tests for trend in binomial proportions: A proposed two-stage procedure, in *Advances in the Statistical Sciences, Vol. V: Biostatistics* (editors I.B. MacNeil and G.J. Umphrey) **1987**, pp. 61-69, Reidel: Boston.

18. Krewski, D.; Smythe, R.T.; Fung, K.Y.; Burnett, R.T. Conditional and unconditional tests with historical controls, Canadian Journal of Statistics **1991**, *19*, 407-423.

19. Milman, H.A.; Weisburger, E.K. (editors) *Handbook of Carcinogen Testing*, Noyes: Park Ridge, NJ, 1985.

20. NTP: National Toxicology Program. *Report of the NTP Ad Hoc Panel on Chemical Carcinogenesis Testing and Evaluation*; U. S. Department of Health and Human Services: Washington, DC, 1984.

21. Prentice; R.L.; Smythe, R.T.; Krewski, D.; Mason, M. On the use of historical control data to estimate dose response trends in quantal bioassay, Biometrics **1992**, *48*, 459-478.

22. Ramgopal, P.; Laud, P.W.; Smith, A.F.M. Nonparametric Bayesian bioassay with prior constraints on the shape of the potency curve, Biometrika **1993**, *80*, 489-98.

23. Ramsey, F.L. A Bayesian approach to bioassay, Biometrics **1972**, *28*, 841-858.

24. Rubin, D.B. Using the SIR algorithm to simulate posterior distributions, in *Bayesian Statistics 3* (editors J.M. Bernardo *et*

al), Oxford University Press, **1988**, pp. 395-402.

25. Shaked, M.; Singpurwalla, N.D. A Bayesian approach for quantile and response probability estimation with applications to reliability, Annals of the Institute of Statistical Mathematics **1990**, *42*, 1-19.

26. Smythe, R.T.; Krewski, D.; Dewanji, A. Robust tests for trend in binomial proportions, in *Probability and Bayesian Statistics* (editor R. Viertl) **1987**, pp. 443-454, Plenum: New York and London.

27. Smythe, R.T.; Krewski, D.; Murdoch, D. The use of historical control information in modelling dose response relationships in carcinogenesis, Statistics and Probability Letters **1986**, *4*, 87-93.

28. Tamura, R.N.; Young, S.S. The incorportation of historical control information in tests of proportions: simulation study of Tarone's procedure, Biometrics **1986**, *42*, 343-349.

29. Tamura, R.N.; Young, S.S. A stabilized moment estimator for the beta-binomial distribution, Biometrics **1987**, *43*, 813-824.

30. Tarone, R.E. The use of historical control information in tests of proportions, Biometrics **1982**, *38*, 215-220.

31. Tarone, R.E.; Gart, J.J. On the robustness of combined tests for trend in proportion, Journal of the American Statistical Association **1980**, *75*, 110-116.

32. Tiwari, R.C.; Zalzikar, J.N. Tests for a trend in proportion based on mixtures of beta distributions incorporating historical controls, Environmetrics **1999**, *10*, 1-22.

33. Wesley, M.N. Bioassay: estimating the mean of the tolerance distribution, Stanford University Technical Report No: 17 (1 R01 GM 21215-01), **1976**.

34. Yanagawa, T.; Hoel, D.G. Use of historical controls in animal experiments, Environmental Health Perspectives **1985**, *63*, 217-224.

35. Yanagawa, T.; Hoel, D.G.; Brooks, G.T. A conservative use of historical data for a trend test in proportions, Journal of the Japan Statistical Society **1989**, *1*, 83-94.

Adaptive Economic Choices under Recurrent Disasters: A Bayesian Perspective

Manish C. Bhattacharjee[*]

Center for Applied Mathematics and Statistics
Department of Mathematical Sciences
New Jersey Institute of Technology
Newark, New Jersey 07102

I. Introduction

Consider a geographical region susceptible to some natural disaster that occurs randomly over time. Assume that the economic impact of such disasters is limited to the destruction of capital and corresponding investments in place, which consequently interrupts the stream of economic returns, and that after a disaster, there is no delay to reinvest in a feasible economic activity from a set of available alternatives.

Given the profiles of economic activities one may choose from and a parametrically specified model for the stochastic occurrences of the natural disaster over time, the basic problem of choosing an economic strategy which maximizes the expected total discounted stream of returns is conceptually simple enough. If the length of time between

---
[*]This article was prepared while the author was on a sabbatical leave at the Indian Statistical Institute.

consecutive disasters depends *neither* on the economic activities undertaken, *nor* on the past history of such disasters; it is not hard to show that an activity which maximizes the expected return between consecutive disasters is optimal for the infinite horizon.

The situation is not so simple however, if the parameters governing the process of disasters are unknown. In such a case, it is natural to assume a prior on the parameter space and then proceed to determine the best economic choices which are typically adaptive to our evolving beliefs about the parameters through successive posterior distributions. The Bayesian paradigm thus provides a natural framework for the evaluation of our economic options, whether one is a Bayesian by choice or, forced to be one, to model our relative ignorance about the parameters.

With the advent of Markov Chain Monte Carlo (MCMC) methods and algorithms, a computationally intensive approach to our problem is feasible. Complementary to such methods are Bayesian models that are analytically tractable, and when appropriate as a model for the natural disaster, provides a deeper insight into the dynamic nature of sequential investment choices to be made and their asymptotics. It should be remarked here that the specific physical character of the natural disaster is, in a broad sense, not fundamental to our analysis, so long as the distributional assumptions about the disaster event (e.g., floods, earthquakes) are appropriate. The work summarized here, based on an unpublished technical report (1968), focuses on such a Bayesian model and its analysis, in which the family of *inverted Beta distributions* play an important role.

II. The Model

A. Economic Activities

An economic activity is an ordered triple $(a, b(\cdot), m)$ such that $a > 0$, $0 \leq b(\cdot)$ on $[0, \infty)$ and $0 < m \leq \infty$. Here $a > 0$ is the set up cost representing the capital needed to build the corresponding physical infrastructure and technology. The function $b(t)$ describes the net rate of benefits at time t accruing from the operating activity, conditional on no disaster since its inception. The parameter m denotes the (possibly infinite) value of the activity's operating time, if reached without interruption by an intervening disaster, when a planned replacement by rebuilding the same activity is scheduled.

The case $m = \infty$ corresponds to these economic activities which have no scheduled replacement, and will be referred to as activities of "type-I". Such activities continue to operate until destroyed by a disaster. On the other hand, an economic activity is of "type-II" if the benefit rate $b(t)$ is such that it is either necessary, or considered desirable to rebuild the corresponding technology after it has continuously operated for a finite time $m > 0$. A "one-hass shay" benefit function $b(t) = b1_{\{t \le m\}}$ where $b > 0$ and $m > 0$ are given constants, and 1_A denotes the indicator function of a set A is an example of the former; while the exponentially decaying benefit $b(t) = b\exp(-et)$, $b > 0$, $c > 0$ is an example of the latter. For an economic activity of type-II, there is typically a positive probability of any number of planned replacements between two consecutive disasters.

At time $t > 0$, the present value of a continuously operating activity, with set up cost $a > 0$ and benefit rate $b(\cdot)$, is

$$W(t) = -a + \int_0^t b(x)\exp(-\rho x)dx, \quad t > 0, \tag{II.1}$$

where $\rho > 0$ is the discount rate. To guarantee finiteness of $W(t)$ for all $t > 0$, assuming $\sup_{t>0} b(t) < \infty$ will suffice. Assuming $b(0+) < \infty$ (a finite benefit rate at inception of the activity) will often be sufficient in practice; since in many situations, the benefit rate is typically nonincreasing ($b(t) \downarrow$). Further, since there is no point in engaging in an economic activity from the investor's point of view unless $W(t) > 0$ for same t onward; we additionally assume without loss of generality that all activities are *productive* in the sense that the corresponding present value function $W(t) = 0$ has a finite solution $t \in (0, \infty)$; equivalently if $W(\infty) := \lim_{t \to \infty} W(t) > 0$, since $W(t)$ is monotone $\uparrow$ on $(0, \infty)$ with

$$-a = W(0+) \le W(t) \uparrow W(\infty) \le -a + \rho^{-1} \sup_{t>0} b(t) < \infty$$

as $t \uparrow \infty$.

Consider an activity of type-II with a replacement time m. The *net present value* (NPV) $W^*(t)$ of income from such an activity continuously operating in $(0, t)$ is

$$W^*(t) = \begin{cases} W(t), & \text{if } 0 \leq t < m \\ \left(\displaystyle\sum_{k=0}^{j-1} \exp(-k\rho m) \right) W(m) \\ \quad + \exp(-j\rho m)W(t - jm), & \text{if } jm \leq t < (j+1)m, \end{cases}$$

where $j = 1, 2, \ldots$. Hence, for a cycle of random length T between two consecutive disasters, this income is

$$W^*(T) = \sum_{j=0}^{\infty} 1_{B_j} \left[\left(\frac{1 - \exp(-j\rho m)}{1 - \exp -(\rho m)} \right) W(m) \right.$$
$$\left. + \exp(-j\rho m)W(T - jm) \right] \qquad (\text{II.2})$$

where 1_{B_j} are indicators of the disjoint events

$$B_j = \{ jm \leq T < (j+1)m \}.$$

Note that if $m = \infty$, the above reduces to $W^*(T) \equiv W(T)$ so that (II.2) subsumes activities of type-I.

The financial impact of an activity $(a, b(\cdot), m)$ described by its contribution to the income stream, subject to renewals between disasters is fully captured by $EW^*(T)$, where T denotes a typical inter-occurrence time of disasters. Thus, the physical basis of an economic activity as parameterized via $(a, b(\cdot), m)$ is, in a mathematical sense, essentially redundant for computing the expected value of an income stream so long as we can specify $EW^*(T)$. Note $EW^*(T)$ is a function of the underlying parameters (hyper-parameters, in the Bayesian case) governing the law of the disaster process.

Assume that there are N *productive* economic activities, one must choose from to invest in after each disaster. The ith activity

$$(a_i, b_i(\cdot), m_i), \quad m_i \leq \infty \quad i = 1, 2, \ldots, N$$

is simply denoted by i, for brevity. The set of available choices is $A = \{1, 2, \ldots, N\}$, where each activity $i \in A$ is either of type-I or, type-II.

To avoid trivialities, we may assume that all available activities in A are *essential* and *relevant*, properties which we now define.

Call two activities i, i' to be *equivalent* if $EW_i^*(T) = EW_{i'}^*(T)$. Clearly, there is no reason to prefer an activity $i \in A$ over others, if any, in A which are equivalent. An activity $i \in A$ is *essential* if there is no other activity $i' \in A$ such that i and i' are equivalent.

An activity $i \in A$ is *relevant* if there does *not* exist an activity $i' \in A$ which dominates i in the sense $EW_{i'}^*(T) \geq EW_i^*(T)$ with strict inequality for some value of the underlying parameter(s) of the disaster process.

A given activity $(a, b(\cdot), m)$ can have many additive decompositions ($\oplus$),

$$(a, b(\cdot), m) := \left(a', b'(\cdot), m\right) \oplus \left(a'', b''(\cdot), m\right),$$

where $a = a' + a''$, $b(\cdot) = b'(\cdot) + b''(\cdot)$ and $m \leq \infty$. Note that the decomposed constituents on the right hand side may be *virtual activities* in that they may not be physically available choices. An example of such a decomposition, which we will find useful in our analysis (Section III) is,

$$(a, b(\cdot), m) = (a, 0, m) \oplus (0, b(\cdot), m).$$

Finally, note that while we assume that a disaster destroys an ongoing economic activity; our framework nevertheless can allow for restarting the same after repairs subsequent to a disaster. If the impact of a disaster on an activity $(a, b(\cdot), m) \in A$ is limited to partial destruction; it may be possible to salvage and restart the same physical activity, usually with a benefit rate $b'(t)$ typically smaller than the old $b(t)$, but also usually with a smaller setup cost $a' < a$. If $b(t)$ is $\downarrow$, then a typical example of such a relatively diminished return function after repair is $b'(t) = b(t + l), t > 0$, where $l > 0$ is a parameter that reflects how effective the repaired technology is. Such a salvaged activity may be included among the available choices in A, so long as it is still *productive, essential* and *relevant*.

B.   The Disaster Process, States and Transitions

Let $T_1, T_2, \ldots$ be the inter-occurrence times between disasters. Assume $\{T_n, n \geq 1\}$ to be conditionally i.i.d. exponential with a rate $\lambda > 0$, given λ. In other words, we assume that there is an environmental variable Λ which describes the proneness of disasters such that $\{T_n, n \geq 1\}$ are conditionally i.i.d exponential with rate λ, given $\Lambda = \lambda$. Since the realized value of Λ is unknown, we adopt a Bayesian posture based on past disasters to assign a suitable prior distribution to Λ which is then revised by successive posterior distributions, as we learn more and more about Λ with subsequent disasters as they are observed.

To do this, we track the number (r) of disasters to date and the calendar time (t) when disasters strike. The ordered pair (r, t) summarizes our experience of disasters and describes the set of possible states, which are

$$\mathcal{S} = \{(r, t) : t \geq 0; \ r = 0, 1, 2, \ldots\}.$$

If $\{T_n, n \geq 1\}$ is a sequence of inter-disaster times beginning at the state (r, t), then T_j is the time spent in the jth cycle between the $(r + j - 1)$st and $(r + j)$th disasters. The corresponding state transition is from the state $(r + j - 1, t + S_{j-1})$ to $(r + j, t + S_j)$, where $S_j = \sum_{k=1}^{j} T_k$, $j \geq 1$, $S_0 = 0$. To describe the probability law of the sequence of inter-disaster times, beginning at (r, t), which are conditionally i.i.d. exponential given Λ; we assign to Λ a gamma-prior $G(\cdot | r_0 + r, t_0 + t) \in \mathcal{G}$ for some $r_0 = 0, 1, 2, \ldots$, and $t_0 \geq 0$, where

$$\mathcal{G} = \{G : G = G(.|r, t); \ r > 0; \ t \geq 0\}$$

is the family of *gamma* distributions

$$G(\lambda | r, t) = \frac{t^r}{\Gamma(r)} \int_0^\lambda \exp(-tx) x^{r-1} dx, \quad \lambda > 0.$$

Let

$$\mathcal{F} = \{F : F = F(\cdot | p, q, b); \ p > 0, \ q > 0, \ b > 0\}$$

be the family of *inverted Beta* distributions,

$$F(y|p,q,b) = \frac{\Gamma(p+q)b^q}{\Gamma(p)\Gamma(q)} \int_0^y \frac{x^{p-1}}{(b+x)^{p+q}}dx, \quad y > 0 \qquad (\text{II.3})$$

which correspond to the familiar Beta distributions (Johnson and Kotz, 1970) on the unit interval in the following way. If X is a r.v. with a standard beta density

$$\text{Const. } x^{p-1}(1-x)^{q-1}, \ 0 < x < 1, \ p > 0, \ q > 0$$

then the r.v. $Y = bX/(1-X)$, $b > 0$ has the inverted Beta distribution $F(\cdot|p,q,b)$ in (II.3. The following facts about the probability law $(\mathcal{L})$ of the sequence $\{T_n, n \geq 1\}$ are now standard.

Lemma II.1. *For the inter-occurrence times $\{T_1, T_2, \ldots\}$ between disasters beginning at the state (r,t), as specified above,*

(i) $\mathcal{L}(T_1) = F(\cdot|1, r_0 + r, t_0 + t)$

$$\mathcal{L}(T_{n+1}|T_1, \ldots, T_n) = F(\cdot|1, r_0 + r + n, \ t_0 + t + S_n), \ n \geq 1$$

from which all finite dimensional distributions can be computed.
(ii) *The posterior of Λ at the state $(r+n, t+S_n)$ is*

$$\mathcal{L}(\Lambda|T_1, \ldots, T_n) = G(\cdot|r_0 + r + n, \ t_0 + t + S_n), \ n \geq 1.$$

(iii) *The unconditional distribution of the total time up to n disasters, is $\mathcal{L}(S_n) = F(\cdot|n, r_0 + r, t_0 + t)$ $n \geq 1$.*

Our activities and corresponding incomes evolve as follows. A typical transition is $(r,t) \to (r+1, t+T)$ where T is the cycle time to the next disaster. At a state (r,t), we can choose any economic activity $i \in A$, the set of available choices, which then operates until the next disaster at $(r+1, t+T)$ when we collect an economic return that has a present value $\exp(-\rho t)W_i^*(T)$. The transition probability $q(\cdot|r,t)$ which specifies the distribution of the next state $(r+1, t+T)$ is a probability measure on the states S such that for every bounded measurable $h : S \to (-\infty, \infty)$,

$$\int_S h(\cdot, \cdot) dq(\cdot, \cdot | r, t)$$

$$= \int_0^\infty h(r+1, t+T) dF(y|1, r_0 + r, t_0 + t)$$

$$= (r_0 + r)(t_0 + t)^{r_0 + r} \int_0^\infty \frac{h(r+1, t+y)}{(t_0 + t + y)^{r_0 + r + 1}} dy.$$

An (adaptive) investment program π is a policy in the sense of Blackwell (1965, 1965) that chooses economic activities in A, or more generally a family of conditional distributions (defining a randomized choice mechanism) on the set A of available activities, possibly depending on our past experience of the system's history.

A (non-randomized) *stationary* investment program is specified by a function $J : S \to A$ such at any state $(r, t) \in S$, we choose the economic activity $J(r, t) \in A$. Such stationary policies are conceptually the simplest and intuitively appealing, as they require only a memory of the current state to choose an economic activity to invest in. Since $A = \{1, 2, \ldots, N\}$ is finite, a stationary investment program is equivalent to specifying a disjoint partition $\{A_1, \ldots, A_N\}$ of S such that for states in A_j, we invest in the economic activity $j \in A$ until the next disaster.

III. Optimal Bayesian Policies

In a series of papers, Blackwell (1962, 1965) and Strauch (1966) articulated a rigorous theoretical framework to accommodate and extend Richard Bellman's (2003) ideas and methods for dynamic programming, in three broad categories: the positive, negative and discounted cases. These respectively correspond to the stepwise income at each transition being either positive (profit maximization problems) or, negative (cost minimization problems) or, discounted by a deterministic factor γ^n at the n-th transition, $0 < \gamma < 1$. Since the income $W^*(T)$ between disasters can be negative or positive and the discount factors $\exp(-\rho S_n)$ are stochastic, the model formulated in Section II is thus a *non-standard* dynamic programming problem.

To put our approach to the solution of the economic choice problem on a rigorous footing, there is thus a need to show that expected incomes of all policies in our setup is finite and then find appropriate technical means to identify an optimal investment policy.

A.   Income of Adaptive Programs

The NPV of the income $i(h)$ associated with any history $h = ((r,t), i_1, (r+1, t+S_1), i_2, (r+2, t+S_2), \ldots)$ of the system, is

$$i(h) = \sum_{n=0}^{\infty} \exp\{-\rho(t+S_n)\} W_{i_n}^*(T_{N+1}) := \exp(-\rho t) i^*(h),$$

where $i^*(h)$ is the *spot-value* of the income stream at the initial state (r,t). A policy π in the sense defined in Section II - specifying the activity choices, induces a probability on the set of histories. The expected income of a policy π starting at (r,t) is clearly of the form

$$U_\pi(r,t) = \exp(-\rho t) U_\pi^*(r,t), \tag{III.1}$$

where

$$U_\pi^*(r,t) = E_\pi i^*(h) = E_\pi \left\{ \sum_{n=0}^{\infty} \exp(-\rho S_n) W_{i_n}^*(T_{n+1}) \right\} \tag{III.2}$$

is the spot-value of π at (r,t). The optimal return function is

$$U(r,t) := \sup_\pi U_\pi(r,t) = \exp(-\rho t) U^*(r,t),$$

using (III.1), where $U^*(r,t) := \sup_\pi U_\pi^*(r,t)$ is the *spot-value of the optimal return*. To show $|U_\pi^*(r,t)| < \infty$ for all π and all $(r,t) \in S$; set

$$m_0 := \min_{i \in A} m_i \leq \infty,$$

$$w_0 := \max_{i \in A} \sup_{t>0} |W_i(t)| < \infty, \tag{III.3}$$

$$w^* := \left[1 + \{1 - \exp(-\rho m_0)\}^{-1} \right] w_0 < \infty.$$

m_0 is thus the smallest replacement time among the available activities, with $m_0 = \infty$ iff all activities in A are type-I. In virtue of (II.2);

for any activity $i \in A$, we have

$$
|W_i^*(T)| \leq \sum_{j=0}^{\infty} 1_{B_j} \left(\frac{1 - \exp\left(-j\rho_{m_i}\right)}{1 - \exp\left(-\rho m_i\right)} + \exp\left(-j\rho m_i\right) \right) w_0
$$

$$
\leq \left\{ \frac{1}{1 - \exp\left(-\rho m_i\right)} + 1 \right\} w_0
$$

$$
\leq w^*.
$$

Accordingly, from (III.2),

$$
|U_\pi^*(r,t)| \leq E_\pi \sum_{n=0}^{\infty} \exp\left(-\rho S_n\right) \left|W_{i_n}^*(T_{n+1})\right|
$$

$$
\leq w^* \sum_{n=0}^{\infty} E\left\{\exp\left(-\rho S_n\right)\right\} < \infty, \qquad \text{(III.4)}
$$

provided the sum converges to a finite value for all $(r,t) \in S$. Note, the expected values in the last step no longer depend on π. The terms

$$
\psi_n(\rho; r, t) := E\left\{\exp\left(-\rho S_n\right)\right\}
$$

$$
= E_\Lambda \left\{E(\exp\left(-\rho(T_1 + \ldots + T_n)\right)|\Lambda)\right\}, \; n \geq 1
$$

$$
= E_\Lambda \left[E^n \left\{\exp\left(-\rho T_1\right)|\Lambda\right\}\right],
$$

since $T_1, T_2, \ldots$ are conditionally i.i.d., given Λ. The random variable

$$
\alpha(\Lambda) := E\left\{\exp\left(-\rho T_1\right)|\Lambda\right\} = \frac{\Lambda}{\Lambda + \rho} \in (0,1),
$$

since $T_1|\Lambda$ has distribution $G(\cdot|1,\Lambda)$, i.e. exponential with rate Λ. Hence,

$$
\begin{aligned}
\sum_{n=0}^{\infty} E\left\{\exp\left(-\rho S_n\right)\right\} &= 1 + \sum_{n=1}^{\infty} \psi_n(\rho; r, t) \\
&= \sum_{n=0}^{\infty} E_\Lambda \left\{\alpha\left(\Lambda\right)\right\}^n \\
&= E_\Lambda \left(1 + \rho^{-1}\Lambda\right) \\
&= \int_0^\infty \left(1 + \rho^{-1}\lambda\right) dF\left(\lambda|1, r_0 + t, t_0 + t\right) \\
&= 1 + \frac{(r_0 + r)}{\rho\,(t_0 + t)} < \infty.
\end{aligned}
$$

B. Rebuilding Costs and Static Policies

In our quest for an optimal policy, it will be useful for technical reasons, to first consider the rebuilding cost associated with a *static policy* that always chooses a fixed activity $(a, b(\cdot), m) \in A$, irrespective of our current state and the corresponding posterior distribution of the environmental variable Λ. Conceptually we may think of such an activity decomposed as the "*sum*" of two imaginary activities

$$(a, b(\cdot), m) = (a, 0, m) \oplus (0, b(\cdot), m)$$

so that the spot-value of a static rebuilding policy always using the fixed activity at left is the sum of the corresponding spot values of the two hypothetical static policies defined by $(a, 0, m)$ and $(0, b(\cdot), m)$. The latter gives the expected spot value of the stream of benefits, while the negative of the spot value of the former represents the expected rebuilding costs. Let $K(r, t; a, m)$ denote the spot value of total expected rebuilding costs of the static policy using $(a, b(\cdot), m)$, beginning at (r, t), i.e. the spot value of the total expected costs associated with $(a, 0, m)$. By (III.4), $0 < K(r, t; a, m) < \infty$.

To describe $K(r, t; a, m)$, first note that the spot value $c(r, t; a, m)$ of the expected rebuilding costs in a typical cycle (r, t) to $(r + 1, t +$

T) between two disasters, is

$$c(r, t; a, m) = a \int_0^\infty \{1 - \exp(-\lambda m)\}^{-1} dG(\lambda | r_0 + r, t_0 + t),$$

$$(\text{III.5})$$

since, if $N(T)$ is the number of planned replacements in $[t, t + T)$; then,

$$E(N(T)|\Lambda) = \sum_{j=0}^\infty j \Pr(jm \leq T < (j+1)m|\Lambda)$$

$$= \exp(-\Lambda m) \{1 - \exp(-\Lambda m)\}^{-1},$$

so that,

$$c(r, t; a, m) = E[a\{1 + N(T)\}]$$

$$= a[1 + E_\Lambda\{E(N(T)|\Lambda)\}]$$

$$= a\left[1 + \int_0^\infty \frac{\exp(-\lambda m)}{1 - \exp(-\lambda m)} dG(\lambda | r_0 + r, t_0 + t)\right],$$

which leads to (III.5). Writing $N'(\cdot) \equiv 1 + N(\cdot)$ for brevity; the spot value of the total stream of random rebuilding costs is

$$a \sum_{n=0}^\infty \exp(-\rho S_n) N'(T_{n+1})$$

$$= aN'(T_1) + \exp(-\rho T_1)$$

$$\times \left[a \sum_{n=2}^\infty \exp\{-\rho(T_2 + \ldots + T_n)\} N'(T_{n+1})\right].$$

The second term within the braces is simply the actual rebuilding costs starting at $(r + 1, t + T_1)$ for the same static policy. Thus averaging both sides,

$$K(r, t; a, m) - c(r, t; a, m)$$

$$= E\{\exp(-\rho T)\} K(r + 1, t + T; a, m) \qquad (\text{III.6})$$

$$= \int_0^\infty \exp(-\rho y) K(r + 1, t + y; a, m) dF(y | 1, r_0 + r, t_0 + t).$$

For a type-I activity, $m = \infty$ and $c(r, t; a, \infty) \equiv a$, as is also clear by letting $m \to \infty$ in (III.5). The corresponding total expected rebuilding costs $K(r,t,a) := \lim_{m\to\infty} K(r,t;a,m)$ satisfy the simpler equation

$$K(r,t;a) = a + E\{\exp(-\rho T)\} K(r+1,t+T;a). \qquad \text{(III.7)}$$

C. Finding An Optimal Policy

Upper bounds for the spot value. The spot value of the expected rebuilding costs associated with any policy π is bounded above by

$$k^*(r,t) = \begin{matrix} K\,(r,t;a^*,m_0)\,, & \text{if } m_0 < \infty \\ K\,(r,t;a^*)\,, & \text{if } m_0 = \infty, \end{matrix} \qquad \text{(III.8)}$$

where $m_0 = \min_{i\in A} m_i \le \infty$ is defined in (III.3) and $a^* = \max_{i\in A} a_i$. Hence, the spot value of total expected income satisfies $U_\pi^* \ge -k^*$ pointwise, for all π. Accordingly,

$$U^*(r,t) := \sup_\pi U_\pi^*(r,t) \ge -k^*(r,t), \quad \text{all } (r,t). \qquad \text{(III.9)}$$

To search for reasonable upper bounds for U^*, we may therefore restrict ourselves to

$$Q = \{Q|Q : \mathcal{S} \to (-\infty,\infty),\ Q \ge -k^* \text{ on } \mathcal{S}\}.$$

Consider the linear operator L carrying measurable functions on Q into itself, such that

$$LQ(r,t) := \max_{i\in A} \int_0^\infty \{W_i^*(y) + \exp(-\rho y)Q(r+1,t+y)\}$$
$$\times dF\,(y|1, r_0+r, t_0+t)$$
$$\equiv \max_{i\in A} E\,\{W_i^*(T) + \exp(-\rho T)Q(r+1,t+T)\}.$$

To check that $Q \in Q$ implies $LQ \in Q$, we argue as follows. Corresponding to the activity $i \in A$ defined by $(a_i, b_i(\cdot), m_i)$, let $(a_i, 0, m_i)$ define a *virtual activity* i' (which need not be in A). Then,

$$EW_i^*(T) \ge EW_{i'}^*(T) = -c(r,t;a_i,m_i) \ge -c(r,t;a^*,m_0)$$

where, the function c is defined in (III.5). Together with $Q \ge -k^*$

pointwise, since $Q \in \mathcal{Q}$; the above implies

$$E\left\{W_i^*(T) + \exp(-\rho T)Q(r+1, t+T)\right\}$$
$$\geq -\left[c(r, t; a^*, m_0) + E\left\{\exp(-\rho T)\right\}k^*(r+1, t+T)\right]$$
$$= -k^*(r, t)$$

for all $i \in A$, the last being true in view of (III.6), (III.7) and (III.8). Hence $LQ \geq -k^*$.

Lemma III.1. $Q \in \mathcal{Q}$, $LQ \leq Q \Rightarrow U^* \leq Q$ *pointwise.*

Proof. Consider a hypothetical problem with the same states and transition probabilities except that each economic activity $i \in A = \{1, 2, \ldots, N\}$ is replaced by an hypothetical *action* i^* such that choosing i^* at (r, t) yields an income with present value $\exp(-\rho t)\widehat{W}_{i^*}(T)$ realized at the next state $(r+1, t+T)$, where

$$\widehat{W}_{i^*}(T) = W_i^*(T) + a^*\left\{1 + N(T, m_0)\right\}, \tag{III.10}$$

where $N(t, m_0)$ is the number of planned replacements between disasters, of an activity with scheduled replacement time m_0; i.e. $N(T, m_0) = n$ on $\{nm_0 \leq T < (n+1)m_0\}$, $n = 0, 1, 2, \ldots$ if $m_0 < \infty$, and $N(T, m_0) = 0$ w.p. 1 if $m_0 = \infty$ (all activities are type - I). In any case, $m_0 \leq m_i \leq \infty$ and $a_i \leq a^* < \infty$, implies

$$\widehat{W}_{i^*}(T) \geq W_i^*(T) + a_i\left\{1 + N(T, m_i)\right\} \geq 0,$$

since the second term, being the actual rebuilding costs of $i \in A$ in the time interval $[t, t+T)$, represents the negative part of $W_i^*(T)$. The hypothetical problem so defined, with A replaced by the set of *virtual activities* $A^* = \{1^*, 2^*, \ldots, N^*\}$, is a *positive dynamic programming problem* in the sense of Blackwell (1965). Clearly, the optimal income's spot value for the "positive" problem has the form

$$\widehat{U}(r, t) = U^*(r, t) + k^*(r, t)$$

where U^* is the optimal spot value function of our original problem and the second term k^* represents the expected contribution of the stream of second terms in (III.10) over entire histories. If $\widehat{L}$ is the operator

$$\widehat{L}h(r,t) := \max_{i^* \in A^*} E\left\{\widehat{W}_{i^*}(T) + \exp(-\rho T)h(r+1, t+T)\right\}$$

carrying non-negative functions h on the (r,t)-plane into itself; then Blackwell's (1965) results imply that any function $h \geq 0$ that satisfies $\widehat{L}h \leq h$ must itself be an upper bound on $\widehat{U}$. If Q satisfies the hypothesis in the statement of Lemma 2, then, $\widehat{Q} := Q + k \geq 0$ is such a function; viz.

$$\widehat{L}\widehat{Q}(r,t) = \max_{i^* \in A^*} E\left[\widehat{W}_{i^*}(T) + \exp(-\rho T)\widehat{Q}(r+1, t+T)\right]$$

$$= \max_{i \in A} E\left[W_i^*(T) + a^*\{1 + N(T, m_0)\}\right.$$

$$\left. + \exp(-\rho T)\{Q(r+1, t+T) + k^*(r+1, t+T)\}\right]$$

$$= LQ(r,t) + E\left[a^*\{1 + N(T, m_0)\}\right.$$

$$\left. + \exp(-\rho T)k^*(r+1, t+T)\right].$$

In view of (III.6), (III.7) and (III.8), the second term equals $k^*(r,t)$. Hence,

$$\widehat{L}\widehat{Q}(r,t) = LQ(r,t) + k^*(r,t) \leq Q(r,t) + k^*(r,t) \equiv \widehat{Q}(r,t).$$

This in term implies $U^* + k^* = \widehat{U} \leq \widehat{Q} = Q + k^*$, or $U^* \leq Q$ pointwise. $\square$

Set,

$$A_j = \left\{(r,t) : EW_j^*(T) = L0(r,t)\right\}, \quad j \in A, \tag{III.11}$$

where

$$L0(r,t) = \max_{i \in A} EW_i^*(T)$$

$$= \max_{i \in A} \int_0^\infty W_i^*(y)dF(y|1, r_0 + r, t_0 + t)$$

$$= \max_{i \in A} \int_0^\infty E(W_i^*(T)|\lambda)\, dG(\lambda|r_0 + r, t_0 + t),$$

is the spot value of the maximal one-cycle income. Without loss of generality, $\{A_1, \ldots, A_N\}$ may be taken as a disjoint partition of S.

Let $g : S \to A = \{1, 2, \ldots, N\}$ such that

$$g(r, t) = j, \quad \text{if } (r, t) \in A_j. \tag{III.12}$$

Theorem III.2. *The stationary adaptive investment policy defined by g is optimal.*

Proof. The spot value function U_g of the stationary policy defined by g in (III.12), obviously satisfies

$$\begin{aligned}
U_g(r, t) &= E\left\{W^*_{g(r,t)}(T) + \exp(-\rho T)U_g(r + 1, t + T)\right\} \\
&= L0(r, t) + E\left\{\exp(-\rho T)\right\} U_g(r + 1, t + T) \\
&= LU_g(r, t).
\end{aligned}$$

Further, $U_g \geq -k^*$ as argued in the remark following (III.8). Hence the optimal spot value $U^* \leq U_g$ by Lemma 2; while $U_g \leq U^*$ necessarily. Thus $U_g(r, t) = U^*(r, t)$ which proves the optimality of g.

$\square$

The optimal Bayesian spot value is

$$\begin{aligned}
&U^*(r, t) \\
&= U_g(r, t) \\
&= \sum_{n=0}^{\infty} E\left\{\exp\left(-\rho S_n\right) W^*_{g(r+n,t+S_n)}(T_{n+1})\right\} \\
&= \sum_{n=0}^{\infty} E\left\{\exp\left(-\rho S_n\right) \sum_{j=1}^{N} W^*_j(T_{n+1}) 1_{A_j}(r + n, t + S_n)\right\} \tag{III.13} \\
&= \sum_{n=0}^{\infty} \sum_{j=1}^{N} \int_0^{\infty} \int_0^{\infty} 1_{A_j}(r + n, t + x) \exp(-\rho x) \\
&\qquad\qquad \times W^*_j(y) dH_1(y|x) dH_2(x)
\end{aligned}$$

where the integrating measures are the *inverted beta* distributions

$$\begin{aligned}
H_1(y|x) &= F\left(y|1, r_0 + r + n, t_0 + t + x\right), \\
H_2(x) &= F\left(x|n, r_0 + r, t_0 + t\right)
\end{aligned}$$

defined in (II.3). Clearly (III.13) shows that U^* cannot in general be evaluated in a closed form. For a given set of economic activities with specified set up costs and benefit rates, first the optimal partition $\{A_j : j = 1, 2, \ldots, N\}$ and then U^* can be numerically evaluated using (III.11)–(III.13).

Static rebuilding programs. These correspond to stationary policies that repeatedly choose a fixed activity in a A, and are typically not optimal, although they are conceptually appealing and among the simplest. Another reason they are useful is that such policies play an important role in the asymptotic behavior of the optimal return $U^*(r, t)$, defined in (III.9), under fairly reasonable conditions on the benefit rates, as we show in Section (IV.

The Bayesian spot value $U_0(r, t)$ of a static policy has a simple representation

$$U_0(r, t) = \int_0^\infty V(\lambda) dG\left(\lambda | r_0 + r, t_0 + t\right), \qquad \text{(III.14)}$$

where $V(\lambda)$ is the corresponding spot value in the non-Bayesian case, i.e. when the environmental variable Λ is fixed at a value $\lambda > 0$. To justify (III.14), suppose the fixed economic activity defining the static policy has a planned replacement time $m \leq \infty$. Since given $\Lambda = \lambda$, the time T between disasters is exponential; the memoryless property of exponential distributions imply that the fixed activity defining the stationary policy is *renewed* after operating for time $\min(T, m)$, so that V satisfies the renewal type equation

$$V = E_\lambda \left[W(\min(T, m)) + \exp\{-\rho \min(T, m)\} V \right], \qquad \text{(III.15)}$$

which gives

$$V(\lambda) = \frac{E_\lambda W(\min(T, m))}{E_\lambda \left[1 - \exp\{-\rho \min(T, m)\} \right]}, \quad \lambda > 0. \qquad \text{(III.16)}$$

For a type-I activity, this takes a simpler form

$$V(\lambda) = \left(1 + \frac{\lambda}{\rho}\right) E_\lambda W(T), \quad \lambda > 0. \qquad \text{(III.17)}$$

While (III.16) is the preferred way to compute V for a type -II activity, it can also be expressed in a form analogues to (III.17), resulting from the equation

$$V = E_\lambda \left\{ W^*(T) + \exp(-\rho T) V \right\}$$

instead of (III.15), where $W^*(T)$ given in (II.2) is the income between consecutive disasters, allowing for possibly multiple planned replacements in between. Thus implies

$$V(\lambda) = \frac{E_\lambda W^*(T)}{1 - \alpha(\lambda)} = \left(1 + \frac{\lambda}{\rho}\right) E_\lambda W^*(T), \tag{III.18}$$

where

$$\alpha(\lambda) := E_\lambda \exp(-\rho T) = \lambda/(\lambda + \rho) \in (0, 1).$$

Equation (III.17) of course corresponds to (III.18), when $m = \infty$. The claim (III.14) is now immediate. By (III.2), a static policy's spot value function is

$$U_0(r, t) = E \sum_{n=0}^{\infty} \exp(-\rho S_n) W^*(T_{n+1})$$

$$= \sum_{n=0}^{\infty} E\left[E\{\exp(-\rho S_n) W^*(T_{n+1})| T_1, \ldots, T_n, \Lambda\}\right]$$

$$= \sum_{n=0}^{\infty} E\left[E(\exp(-\rho S_n)| \Lambda)\right.$$

$$\left. \times E\{W^*(T_{n+1})| T_1, \ldots, T_n, \Lambda\}\right].$$

Since T_n are conditionally i.i.d. given Λ; the nth term above, conditional on $\Lambda = \lambda$, is

$$E_\lambda^n\{\exp(-\rho T)\} E_\lambda W^*(T) \equiv \alpha^n(\lambda) E_\lambda W^*(T), \quad n \geq 0$$

which implies

$$U_0(r, t) = \sum_{n=0}^{\infty} E[\alpha^n(\Lambda) E(W^*(T)|\Lambda)]$$

$$= E\left\{\frac{E(W^*(T)|\Lambda)}{1 - \alpha(\Lambda)}\right\}$$

$$= \int_0^\infty \frac{E_\lambda W^*(T)}{1 - \alpha(\lambda)} dG(\lambda|r_0 + r, t_0 + t).$$

Together with (III.18), this proves (III.14).

As remarked in Section I, in the non-Bayesian case, an optimal policy is always static. McGuire et al. (1972) consider some illustrative examples of benefit rates and the spot values $V(\lambda)$ for the corresponding static policies when the disasters are assumed to be a Poisson process with a known rate λ. The following is a summary of their results, which we then use to evaluate the respective Bayesian spot value functions, by exploiting the representation (III.14).

Activities

No decay	$(a, b(t) \equiv b, \infty)$, $b > 0$
Delayed benefits	$(a, b(t) = b1_{\{t>l\}}, \infty)$, $b > 0$
One-hoss shay	$(a, b(t) = b1_{\{t<m\}}, m)$ $b > 0, l > 0$
Exponential decay	$(a, b(t) = b\exp(-\delta t), m)$ $b > 0, m > 0, \delta > 0$

The spot value functions $U_0(r,t)$, using (III.14), (III.16)–(III.17) are

No decay:

$$V(\lambda) = -a(1 + \rho^{-1}\lambda) + \rho^{-1}b,$$

$$U_0(r,t) = EV(\lambda) = \frac{b}{\rho} - a\left(1 + \frac{r_0 + r}{\rho(t_0 + t)}\right).$$

Delayed benefits:

$$V(\lambda) = \rho^{-1}b\exp\{-(\lambda + \rho)l\} - a\left(1 + \rho^{-1}\lambda\right),$$

$$U_0(r,t) = EV(\lambda) = \frac{b}{\rho}\left(\frac{t_0 + t}{t_0 + t + l}\right)^{r_0+r}\exp(-\rho l)$$
$$-a\left(1 + \frac{r_0 + r}{\rho(t_0 + t)}\right).$$

As $l \to 0$, this reduces to 'no decay'.
One-hoss shay:

$$V(\lambda) = \frac{b}{\rho} - \frac{a(\lambda + \rho)}{\rho[1 - \exp\{-(\lambda + \rho)m\}]}.$$

With the prior $G(\cdot|k,c)$ of λ, where $k \equiv r_0 + r$, $c \equiv t_0 + t$; we have

$$
E_\Lambda \left[\frac{\Lambda + \rho}{1 - \exp\{-(\Lambda + \rho)m\}} \right]
$$

$$
= \frac{c^k}{\Gamma(k)} \sum_{j=0}^{\infty} \int_0^{\infty} (\lambda + \rho) \exp\{-j(\lambda + \rho)m\} \exp(-c\lambda)\lambda^{k-1} d\lambda
$$

$$
= \frac{c^k}{\Gamma(k)} \sum_{j=0}^{\infty} \exp(-j\rho m)
$$

$$
\times \int_0^{\infty} (\lambda + \rho) \exp\{-(c + jm)\lambda\} \lambda^{k-1} d\lambda
$$
(III.19)

$$
= c^k \sum_{j=0}^{\infty} \exp(-j\rho m) \left\{ \frac{k}{(c+jm)^{k+1}} + \frac{\rho}{(k+jm)^k} \right\}
$$

$$
= c^k \sum_{j=0}^{\infty} \frac{\exp(-j\rho m)}{(c+jm)^k} \left(\rho + \frac{k}{c+jm} \right).
$$

Accordingly, the Bayesian spot-value is

$$
U_0(r,t) = \frac{b}{\rho} - \frac{a}{\rho}(t_0 + t)^{r_0+r} \sum_{j=0}^{\infty} \frac{\exp(-j\rho m)}{(t_0 + t + jm)^{r_0+r}}
$$

$$
\times \left(\rho + \frac{r_0 + r}{t_0 + t + jm} \right).
$$

As $m \to \infty$, we get

$$
U_0(r,t) \to \frac{b}{\rho} - \frac{a}{\rho} \left(\rho + \frac{r_0 + r}{t_0 + t} \right),
$$

the case of 'no-decay'.

For the case of exponential decay, we have

$$
V(\lambda) = \left(1 + \frac{\lambda}{\rho}\right) \left[-a + \frac{b[1 - \exp\{-(\lambda + \rho + \delta)m\}]}{\lambda + \rho + \delta} \right]
$$

$$
\Big/ [1 - \exp\{-(\lambda + \rho)m\}]^{-1}.
$$

The Bayesian spot value $EV(\Lambda)$ is rather messy. For exponential decay with no planned replacement ($m = \infty$),

$$
U_0(r,t) = E \left\{ \left(1 + \frac{\Lambda}{\rho}\right) \left(-a + \frac{b}{\Lambda + \rho + \delta}\right) \right\}
$$

can be evaluated by computing integrals along the lines of (III.19).

IV. Asymptotic Behavior

As we have seen, whenever the set of activities contains two or more choices, the return of the optimal Bayesian policy is usually not readily computable in a simple way. An investigation of the asymptotic behavior of the optimal spot value shows that with type-I activities, an investor, who pretends to behave as if Λ is known by setting the value of Λ as the Bayes' estimator with squared error loss will be close to optimal, for initial states which are suitably "large". This nearly Bayesian optimal behavior is a static rebuilding policy.

Let $V_j(\lambda)$ be the non-Bayesian spot value of the static policy using activity $j \in A; j = 1, 2, \ldots, N$, and let

$$\widetilde{V}(\lambda) = \max_{j \leq N} V_j(\lambda), \quad \lambda > 0$$

be their envelope.

Theorem IV.1. *If all available activities are of type-I with set up costs and benefit rates $(a_j, b_j(\cdot))$, such that each $b_j(t)$ is nonincreasing and differentiable with $\sup_{t>0} b_j(t) < \infty$, $j = 1, 2, \ldots, N$; then*

$$\left| U^*(r,t) - \widetilde{V}(\beta) \right| \to 0$$

as $r, t \to \infty$ along the paths $r \sim \beta t$ (i.e. along any locus on the (r,t)-plane, such that $r/t \to \beta > 0$).

To prove our claim, we will exploit the following technical results. Recall, $W(t)$ of (II.1) is the cumulative return at time t from an uninterrupted activity in $(0, t)$ with expected value $g(\lambda) := E_\lambda W(T)$. Let $H(\theta) := g(\theta^{-1})$ be this expected value as a function of the mean time θ between two disasters.

Lemma IV.2. *(a) If $b(t)$ is differentiable a.e., $\sup_{t>0} b(t) < \infty$, then*
i) $g(\lambda)$ is convex and nonincreasing.
ii) W and H both satisfy the Lipschitz condition of order 1.
(b) If $b(t)$ is nonincreasing, then $W(t)$ is concave.

Lemma IV.3. *The Bayesian optimal spot value U^* satisfies*

$$U^*(r,t) \leq \int_0^\infty \tilde{V}(\lambda)dG\left(\lambda|r_0 + r, \ t_0 + t\right). \qquad \text{(IV.1)}$$

Proof of Lemma 3. (a). By (II.1) and interchanging the order of integration

$$g(\lambda) = E_\lambda W(T) = -a + \int_0^\infty \int_0^t \exp(-\rho y)b(y)\lambda \exp(-\lambda t)dydt$$

$$= -a + \int_0^\infty \exp\left\{-(\lambda + \rho)y\right\}b(y)dy$$

which gives,

$$\dot{g}(\lambda) = -\int_0^\infty \exp\left\{-(\lambda + \rho)y\right\}yb(y)dy \leq 0,$$

$$\ddot{g}(\lambda) = \int_0^\infty \exp\left\{-(\lambda + \rho)y\right\}y^2 b(y)dy \ \geq 0,$$

for all $\lambda > 0$. For any $t, t' > 0$; setting $t_0 = \min(t, t')$, $t_1 = \max(t, t')$; again from (II.1),

$$|W(t) - W(t')| = \left|\int_{t'}^t b(y)\exp(-\rho y)dy\right|$$

$$= |t - t'|\, b(\xi)\exp(-\rho\xi), \ \text{some} \ \xi \in (t_0, t_1)$$

$$\leq b_0\, |t - t'|,$$

where $b_0 = \sup_{t>0} b(t) < \infty$. To show H also satisfies the Lipschitz condition, write

$$g(\lambda) = \lambda \int_0^\infty W(t)\exp(-\lambda t)dt = \int_0^\infty W(y/\lambda)\exp(-y)dy,$$

so that for any positive θ, θ'

$$|H(\theta) - H(\theta')| \equiv \left|g\left(\frac{1}{\theta}\right) - g\left(\frac{1}{\theta'}\right)\right|$$

$$\leq \int_0^\infty |W(\theta y) - W(\theta' y)|\exp(-y)dy \qquad \text{(IV.2)}$$

$$\leq b_0 \int_0^\infty |\theta - \theta'|\, y\exp(-y)dy$$

$$= b_0\, |\theta - \theta'|$$

(b) Finally note, if $0 \leq b(t)$ is decreasing then (II.1) implies

$$\dot{W}(t) = \exp(-\rho t) b(t) \geq 0$$

is decreasing, since $b(t)$ is. Hence W is concave. Note that $0 \leq b(t) \downarrow$ is sufficient for W to be concave, and $b(t)$ need not be differentiable. □

Proof of Lemma 4. We show, the function $\tilde{Q}(r,t) := E\tilde{V}(\Lambda)$ defined in the righthand side of (IV.1), satisfies the hypothesis of Lemma 2, which will prove our claim. If $U_j(r,t) := EV_j(\Lambda)$ denotes the spot value at (r,t) of the static policy that always uses activity $j \in A$, then

$$\tilde{Q}(r,t) \geq E\tilde{V}(\Lambda) = EV_j(\Lambda) = U_j(r,t) \geq -k^*(r,t),$$

where the last inequality holds in virtue of the argument preceding (III.9). To prove $L\tilde{Q} \leq \tilde{Q}$, consider $E\{W_j^*(T) + \exp(-\rho T)\tilde{Q}(r + 1, t + T)\}$. The second term, by appealing to Lemma 1, and writing $k = r_0 + r, c = t_0 + t$ for brevity, is

$$E\left\{\exp(-\rho T) \int_0^\infty \tilde{V}(\lambda) dG\left(\lambda | r_0 + r + 1, t_0 + t + T\right)\right\}$$

$$= \int_0^\infty \int_0^\infty \exp(-\rho y)\tilde{V}(\lambda) dG\left(\lambda | r_0 + r + 1, t_0 + t + y\right)$$
$$\times dF\left(y | 1, r_0 + r, t_0 + t\right)$$

$$= \int_0^\infty \int_0^\infty \exp(-\rho y)\tilde{V}(\lambda) \frac{(c+y)^{k+1}}{\Gamma(k+1)} \exp\left\{-(c+y)\lambda\right\}$$
$$\times \frac{k(\lambda c)^k}{(c+y)^{k+1}} d\lambda$$

$$= \int_0^\infty \left(\int_0^\infty \exp\left\{-(\lambda + \rho)y\right\} dy\right) \tilde{V}(\lambda) \frac{(c\lambda)^k}{\Gamma(k)} \exp\left(-c\lambda\right) d\lambda$$

$$= \int_0^\infty \alpha(\lambda)\tilde{V}(\lambda) dG(\lambda | k, c),$$

where $\alpha(\lambda) = E_\lambda \exp(-\rho T) = \lambda/(\lambda + \rho)$. Hence, for any $j \in A$, the above and an appeal to (III.18) gives,

$$E\left\{W_j^*(T) + \exp\left(-\rho T\right)\tilde{Q}(r+1, t+T)\right\}$$

$$= \int_0^\infty E_\lambda W_j^*(T)dG\left(\lambda|r_0 + r, t_0 + t\right)$$

$$+ E\left\{\exp(-\rho T)\int_0^\infty \tilde{V}(\lambda)dG\left(\lambda|r_0 + r + 1, t_0 + t + T\right)\right\}$$

$$= \left[\{1 - \alpha(\lambda)\}V_j(\lambda) + \alpha(\lambda)\tilde{V}(\lambda)\right]dG\left(\lambda|r_0 + r, t_0 + t\right)$$

$$\leq \int_0^\infty \tilde{V}(\lambda)dG\left(\lambda|r_0 + r, t_0 + t\right) \equiv \tilde{Q}(r, t)$$

since $0 < \alpha(\lambda) < 1$ and $V_j(\lambda) \leq \tilde{V}(\lambda)$ all $\lambda > 0$, and all $j \in A$. Thus $L\tilde{Q} \leq \tilde{Q}$. □

Proof of Theorem 2. Let $U_j(r, t)$ denote the Bayesian spot value of the static policy using activity $j \in A$. For fixed $\lambda > 0$, the expected income $g_j(\lambda)$ of type-I activity j between two disasters, satisfies

$$g_j(\lambda) \geq g_j(\beta) + (\lambda - \beta)\dot{g}_j(\beta), \quad \text{all } \lambda > 0, \ \beta > 0$$

where the slope $\dot{g}_j \leq 0$ since $g_j(\lambda)$ is convex and nonincreasing in λ (lemma 3) under the stated assumptions. Hence, for the corresponding static policy,

$$V_j(\lambda) \geq \left(1 + \rho^{-1}\lambda\right)g_j(\beta) + \left\{\rho^{-1}\lambda^2 + \lambda\left(1 - \rho^{-1}\beta\right) - \beta\right\}\dot{g}_j(\beta)$$

by a reference to (III.17), so that (III.14) then implies

$$U_j(r, t) = \int_0^\infty V_j(\lambda)dG\left(\lambda|r_0 + r, t_0 + t\right)$$

$$\geq \left(1 + \frac{r_0 + r}{\rho(t_0 + t)}\right)g_j(\beta) + \dot{g}_j(\beta)I(r, t; \rho, \beta), \tag{IV.3}$$

where

$$I(r, t; \rho, \beta) := E_\lambda\left\{\rho^{-1}\Lambda^2 + \Lambda(1 - \rho^{-1}\beta) - \beta\right\}$$

$$= \frac{(r_0 + r)(r_0 + r + 1)}{\rho(t_0 + t)^2} + \frac{r_0 + r}{t_0 + t}\left(1 - \frac{\beta}{\rho}\right) - \beta. \tag{IV.4}$$

Note, since each benefit rate $b_j(t)$ is nonincreasing, we have

$$0 \le -\dot{g}_j(\beta) = \int_0^\infty \exp\{-(\beta + \rho)t\} \, t b_j(t) dt \le \frac{b^*}{(\beta + \rho)^2} < \infty,$$

for all $j \le N$, where

$$b^* = \max_{j \le N} \sup_{t > 0} b_j(t) = \max_{j \le N} b_j(0+) < \infty. \tag{IV.5}$$

Consider any locus on the (r, t)-plane such that $r \sim \beta t$, i.e. such that $r = \beta t + o(t)$. Denote by $\liminf_{r \sim \beta t}$ ($\limsup_{r \sim \beta t}$, respectively) the operation $\liminf_{t \to \infty}$ ($\limsup_{t \to \infty}$, respectively), as $t \to \infty$, along any path on the (r, t)-plane for which $r \sim \beta t$. Then from (IV.3)–(IV.4),

$$\liminf_{r \sim \beta t} U_j(r, t) = (1 + \rho^{-1}\beta) \, g_j(\beta) + \dot{g}_j(\beta) \lim_{t \to \infty} I(\beta t, t; \rho, \beta)$$
$$= V_j(\beta) + \dot{g}_j(\beta) \left\{ \rho^{-1}\beta^2 + \beta(1 - \rho^{-1}\beta) - \beta \right\}$$
$$= V_j(\beta).$$

As the optimal spot value U^* clearly satisfies $U^*(r, t) \ge U_j(r, t)$, for all $j \in A$, we get,

$$\liminf_{r \sim \beta t} U^*(r, t) \ge \liminf_{r \sim \beta t} U_j(r, t) \ge V_j(\beta), \quad \text{all } j \in A$$

so that,

$$\liminf_{r \sim \beta t} U^*(r, t) \ge \max_{j \le N} V_j(\beta) = \tilde{V}(\beta). \tag{IV.6}$$

On the other hand, by Lemma 3(a) -ii), viz. (IV.2),

$$|g_j(\lambda) - g_j(\beta)| \le b^* \left| \lambda^{-1} - \beta^{-1} \right|, \quad \text{all } \lambda > 0, \beta > 0$$

for each activity j, where b^* is given by (IV.5). Thus,

$$\tilde{V}(\lambda) = (1 + \rho^{-1}\lambda) \max_j g_j(\lambda)$$
$$\le (1 + \rho^{-1}\lambda) \max_j g_j(\beta) + b^* (1 + \rho^{-1}\lambda) \left| \lambda^{-1} - \beta^{-1} \right|.$$

Hence, an appeal to Lemma 4 yields,

$$U^*(r, t) \le \left(1 + \frac{r_0 + r}{\rho(t_0 + t)} \right) \max_j g_j(\beta) + b^* \Delta(r, t). \tag{IV.7}$$

where

$$0 < \Delta(r,t) = \int_0^\infty \left(1 + \rho^{-1}\lambda\right) \left|\lambda^{-1} - \beta^{-1}\right| dG\left(\lambda|r_0 + r, t_0 + t\right) \to 0,$$

as $r, t \to \infty$ such that $r \sim \beta t$, since

$$\Delta^2(r,t)$$
$$= E^2 \left\{ (1 + \rho^{-1}\Lambda) \left|\Lambda^{-1} - \beta^{-1}\right| \right\}$$
$$\leq E \left\{ (1 + \rho^{-1}\Lambda)^2 \right\} E \left\{ \left(\Lambda^{-1} - \beta^{-1}\right)^2 \right\}$$
$$= \left\{ 1 + \frac{2}{\rho} \frac{r_0 + r}{t_0 + t} + \frac{1}{\rho^2} \frac{(r_0 + r)(r_0 + r + 1)}{(t_0 + t)(t_0 + t + 1)} \right\}$$
$$\times \left\{ \frac{1}{\beta^2} - \frac{2}{\beta} \frac{t_0 + t}{r_0 + r - 1} + \frac{(t_0 + t)^2}{(r_0 + r - 1)(r_0 + r - 2)} \right\}$$

implies

$$\lim_{r \sim \beta t} \Delta^2(r,t) = \lim_{t \to \infty} \Delta^2(\beta t, t)$$
$$= (1 + \rho^{-1}\beta)^2(\beta^{-2} - 2\beta^{-2} + \beta^{-2})$$
$$= 0.$$

By (IV.7), we then have

$$\limsup_{r \sim \beta t} U^*(r,t) \leq \left(1 + \rho^{-1}\beta\right) \max_j g_j(\beta) + b^* \lim_{t \to \infty} \Delta(\beta t, t)$$
$$= \tilde{V}(\beta).$$

Together with (IV.6), this yields the desired conclusion. □

When available choices include activities of type-II, the asymptotic behavior of the optimal Bayesian policy in Theorem 2, remains an open question as of this writing.

References

1. Bellman, R.E. *Dynamic Programming*; Dover Publications, 2003.
2. Bhattacharjee, M.C. Optimal investments in the face of floods,

Working Paper # 259, Center for Research in Management Science, University of California, Berkeley, 1968.

3. Blackwell, D. Discounted dynamic programming, *Annals of Mathematical Statistics* **1965**, *36*, 226-235.

4. Blackwell, D. Positive dynamic programming, in *Proceedings of the Fifth Berkeley Symposium on Mathematical Statistics and Probability* **1965**, volume 1, pp. 415-418, University of California Press.

5. Johnson, N.L.; Kotz, S. *Distributions in Statistics: Continuous Univariate Distributions* (volume 2); John Wiley and Sons: New York, 1970.

6. McGuire, C.B.; Brown, J.P.; Contini, B. An economic model of flood plain use, *Water Resources Research* **1972**.

7. Strauch, R.E. Negative dynamic programming, *Annals of Mathematical Statistics* **1966**, *37*, 871-890.

Probabilistic Analysis of Machine Tool Errors Using Beta Distribution

Kyoung-Gee Ahn

Department of Mechanical Engineering
University of Michigan
2250 Hayward, Ann Arbor, Michigan 48109

I. Introduction

Theoretical and empirical analyses of uncertainty in manufacturing processes are very important in designing and manufacturing engineering products on a rational basis. Uncertainty signifies a degree of discrepancy between an idealized object and its physical realization. Such discrepancy inevitably creeps into our product realization processes because of practical cost considerations or our inability to fully control manufacturing processes. Uncertainty in machine tools comes from many sources. Estimating the total uncertainty is, however, difficult, due to the fact that the uncertainties vary with the task being performed, the environment, the operator, the chosen processes, etc. Here we refer to the sources of uncertainty caused by errors inherent to the design of the machine, its geometry, its dynamics, the environment in which it is placed, machine temperatures and vibration, elastic deformation of the tool and workpiece due to cutting forces, workpiece loading errors, clamping effects, the

uncorrected parts of the 21 parametric errors, and random components of these parametric errors. This research crucially addresses
the stochastic nature of errors and models it systematically. Individual machine tool errors are divided into two categories, systematic
errors and random errors, which can be interpreted as deterministic
values and probabilistic distribution of random variation, respectively (Slocum, 1992; Weck, 1980). However, the variation is small
compared with the error itself. It is therefore convenient to divide
the error into a constant part and a variable part. We may use the
mean value of the error as the constant component of an error. Based
on this convention, the error E may be written as:

$$E = E_d + E_s,$$

where E_d and E_s are the deterministic and stochastic parts of the
error E, respectively. For many production machines, the stochastic
part of the error is defined as repeatability that accounts for a significant part of the total error (Shin et al., 1991), and two or three
standard deviations of the repeat measurement error have been used
as the random error (Slocum, 1992; Pahk, 1990; Weck, 1980).

Stochastic errors become more significant as higher positional accuracy is required or smaller parts are to be machined. When machining small parts, the repeatability, determined by stochastic error
components, may affect the final tolerance more than the mean positional accuracy. The deterministic part, if known, can be compensated for, but the stochastic part cannot be corrected. Since stochastic errors like deterministic parts vary from one position to another, a
working volume must be carefully selected instead so that repeatability within that volume has an envelope lower than required tolerance
(Shin et al., 1992). This consequently requires a complete analytical
modeling of stochastic errors. Figure 1 shows the importance of modeling stochastic errors. In Figure 1a the gray hexahedron represents
the stochastic error volume of a machine tool before compensating
for deterministic errors and the white hexahedron indicates the desired tolerance volume. Figure 1b shows that the deterministic error
can be eliminated if the error can be predicted and compensated
for, and if the stochastic variation of the errors is within the desired

tolerance volume. If, however, the random deviation of the errors is too large, then only a certain number of the parts produced on this machine will meet the specified tolerance requirement and the remaining parts will exceed the allowable tolerance. This can be represented by the remaining gray volume when the white volume is removed in Figure 1b. This illustrates the importance of stochastic deviation of errors.

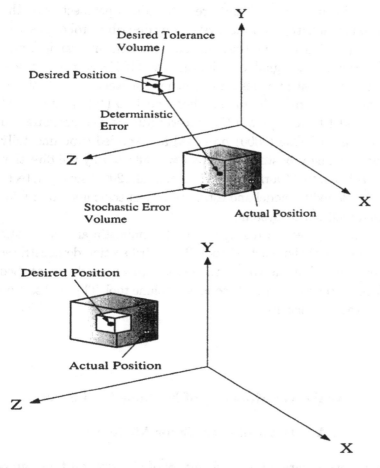

Figure 1. Comparison between before and after compensation of deterministic error, the top figure represents before compensation (Figure 1a) while the bottom represents after compensation (Figure 1b).

In recent years, researchers have become interested in the uncertainty analysis of errors. Shen and Duffie (1991, 1995) presented the uncertainty analysis method for coordinate referencing, and used the uncertainty interval concept to describe the essential characteristics of uncertainty sources in coordinate referencing and coordinate transformation relationships. Shin and Wei (1992) placed great importance on the random error of machine tools and tried to express it. Qian and Kazerounian (1996) presented a new perspective on the calibration of industrial robotics systems by describing robot position errors as variation in time that can be separated into random system variation and assignable variation. Yau (1997) proposed a new rotation vector that provides a more general mathematical basis for representing vectorial tolerances. Chen and Rao (1997) introduced a fuzzy model to manipulate the uncertainties in the optimization process. Ahn and Cho (2000) proposed and verified experimentally that the machine tool stochastic error was affected by the direction of approach of a tool/workpiece. Wilhelm et al. (2001) surveyed techniques developed to model and estimate task-specific uncertainty for coordinate measuring systems.

The following section discusses the deterministic and stochastic error modeling of the machine tool. The third section demonstrates the probability that the volumetric errors are within the prescribed tolerances in the case of the three-axis machine tool. The final section presents the conclusions.

II. Analytical Modeling of Machine Tool Errors

A. Deterministic Error Modeling

Twenty-one separate geometric errors of a horizontal type three-axis machine tool were assessed using a laser measurement system. The rigid body kinematics model and homogeneous transformation were used for this purpose. The mechanical structure and assigned coordinate systems of the machine tool are shown in Figure 2.

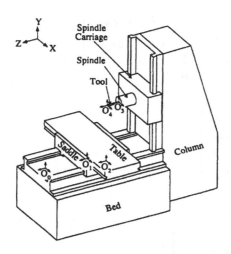

Figure 2. The mechanical structure and the assigned coordinate systems of the machine tool.

Using the associated transformations and corresponding error matrices, the actual position of the cutting tool with respect to the reference frame is represented by the following matrix multiplication:

$$
{}^{0}T_{1}^{a} = \begin{bmatrix} 1 & -\epsilon_{zz} & \epsilon_{yz} & \delta_{xz} - S_{xz}z \\ \epsilon_{zz} & 1 & -\epsilon_{xz} & \delta_{yz} - S_{yz}z \\ -\epsilon_{yz} & \epsilon_{xz} & 1 & z + \delta_{zz} \\ 0 & 0 & 0 & 1 \end{bmatrix},
$$

$$
{}^{1}T_{2}^{a} = \begin{bmatrix} 1 & -\epsilon_{zx} & \epsilon_{yx} & x + \delta_{xx} \\ \epsilon_{zx} & 1 & -\epsilon_{xx} & \delta_{yx} \\ -\epsilon_{yx} & \epsilon_{xx} & 1 & z \\ 0 & 0 & 0 & 1 \end{bmatrix},
$$

$$
{}^{0}T_{3}^{a} = \begin{bmatrix} 1 & -\epsilon_{zy} & \epsilon_{yy} & \delta_{xy} - S_{xy}y \\ \epsilon_{zy} & 1 & -\epsilon_{xy} & y + \delta_{yy} \\ -\epsilon_{yy} & \epsilon_{xy} & 1 & \delta_{zy} \\ 0 & 0 & 0 & 1 \end{bmatrix},
$$

$$
{}^{3}T_{4} = \begin{bmatrix} 1 & 0 & 0 & 0 \\ 0 & 1 & 0 & 0 \\ 0 & 0 & 1 & Z_{t} \\ 0 & 0 & 0 & 1 \end{bmatrix},
$$

$$
T_{w} = \begin{bmatrix} 1 & 0 & 0 & -x \\ 0 & 1 & 0 & y \\ 0 & 0 & 1 & -z + Z_{t} \\ 0 & 0 & 0 & 1 \end{bmatrix},
$$

$$
{}^{ref}T_{tool}^{a} = \left[{}^{ref}T_{spindle}^{a} \right] \left[{}^{spindle}T_{tool} \right]
$$
$$
= \left[{}^{0}T_{3}^{a} \right] \left[{}^{3}T_{4} \right],
$$

$$
{}^{ref}T_{workpiece}^{a} = \left[{}^{ref}T_{spindle}^{a} \right] \left[{}^{saddle}T_{table}^{a} \right] \left[{}^{table}T_{workpiece} \right]
$$
$$
= \left[{}^{0}T_{1}^{a} \right] \left[{}^{1}T_{2}^{a} \right] \left[T_{w} \right],
$$

$$
E_{total} = \left[{}^{ref}T_{workpiece}^{a} \right]^{-1} \left[{}^{ref}T_{tool}^{a} \right]
$$
$$
= \left[\left[{}^{0}T_{1}^{a} \right] \left[{}^{1}T_{2}^{a} \right] \left[T_{w} \right] \right]^{-1} \left[\left[{}^{0}T_{3}^{a} \right] \left[{}^{3}T_{4} \right] \right]. \tag{II.1}
$$

After expanding the terms in Equation (II.1), E_{total} can be rewritten as

$$
E_{total} = \begin{bmatrix} e_{11} & e_{12} & e_{13} & P_{x} \\ e_{21} & e_{22} & e_{23} & P_{y} \\ e_{31} & e_{32} & e_{33} & P_{z} \\ 0 & 0 & 0 & 1 \end{bmatrix}. \tag{II.2}
$$

Since the error terms in machine tools are usually small, only the first, or sometimes the second order terms are used to model machine tool errors. Using the first order terms, the elements of the positional error vectors in Equation (II.2) can be written as follows:

$$P_x = \delta_{xx} - \delta_{xy} + \delta_{xz} - z\epsilon_{yx} + Z_t\epsilon_{yx} - y\epsilon_{zx} - Z_t\epsilon_{yy} - z\epsilon_{yz}$$
$$+ Z_t\epsilon_{yz} - y\epsilon_{zz} + yS_{xy} - zS_{xz}, \qquad (II.3)$$

$$P_y = -\delta_{yx} + \delta_{yy} - \delta_{yz} - z\epsilon_{xx} + Z_t\epsilon_{xx} + x\epsilon_{zx} - Z_t\epsilon_{xy} - z\epsilon_{xz}$$
$$+ Z_t\epsilon_{xz} + zS_{yz},$$

$$P_z = \delta_{zx} - \delta_{zy} + \delta_{zz} + y\epsilon_{xx} + x\epsilon_{yx} + y\epsilon_{xz}.$$

B. Stochastic Error Modeling Using Beta Distribution

   The stochastic part of the volumetric error is the sum of the individual stochastic errors. Since the individual errors are random variables, their sum also has to be random. The parameters describing the statistical behavior of the sum of the random variables have to be calculated from the statistical parameters of the individual random variables. In the error matrix transformation, each element contains both deterministic and stochastic error components. Therefore, the transformation must be performed on both the distribution function and the corresponding mean error. The positional error vector is represented by

$$\vec{P} = \vec{P}_d + \vec{P}_s,$$

$$\vec{P} = P_x\hat{i} + P_y\hat{j} + P_z\hat{k},$$

$$\vec{P}_d = P_{xd}\hat{i} + P_{yd}\hat{j} + P_{zd}\hat{k},$$

$$\vec{P}_s = P_{xs}\hat{i} + P_{ys}\hat{j} + P_{zs}\hat{k}.$$

   From Equation (II.3), the expectation and variance of P_x denoted by the statistical parameters $E(P_x)$ and $Var(P_x)$, respectively, can be calculated as

$$
\begin{aligned}
E\left(P_{x}\right) &= \mu_{x} \\
&= \mu_{\delta_{xx}} - \mu_{\delta_{xy}} + \mu_{\delta_{xz}} - z\mu_{\epsilon_{yx}} + Z_{t}\mu_{\epsilon_{yx}} - y\mu_{\epsilon_{zx}} \\
&\quad - Z_{t}\mu_{\epsilon_{yy}} - z\mu_{\epsilon_{yz}} + Z_{t}\mu_{\epsilon_{yz}} - y\mu_{\epsilon_{zz}} + y\mu_{S_{xy}} \\
&\quad - z\mu_{S_{xz}},
\end{aligned}
\tag{II.4}
$$

$$
\begin{aligned}
Var\left(P_{x}\right) &= \sigma_{x}^{2} \\
&= \sigma_{\delta_{xx}}^{2} + \sigma_{\delta_{xy}}^{2} + \sigma_{\delta_{xz}}^{2} + z^{2}\sigma_{\epsilon_{yx}}^{2} + Z_{t}^{2}\sigma_{\epsilon_{yx}}^{2} + y^{2}\sigma_{\epsilon_{zx}}^{2} \\
&\quad + Z_{t}^{2}\sigma_{\epsilon_{yy}}^{2} + z^{2}\sigma_{\epsilon_{yz}}^{2} + Z_{t}^{2}\sigma_{\epsilon_{yz}}^{2} + y^{2}\sigma_{\epsilon_{zz}}^{2} + y^{2}\sigma_{S_{xy}}^{2} \\
&\quad + z^{2}\sigma_{S_{xz}}^{2}
\end{aligned}
\tag{II.5}
$$

The expectations and variances of P_{y} and P_{z} can be expressed in a similar manner. The calculations of the expectations and variances of the sum of the random variables may be performed without knowledge of the distributions of the individual random variables involved. On the other hand, when we are going to draw conclusions about the confidence of tolerances of the sum of the random variables, we have to know the distribution of the sum. An exact calculation of the distributions of the sum of the random variables is quite difficult, due both to lack of knowledge of the distributions of the individual dimensions, and to the calculation involving the time consuming process termed convolution. As a consequence of this, we have to establish approximate methods for calculation of the distributions of the sum of the random variables. The method most commonly used is based on the central limit theorem. This theorem implies that the distribution of the sum of the random variables is asymptotically normal, independent of the distributions of the individual random variables, as long as the number of random variables in the sum is large enough. Although the normal distribution model is most commonly used in many fields, there are disadvantages due to the stiffness of a normal distribution; that is, the number of parameters available to modify the distribution. A more flexible model is the beta distribution model. Beta distributions are defined by four parameters compared to the two parameters defining a normal distribution. Furthermore, in engineering applications of probability theory, it is occasionally helpful to have a family of distributions whose set of possible values is a finite interval. One such family is the beta family of distributions, and what's more, a beta distribution can approximate better the real distribution of the sum of the random variables because a beta distribution can cover confidence levels up to 100%, and the distribution of the sum of the random variables may have as few as two random variables (Bj⊥rke, 1978). Figure 3 shows that a beta distribution can cover various kinds of distribution including a rectangular one, which is a worst case estimate of the distribution given by a process, thus making tolerance calculations possible without process knowledge (Ochi, 1990; Vardeman, 1994). Therefore, the beta distribution was applied to the distribution modeling of the stochastic part of the volumetric error.

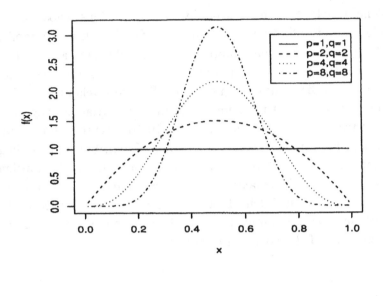

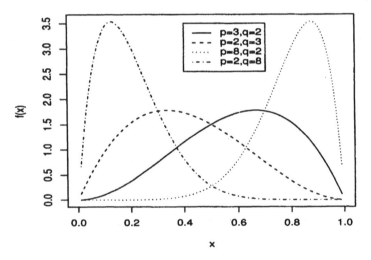

Figure 3. Unit beta probability density functions.

The probability density functions of P_{xs}, P_{ys}, P_{zs} can be approximated by a beta distribution as

$$f_x\left(P_{xs}\right) = \frac{\left(P_{xs} - a_x\right)^{p_x-1} \left(b_x - P_{xs}\right)^{q_x-1}}{B\left(p_x, q_x\right) \left(b_x - a_x\right)^{p_x+q_x-1}} P_{xs}^{p_x-1} \left(1 - P_{xs}\right)^{q_x-1},$$
$$a_x \leq P_{xs} \leq b_x, \quad p_x > 0, q_x > 0, \qquad (\text{II.6})$$

$$f_y\left(P_{ys}\right) = \frac{\left(P_{ys} - a_y\right)^{p_y-1} \left(b_y - P_{ys}\right)^{q_y-1}}{B\left(p_y, q_y\right) \left(b_y - a_y\right)^{p_y+q_y-1}} P_{ys}^{p_y-1} \left(1 - P_{ys}\right)^{q_y-1},$$
$$a_y \leq P_{ys} \leq b_y, \quad p_y > 0, q_y > 0, \qquad (\text{II.7})$$

$$f_z\left(P_{zs}\right) = \frac{\left(P_{zs} - a_z\right)^{p_z-1} \left(b_z - P_{zs}\right)^{q_z-1}}{B\left(p_z, q_z\right) \left(b_z - a_z\right)^{p_z+q_z-1}} P_{zs}^{p_z-1} \left(1 - P_{zs}\right)^{q_z-1},$$
$$a_z \leq P_{zs} \leq b_z, \quad p_z > 0, q_z > 0. \qquad (\text{II.8})$$

It is fairly clear from Equations (II.6)–(II.8) that a_x, b_x, a_y, b_y, a_z and b_z control the interval with a positive probability, while p_x, q_x, p_y, q_y, p_z and q_z control the shape of the distribution over that interval. $f_x(P_{xs})$, $f_y(P_{ys})$ and $f_z(P_{zs})$ are independent, and the probability density function of $\overrightarrow{P}_s$ can be obtained as a joint function of f_x, f_y and f_z. That is, the probability density function of $\overrightarrow{P}_s$ can be written as

$$f\left(\overrightarrow{P}_s\right) = f_x\left(P_{xs}\right) f_y\left(P_{ys}\right) f_z\left(P_{zs}\right).$$

Thus the probability of the tool tip being within an allowable error volume is given by

$$\Pr\left\{P_{xs} \leq |t_x|, P_{ys} \leq |t_y|, P_{zs} \leq |t_z|\right\}$$
$$= \iiint\limits_{P_{xs}\leq|t_x|,P_{ys}\leq|t_y|,P_{zs}\leq|t_z|} f_x\left(P_{xs}\right) f_y\left(P_{ys}\right) f_z\left(P_{zs}\right) dP_{xs}dP_{ys}dP_{zs}$$
$$= \int_{-t_z}^{t_z} \int_{-t_y}^{t_y} \int_{-t_x}^{t_x} f_x\left(P_{xs}\right) f_y\left(P_{ys}\right) f_z\left(P_{zs}\right) dP_{xs}dP_{ys}dP_{zs} \qquad (\text{II.9})$$

Equation (II.9) can be rewritten by substituting Equations (II.6)–

(II.8) into Equation (II.9) to give

$$\Pr\left\{P_{xs} \leq |t_x|, P_{ys} \leq |t_y|, P_{zs} \leq |t_z|\right\}$$
$$= \alpha\beta\gamma \int_{-t_z}^{t_z} \int_{-t_y}^{t_y} \int_{-t_x}^{t_x} \xi\psi\eta\, dP_{xs}\, dP_{ys}\, dP_{zs},$$

where

$$\alpha = \frac{(b_x - a_x)^{1-p_x-q_x}}{B(p_x, q_x)},$$

$$\beta = \frac{(b_y - a_y)^{1-p_y-q_y}}{B(p_y, q_y)},$$

$$\gamma = \frac{(b_z - a_z)^{1-p_z-q_z}}{B(p_z, q_z)},$$

$$\xi = (P_{xs} - a_x)^{p_x-1}(b_x - P_{xs})^{q_x-1},$$

$$\psi = (P_{ys} - a_y)^{p_y-1}(b_y - P_{ys})^{q_y-1}$$

and

$$\eta = (P_{zs} - a_z)^{p_z-1}(b_z - P_{zs})^{q_z-1}.$$

Now we can calculate the probability that the volumetric errors are within the prescribed tolerances.

III.   Volumetric Uncertainty Errors

The working volume was set to measure the machine tool errors as follows: X-axis: 10-300mm; Y-axis: 10-200mm; Z-axis: 10-300mm, and the individual errors were measured along the boundaries of the working volume to construct the error models. Each of the 21 individual errors of the machine tool was assessed seven times using a laser measurement system. The means and variances of the individual errors were obtained at each measuring point. The mean and variance of the volumetric error in the x-direction in the working volume were calculated by substituting the mean and variance of the individual errors in Equations (II.4) and (II.5). Those in the y- and z-directions were calculated in a similar way. The mean and variance of the volumetric error at an arbitrary point in the working volume can then be obtained by interpolation. The probability that the volumetric error is within the given or allowable tolerances can be calculated at each corresponding position in the working volume.

Figures 4–6 show the probabilities that errors exist within these tolerances in a three-dimensional graphic form when the prescribed tolerances are set at $x = \pm 4\mu$m, $y = \pm 4\mu$m, $z = \pm 4\mu$m. The right hand side gray-scale bar indicates the probability scale, which becomes darker as the probability decreases. Figure 4 shows the probabilities that errors exist in the yz-plane at $x = 10, 100, 200$ and 300mm. We can note that the probabilities along the x-axis in this figure show a similar tendency throughout. Along the y- and z-axes, the probabilities decrease, making C-curves, and have similar values in each specified yz-plane.

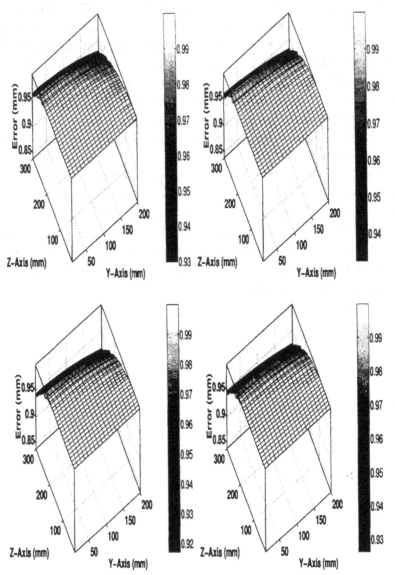

Figure 4. Probabilities that the errors are within the prescribed tolerances on the yz-plane along the x-axis (tolerances: $x = \pm 4\mu m$, $y = \pm 4\mu m$, $z = \pm 4\mu m$). For the top two figures, $x = 10$mm and $x = 100$mm while for the bottom two $x = 200$mm and $x = 300$mm.

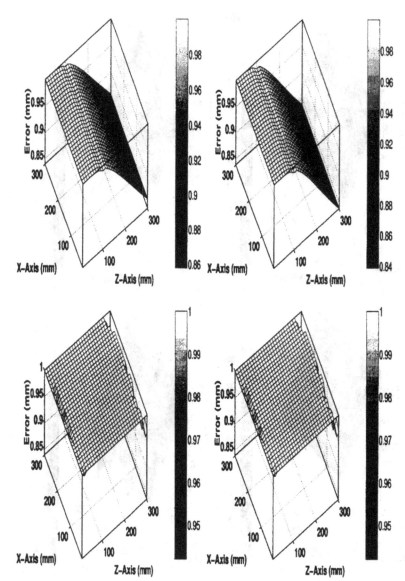

Figure 5. Probabilities that the errors are within the prescribed tolerances on the xz-plane along the y-axis (tolerances: $x = \pm 4\mu$m, $y = \pm 4\mu$m, $z = \pm 4\mu$m). For the top two figures, $x = 10$mm and $x = 70$mm while for the bottom two $x = 140$mm and $x = 200$mm.

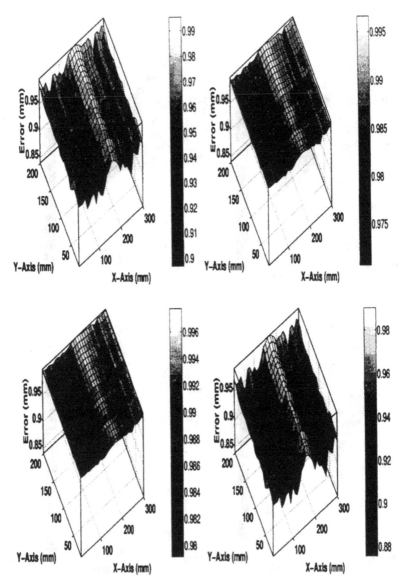

Figure 6. Probabilities that the errors are within the prescribed tolerances on the xy-plane along the z-axis (tolerances: $x = \pm4\mu$m, $y = \pm4\mu$m, $z = \pm4\mu$m). For the top two figures, $x = 10$mm and $x = 100$mm while for the bottom two $x = 200$mm and $x = 300$mm.

Figure 5 shows the probabilities that errors exist in the xz-plane at $y = 10, 70, 140$ and 200mm. The probabilities along the y-axis are uniform along the x-axis, while along the z-axis the probabilities decrease as the z-axis position increases after around the 50mm z-position at $y = 10$ and 70mm. The probabilities at $y = 140$ and 200mm also show very similar tendencies; the wide middle areas show very high probabilities, while the probabilities in the end of the z-axis decrease sharply.

Figure 6 shows the probabilities that errors exist within the xy-plane at $z = 10$, 100, 200 and 300mm. The probabilities in the y-direction are almost uniform except that there is a little fluctuation, while there is considerable fluctuation in the x-direction at $z = 10$ and 300mm. The probabilities are relatively very high, and uniform in both directions at $z = 100$ and 200mm.

We can select the best working position to minimize the volumetric uncertainty error from the probability data. For example, in the x-direction, the probability decreases as the positions along the y- and z-axes increase. Therefore, the region with small coordinate values in the y- and especially z-directions is favorable for reducing uncertainty error in the x-direction. This means that using the rear and lower parts of the working volume will reduce the x-direction uncertainty error. In the y-direction, the probability is uniform in x-direction, with no relation to the x-axis position, and the probability is much higher and more stable at $y = 140$ and 200mm than that at $y = 10$ and 70. Therefore, to reduce y-direction random uncertainty, machining is allowable at the middle area above the half of the y-axis. In the z-direction, the probability is pretty high and stable at $z = 100$ and 200mm, while it is low and unstable at $z = 10$ and 300. Thus the middle of the z-axis is recommended as a work area to reduce the uncertainty error in z-direction.

IV.   Conclusions

Traditionally, dimensional error or tolerance has been investigated using various statistical techniques after a product is fabricated.

However, conventional approaches give rise to great difficulty when the cause of errors must be inferred after a part is made, since the performance of the machine is unknown. The proposed model allows us to predict the tolerances of products of various sizes caused by a machine tool itself on a given machine without going through a very costly and time-consuming production evaluation. It was shown that the probability of volumetric errors within a prescribed tolerance can be predicted. This stochastic analysis provides a scientific basis for controlling product quality, thus improving product quality.

Notation

E: error

E_d, E_s: deterministic and stochastic parts of the error, E, respectively.

${}^0T_1^a = {}^{ref}T_{saddle}^a$: actual HTM (Homogeneous Transformation Matrix) of the saddle in the reference frame.

${}^1T_2^a = {}^{saddle}T_{table}^a$: actual HTM of the table in the saddle frame.

${}^0T_3^a = {}^{ref}T_{spindle}^a$: actual HTM of the spindle in the reference frame.

${}^3T_4^a = {}^{spindle}T_{tool}^a$: HTM of the cutting tool in the spindle frame.

$T_w = {}^{table}T_{workpiece}$: HTM of the workpiece in the reference frame.

Z_t: ideal tool dimension in the z direction.

x, y, z: joint displacements along the x, y, and z axes respectively.

δ_{ij}: positional error in the i-th axis direction along the j-th axis $(i, j = x, y, z)$.

ϵ_{ij}: angular errors, where the first subscript represents which axis the rotation error is around, and the second subscript refers to the moving direction of the slide $(i, j = x, y, z)$.

S_{xy}, S_{xz}, S_{yz}: squareness errors between the axis pairs.

E_{total}: machine tool error matrix of the tool tip in the reference frame.

P_x, P_y, P_z: positional errors in the x, y, and z directions, respectively.

$\overrightarrow{P}$: positional error vector.

$\overrightarrow{P}_d$: deterministic positional error vector.

P_{xd}, P_{yd}, P_{zd}: elements of the vector, $\vec{P}_d$ in the x, y, and z directions, respectively.

$\vec{P}_s$: stochastic positional error vector.

P_{xs}, P_{ys}, P_{zs}: elements of the vector, $\vec{P}_s$ in the x, y, and z directions, respectively.

$\hat{i}$, $\hat{j}$, $\hat{k}$: unit vectors in the x, y, and z directions, respectively.

$E(P_x)$, $E(P_y)$, $E(P_z)$: expectations of P_x, P_y and P_z, respectively.

$Var(P_x)$, $Var(P_y)$, $Var(P_z)$: variances of P_x, P_y and P_z, respectively.

$\mu_{\delta_{xx}}$, $\mu_{\delta_{yy}}$, $\mu_{\delta_{zz}}$: expected linear displacement errors along the x, y, and z axes, respectively.

$\mu_{\delta_{yx}}$, $\mu_{\delta_{zx}}$, $\mu_{\delta_{xy}}$, $\mu_{\delta_{zy}}$, $\mu_{\delta_{xz}}$, $\mu_{\delta_{yz}}$: expected straightness errors, where the first subscript represents the error direction and the second subscript refers to the direction in which the slide is moving.

$\mu_{\epsilon_{xx}}$, $\mu_{\epsilon_{yx}}$, $\mu_{\epsilon_{zx}}$, $\mu_{\epsilon_{xy}}$, $\mu_{\epsilon_{yy}}$, $\mu_{\epsilon_{zy}}$, $\mu_{\epsilon_{xz}}$, $\mu_{\epsilon_{yz}}$, $\mu_{\epsilon_{zz}}$: expected angular errors where the first subscript represents which axis the rotation error is around, and the second subscript refers to the direction in which the slide is moving.

$\mu_{S_{xy}}$, $\mu_{S_{xz}}$, $\mu_{S_{yz}}$: expected squareness errors between the pairs of axes.

$\sigma^2_{\delta_{xx}}$, $\sigma^2_{\delta_{yy}}$, $\sigma^2_{\delta_{zz}}$: variances of the linear displacement error along the x, y, and z axes, respectively.

$\sigma^2_{\delta_{yx}}$, $\sigma^2_{\delta_{zx}}$, $\sigma^2_{\delta_{xy}}$, $\sigma^2_{\delta_{zy}}$, $\sigma^2_{\delta_{xz}}$, $\sigma^2_{\delta_{yz}}$: variances of the straightness errors, where the first subscript represents the error direction and the second subscript refers to the direction in which the slide is moving.

$\sigma^2_{\epsilon_{xx}}$, $\sigma^2_{\epsilon_{yx}}$, $\sigma^2_{\epsilon_{zx}}$, $\sigma^2_{\epsilon_{xy}}$, $\sigma^2_{\epsilon_{yy}}$, $\sigma^2_{\epsilon_{zy}}$, $\sigma^2_{\epsilon_{xz}}$, $\sigma^2_{\epsilon_{yz}}$, $\sigma^2_{\epsilon_{zz}}$: variances of the angular errors, where the first subscript represents which axis the rotation error is around, and the second subscript refers to the direction in which the slide is moving.

$\sigma^2_{S_{xy}}$, $\sigma^2_{S_{xz}}$, $\sigma^2_{S_{yz}}$: variances of the squareness errors between the pairs of axes.

f_x, f_y, f_z: probability density functions of P_{xs}, P_{ys}, and P_{zs}, respectively.

$B(p_i, q_i)$: beta function ($i = x, y, z$).

$[a_i, b_i]$: intervals of the probability density functions ($i = x, y, z$).

t_x, t_y, t_z: allowable error limits in x, y, and z directions, respectively.

References

1. Ahn, K.G.; Cho, D.W. An analysis of the volumetric error uncertainty of a three-axis machine tool by beta distribution, International Journal of Machine tools and Manufacture **2000**, *40*, 2235-2248.

2. Bj⊥rke, ⊥. *Computer-Aided Tolerancing*; Tapir Publishers, 1978.

3. Chen, L.; Rao, S.S Manipulation of uncertainties in the determination of optimal machining conditions under multiple criteria, Transactions of the ASME, Journal of Manufacturing Science and Engineering **1997**, *119*, 186-192.

4. Ochi, M.K. *Applied Probability and Stochastic Processes in Engineering and Physical Sciences*; John Wiley and Sons, 1990.

5. Pahk, H.J. Computer aided volumetric error calibration of coordinate measuring machine, Ph.D. Thesis, University of Manchester, 1990.

6. Qian, G.Z.; Kazerounian, K. Statistical error analysis and calibration of industrial robots for precision manufacturing, International Journal of Advanced Manufacturing Technology **1996**, *11*, 300-308.

7. Shen, Y.L.; Duffie, N.A. Uncertainties in the acquisition and utilization of coordinate frames in manufacturing systems, Annals of the CIRP **1991**, *40*, 527-530.

8. Shen, Y ; Duffie, N.A. An uncertainty analysis method for coordinate referencing in manufacturing systems, Transactions of the ASME, Journal of Engineering for Industry **1995**, *117*, 42-48.

9. Shin, Y.C.; Chin, H.; Brink, M.J. Characterization of CNC machining centers, Journal of Manufacturing Systems **1991**, *10*, 407-421.

10. Shin, Y.C.; Wei, Y. A statistical analysis of positional errors of a multiaxis machine tool, Precision Engineering **1992**, *14*, 139-146.

11. Slocum, A.H. *Precision Machine Design*; Prentice-Hall, 1992.

12. Vardeman, S.B. *Statistics for Engineering Problem Solving*;

IEEE Press, PWS Publishing Company: Boston, 1994.

13. Weck, M. *Handbook of Machine Tools: Metrological Analysis and Performance Test* (volume 4); John Wiley and Sons, 1980.

14. Wilhelm, R.G.; Hocken, R.; Schwenke, H. Task specific uncertainty in coordinate measurement, Annals of the CIRP **2001**, *50*, 553-563.

15. Yau, H.T Evaluation and uncertainty analysis of vectorial tolerances, Precision Engineering **1997**, *20*, 123-137.

An Attempt to Model Distributions of Part Dimensions by using Beta Distribution in Quality Control

Can Cogun

Department of Mechanical Engineering
Gazi University
Maltepe, 06570 Ankara
Turkey

I. Introduction

Workpart dimensions are random variables and statistical frequency distributions of their dimensions vary from process to process. Quality control is a professional field that deals with these variations in an effort to provide quality production at minimum cost. The point of quality control is to study ongoing processes, which involves analysis of the characteristics of the population output by inference of the sample output. The trends those are detected result from assignable causes as opposed to random causes those are inherent in the manufacturing processes. The main tools used for identifying assignable causes of variation are control charts, of which the x-bar chart and the p-chart are prominent.

The control charts or quality control technique always assumes 'normal dispersion (distribution) of dimensions'. Some of the stud-

507

ies conducted in the field have shown that the normal distribution, which assumes the symmetry of distributions, may not properly model the dimension and tolerance distributions due to the existence of skewness in their shape. For the dimension and tolerance distributions limited number of researchers proposed different models than normal, like right-skewed normal (Shainin, 1949), semi circle (Gibson, 1951), uniform (Crafts, 1952; Fortini, 1956), triangular (Doyle, 1951; Mansoor, 1960; Mansoor, 1964), moving normal (Gladman, 1959), beta (He, 1991) and sinus (Burr, 1958; Mansoor, 1960; Mansoor, 1964). In all of these works, very few distribution functions (mostly only one) are tested with a rather limited number of dimension or tolerance frequency distributions (samples). Some others have also raised the need for a different model than the normal to reflect the behavior of dimension and tolerance distributions (Bennet, 1964; Fortini, 1967; Bjorke, 1978; Gladman, 1980; Nelson, 1984; Zhang, 1992). Although, some attempts have been made to represent the dimension distributions by another model than the normal, no attempt has been made to reconstruct x-bar control charts, according to the new proposed models. Some published works (Shilling, 1976; Nelson, 1979; Lashkari, 1982; Balakrishnan, 1986; Choobineh, 1987; Chan, 1988; Padget, 1990; Bai, 1995; Haridy, 1996; Shore, 1998; Borror, 1999) have provided some motivation for normal based control charts that deal with data that is not symmetric.

One of the purposes of this study is to propose some other statistical probability density functions (models) than the normal which would reflect the statistical behavior of the frequency distribution of dimensions better than the normal model. The beta model (distribution function) is found to be the best model among the proposed seven distribution functions to reflect the shape behavior of dimension (frequency) distributions. An attempt is made for the adoption of the unit beta model to x-bar charts of quality control in manufacturing. The changes those should be made in the construction of x-bar charts in the use of the beta model are proposed.

II. Research Model

A. Statistical Modeling of Dimension Distributions

i. Data Sets Used in the Study

The data sets (set of dimensions) are collected from parts, which are produced by well-known machine component producers in Turkey. Special attention is paid to choose functionally different parts with different sizes, shapes, tolerances and manufacturing processes to eliminate concerns those could be raised from results due to similar workparts (functionally) produced by similar manufacturing techniques and dimensions.

Although, a large number of sets of dimensional measurements (data sets) are collected from various workparts (more than 100 sets of data), a limited number of them is presented in this paper. The information about the data sets used in the study is given in Table 1. The first 5 characters of the code of the data set is for the short description of manufacturing process, part name and dimension information. The letters after the dot (.) in the code, namely, HMA, ORS, TS, ASE, MKE, HE and MAN, are the abbreviations for HEMA Gear Company (gear manufacturer), ORS Bearing Company (bearing manufacturer), Konya Trigger and Valve Company (valve manufacturer), ASELSAN (Military Electronics Industries), MKEK Machinery and Chemicals Industry and data sets taken from published works of He (1991) and Mansoor (1964), respectively. The data sets used in other published works for reviewing the probability density functions are not included in this study due to lack of information about the dimensions and parts.

Table 1 Summary of the Information about the Measurement Sets (Data Sets)

Code of the Data Set	Part Definition	Production Method	Nominal Size-Size Limits USL/LSL (mm or inch(+))	Sample Size	Number of Samples	Number of Data in the Set	Measurement Device	Reading Accuracy (mm)
TAM1F.HMA	475 Massey Ferguson Engine gearbox outlet flange diameter	grinding	53,955/53,995	5	9	45	Micrometer	0.001
TAM2F.HMA	Gearbox outlet flange length	"	28,53/28,63	5	9	45	"	0.01
PRTSD.HMA	Rear axis gear tooth thickness	milling	6,82/6,86	5	25	125	"	0.01
TOISD.HMA	Flatness of Maltese cross	turning	0.07	5	24	120	Dial Gage	0.01
TOKMC.HMA	FIAT Engine rear axis outlet dia.	"	37,25/37,45	5	15	75	Micrometer	0.01
TAPDK.HMA	Gearbox rear axis oil sealent dia.	grinding	39,662/39,713	5	10	50	"	0.001
TAR1D.ORS	Roller bearing (6000 series) outer ring diameter	"	25,994/25,999	5	20	100	"	0.001
TAR3D.ORS	Roller bearing (6308 series) outer ring diameter	"	89,991/89,998	5	20	100	"	0.001
TAR6Y.ORS	Roller bearing (6002 series) inner ring rollway diameter	"	13,233/13,245	5	20	100	Dial Gage	0.001
TAR2E.ORS	Roller bearing (6307 series) outer ring width	"	20,920/20,980	5	20	100	Micrometer	0.001
TOR4C.ORS	Roller bearing (6202 series) outer ring inner diameter	turning	29,00/29,10	5	18	87 *	"	0.01
TOREN.ORS	Roller bearing (6202 series) outer ring width	"	11,50/11,75	5	16	80	"	0.01
TAEKB.TS	UAZ engine exhaust valve length	grinding	116,978/117,000	3	10	30	"	0.001
TAEMO.TS	UAZ engine inlet valve seat length	"	4,07/4,57	3	11	32 *	"	0.01
TAEKM.TS	UAZ engine exhaust valve cam side diameter	"	8,921/8,930	3	11	33	"	0.001
DESA1.ASE	Solenoid valve center axis hole center	"	0,115/0,135(+)	5	14	70	"	0.001
DESA2.ASE	Solenoid valve center axis distance	"	0,115/0,135(+)	5	28	140	"	0.001
DESA3.ASE	Solenoid valve center axis distance	"	0,415/0,435(+)	5	14	70	"	0.001
OKT1C.MKE	Galvanized wire diameter	coating	1,93/2,07	5	24	120	"	0.01
OKT2C.MKE	Galvanized wire diameter	"	2,43/2,57	5	19	95	"	0.01
TAEMM.HU	Electric motor rotor diameter	grinding	0,22/0,24	-	-	72	"	0.001
TOSBB.MAN	Valve ring diameter (+)	turning	A:1,000/1,002 B:1,065/1,070	-	-	300	"	0.005
DIYPB.MAN	Oil pump vane clearance (+)	drilling	A:0,3775/0,3825 B:0,3675/0,3725	-	-	240	"	0.005

(*) Insufficient number of measurements

Micrometers and dial gages with different accuracies (Table 1) were used in the measurements. The measurements were performed by the quality control personnel of the companies.

ii.   Distribution Functions Used in This Study

In this study, normal, log-normal, triangular, uniform, Weibull, Erlang and beta probability density functions are tried for the fit of behavior of frequency distributions of part dimensions collected from various manufacturing companies. Weibull, Erlang, beta and log-normal distributions could be symmetric or non-symmetric (right- or left-skewed) in shape depending on the values of model parameters. The distribution functions, estimation of model parameters and shape variations of the models could be found in statistics books (Bury, 1975; Bain, 1978). The shapes of distribution functions of these models are given in Figure 1. The use of the beta distribution function requires long computations involving its four model parameters. A useful and practical form of the beta distribution which requires less and simple computations is the unit beta distribution

and is used in this study. Brief information about the unit beta model
is given in Appendix.

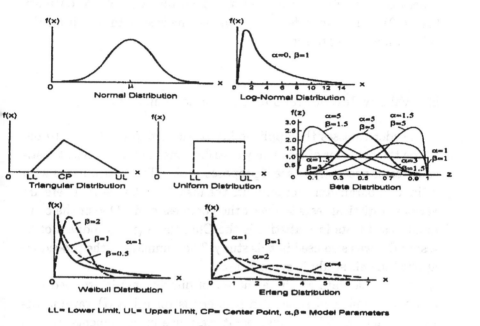

LL= Lower Limit, UL= Upper Limit, CP= Center Point, α,β= Model Parameters

Figure 1 The Distribution Functions Used in the Study

Since the Weibull, Erlang and unit beta distributions starts from
zero (0) and unit beta function outside the interval (0, 1) gives probability density zero, all the collected measurements (dimensions) are
normalized by using the formula

$$z_i = (x_i - a)/(b - a) \tag{II.1}$$

where a and b are the smallest and largest measurements (lower and
upper limits of the distribution), respectively, x_i is the measured dimension (variable) and z_i is the normalized value of x_i (unit dimension). After normalization, the measured variables are distributed
between the values 0 and 1. For normalized Weibull and Erlang
distributions, the model parameter estimators give more accurate
results than non-normalized variables. Unit dimensions (and their
distributions) can be used successfully in normal, log-normal, triangular and uniform distribution functions since the frequency and

shape characteristics of the distributions are not affected by normalization. In this study, equation (II.1) is used for normalizing the collected data sets. The first letter 'Z' in the code of the data set (Table 3) is used to differentiate the normalized data sets from the non-normalized data sets.

iii. Validity Test of Proposed Distribution Functions

In order to asses the applicability of the proposed models to describe the behavior of frequency distributions of dimensions, statistically test is necessary to see if proposed probability distributions with estimated parameters actually fits the measured data sets. There are various statistical tests to check the goodness of fit. One of the commonly used tests in statistics is the Chi-square (χ^2) goodness of fit test and it was also used in this study. The summary of the procedure applied in this study is given below.

i) The sample data x^* (here the set of dimensions) is grouped into a proper number of equal width intervals (or cells) (l) varying between 5 and 30. There must be at least five measurements in each cell. For the manufacturing applications, Ishikawa (1976) suggests that the test give reliable results for 5-7 intervals if the measurements are less than 50. He recommends 6-10 intervals for 50-100 measurements, 7-12 intervals for 100-250 measurements and 10-20 intervals for measurements above 250 for the Chi-square test.

ii) From the available family of statistical distributions (in this study, normal, log-normal, triangular, uniform, Weibull, Erlang and beta probability density functions), a model distribution function $F_o(x)$ is hypothesized to represent and fit the sample data.

iii) The parameters of the hypothesized model ($F_o(x)$) are estimated from the data by using estimation techniques (estimators).

iv) The following statistic is calculated from the observed (measured) and expected (model) frequencies.

$$\chi^2 = \sum_{i=1}^{l} \frac{(O_i - E_i)^2}{E_i}$$

where O_i is the number of observed data in cell i, E_i is the number of expected data in cell i and l is the number of cells. Values of E_i are calculated by using the postulated distribution function, $F_0(x)$. It is known from the probability theory that the statistic χ^2 is distributed as a Chi-square variable with $\nu = l - k - 1$ degrees of freedom. Here k is the number of model parameters, ν is the degrees of freedom of the χ^2 distribution. The k value for triangular distribution is three and for the other distribution functions used in this study is two.

v) The terms of the χ^2 statistic above measure the discrepancy between the observed and postulated (theoretical) class frequencies. Smaller χ^2 values indicate better fit of the distribution. From the standard χ^2 tables, the level of significance (α) can be found by using χ^2 value and degrees of freedom (ν). For a given significance level α and degrees of freedom ν, the critical value (χ_c^2) is obtained from theoretical chi-square tables. The postulated model $F_0(x)$ and the sample data give rise to single value (χ^2) of the test statistic. If, $\chi_c^2 > \chi^2$ the hypothesis, that $F_0(x)$ is the underlying measurement is accepted; otherwise it is rejected. If two different hypothesized models are tested for the same experimental data, the model, which gives the smaller chi-square value (or bigger level of significance) for the same degree of freedom, indicates the better fit.

In this study, for the same set of dimensions, chi-square test is applied to normal, log-normal, triangular, uniform, Weibull, Erlang and unit beta distribution functions, and χ^2 values and level of significances are found for the same degrees of freedom. The STATGRAF (1993) software package is used for the χ^2 test of the models.

B.   Results and Discussion

In the study, normal, log-normal, uniform, triangle and Weibull models are tested for the 23 sets of data given in Table 1. The Erlang and beta models are only tested for TOISD.HMA set, which has variables between 0 and 1. Although DESA1.ASE, DESA2.ASE, DESA3.ASE, TAEMM.HE, TDSBB.MAN and DIYPB.MAN sets have variables in between 0 and 1, the parameter estimates of beta and Erlang distributions were not computable due to the very close values of the distribution variables. The chi-square test parameters and results (l, k, ν, χ^2, α) for sample hypothesized models are given in Table 2 for the 23 sets of dimensions. In the last column of the Table 1, the result of the χ^2 test is summarized by giving the first letter of the names of the models in the order of increasing values of χ^2 (or decreasing level of significance). For TAM1F.HMA set in Table 2, the N, L, U, T$^+$ listing is given which indicates that the model which gives the minimum χ^2 value is 'normal' (shortly N) and triangular (shortly T) is in the 4th position. The '+' sign placed as a superscript on the letter is the indication of significance level less than 0.05 for the model. The best four distribution functions given in Table 2 are plotted on the distributions of the measurement sets. Sample plots are given in Figure 2. The figure indicates clearly that it is impossible to decide the best-fit model by visual inspection. This verifies the strong need for the chi-square test.

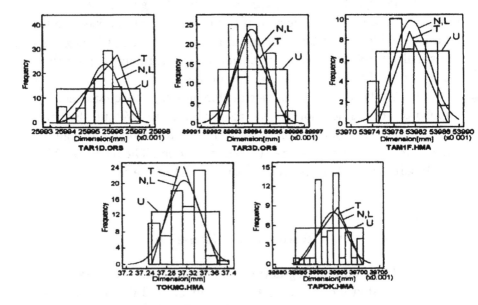

Figure 2 Frequency Distributions of Sample Measurement Sets and the Plot of the Best Three Distribution Functions (N=Normal, L=Log-normal, T=Triangular, U=Uniform, W=Weibull, E=Erlang, B=Beta)

Table 2 Chi-square Test Results of Sample Distribution Functions for Measurement Sets

Code of the Data Set	Number of Data	NORMAL							LOG-NORMAL							UNIFORM							FITNESS OF THE MODELS
		Model Parameters		Chi-square Test					Model Parameters		Chi-square Test					Model Parameters		Chi-square Test					
		$\bar{x}$	s	l	k	u	χ^2	SL	$\bar{x}$	s	l	k	u	χ^2	SL	LL	UL	l	k	u	χ^2	SL	
TAM1F.HMA	45	53.981	0.0033	6	2	3	2.03	0.564	53.981	0.0033	6	2	3	2.03	0.564	53.975	53.988	6	2	3	5.21	0.156	N,L,U,T+
TAM2F.HMA	45	28.587	0.0011	5	2	2	15.9	0.0003	28.587	0.0011	5	2	2	15.9	0.0003	28.57	28.61	5	2	2	7.78	0.0203	U+,N+,L+,T+
FRTSD.HMA	125	6.837	0.010	7	2	4	75.3	1.7E-15	6.837	0.010	7	2	4	75.2	1.8E-15	6.82	6.86	7	2	4	78.13	4E-06	U+,N+,L+,T+
TOISD.HMA	120	0.046	0.023	8	2	5	5.52	0.354	0.0471	0.0276	8	2	5	23	0.0003	0.01	0.14	8	2	5	158.2	0.0	N,W,B+,L+
TOKMC.HMA	75	37.312	0.0122	6	2	3	17.0	0.0007			6	2	3	17	0.00069	37.25	37.38	6	2	3	21.68	0.00007	N+,L+,U+,T+
TAPDK.HMA	50	39.693	0.0035	6	2	3	18	0.0004	39.693	0.0035	6	2	3	18	0.0004	39.685	39.701	6	2	3	30.0	1.3E-06	N+,L+,T+,U+
TAR1D.ORS	100	25.996	0.00075	6	2	3	8.09	0.044	25.9958	0.00075	6	2	3	8.09	0.044	25.994	25.9975	7	2	4	49.62	4.3E-10	N+,L+,T+,U+
TAR3D.ORS	100	89.994	0.00091	7	2	4	25.3	3.9E-05			7	2	4	25.5	0.000039	89.992	89.996	7	2	4	35.46	3.7E-07	N+,L+,T+,U+
TAR6Y.ORS	100	13.23	0.0027	8	2	5	65.7	7.8E-13	13.23	0.0027	7	2	4	17.2	5.3E-06	13.233	13.243	8	2	5	47.87	3.4E-09	L+,T+,U+,N+
TAR2E.ORS	100	20.973	0.0023	8	2	5	18.6	0.0023			8	2	5	18.6	0.0023	20.970	20.979	8	2	5	43.71	2E-06	N+,L+,T+,U+
TOR4C.ORS	87	29.063	0.016	7	2	4	6.98	0.136	29.063	0.016	7	2	4	6.99	0.136	29.02	29.10	7	2	4	27.57	1.5E-05	N,L,T,U+
TOREN.ORS	80	11.731	0.0047	6	2	3	26.8	6.5E-06	11.731	0.0047	6	2	3	27.2		11.5	11.9	6	2	3	43.8	1.6E-09	N,U+,L+,T+
TAEKB.TS	30	116.991	0.0075	5	2	2	1.50	0.472	116.991	0.0075	5	2	2	1.50	0.472	116.98	117.0	5	2	2	5.86	0.053	N,L,T,U
TAEMO.TS	32	4.367	0.149	5	2	2	1.25	0.536	4.367	0.149	5	2	2	1.23	0.541	4.15	4.62	5	2	2	2.43	0.296	L,N,U,T+
TAEKM.TS	33	8.925	0.0042	5	2	2	0.5	0.779	8.925	0.0042	5	2	2	0.500	0.778	8.916	8.932	5	2	2	4.526	0.104	N,L,U,T+
DESA1.ASE	70	0.129	0.0019	6	2	3	5.84	0.119	0.129	0.0019	6	2	3	5.90	0.116	0.125	0.134	6	2	3	15.99	0.001	N,L,T,U+
DESA2.ASE	140	0.130	0.0039	7	2	4	13	0.011	0.130	0.0039	7	2	4	13.2	0.0102	0.121	0.137	8	2	5	24.87	0.00015	N+,L+,T+,U+
DESA3.ASE	70	0.428	0.0024	7	2	4	10.2	0.038	0.428	0.0024	7	2	4	10.3	0.036	0.422	0.433	7	2	4	29.42	6E-06	N+,L+,T+,W+
OKT1C.MKE	120	2.006	0.0298	7	2	4	30.5	3.8E-06	2.0055	0.0299	7	2	4	30.7	3.4E-06	1.95	2.05	7	2	4	16.24	0.0027	U+,L+,N+,T+
OKT2C.MKE	95	2.501	0.032	6	2	3	1.64	0.65	2.501	0.032	6	2	3	1.54	0.67	2.40	2.57	7	2	4	36.86	1.6E-07	L,N,T+,U+
TAEMM.HE	72	0.228	0.0050	8	2	5	4.76	0.444	0.228	0.0050	8	2	5	4.51	0.477	0.22	0.24	8	2	5	14.28	0.014	N,L,T,U+
TOSBB.MAN	300	0.011	0.00097	9	2	6	15	0.0205	0.011	0.00097	10	2	7	11.9	0.105	0.009	0.014	10	2	7	1343	0	L,N+,T+,W+
DIYFB.MAN	240	0.00625	0.00107	9	2	6	10.2	0.116	0.00625	0.00107	9	2	6	10.6	0.103	0.004	0.009	9	2	6	120	0	U,N,L,W+

(+): SL less than 0.05

SL: Significance Level

The results obtained from the chi-square tests for measurement sets (non-normalized) (Table 2) can be summarized as follows:

1. In 14 of the 23 measurement sets, the normal model gives the best fit. For 7 of these 14 sets of data, the level of significance is less than 0.05.
2. In the 15 measurement sets log-normal model gives the same level of significance with normal and in 5 sets log-normal gives better fit (higher significance level) than normal model. So, the log-normal model gives much better fit than normal model when all the measurement sets are considered.
3. Uniform distribution function models 3 data sets better than the other models. This model is generally placed in the 3rd or 4th places after normal and log-normal models.
4. Triangular distribution models only one data set better than the other distribution functions. In all the data sets, it is placed in the 3rd and 4th position.
5. Weibull model parameters (estimated) and χ^2 values are calculated only for 7 sets of measurements. In these 7 sets, the model is in the 2nd place once and 4th place three times.

According to the above results, it is found that the log-normal and
normal distribution functions model the measurement sets better
than the other distribution functions. In modeling, the log-normal
distribution gives better results than normal due to the advantage of
modeling left-skewness of the data sets. Uniform distribution could
be considered as the 3rd best model in modeling the measurement
sets.

In the above analysis Erlang, Weibull and beta distribution functions couldn't be used in modeling due to the non-normalized data
sets. In the second part of the study, 23 data sets are normalized according to equation (II.1). The chi-square test results $(l, k, \nu, \chi^2, \alpha)$
for sample hypothesized models are given in Table 3 for the 23 sets
of normalized dimension measurements. The best four distribution
functions given in Table 3 are plotted on the frequency distributions
of the normalized measurement sets. Sample plots showing the best
three distributions are given in Figure 3. When Table 2 and Table
3 are analyzed together, it can be easily deduced that the goodness of fit of the models (last column of the Table), except Erlang,
Weibull and beta distributions, are same. As an example for the set,
TAM1F.HMA the goodness of fit list is in the order of N, L, U, T in
Table 2. In the normalized set, ZTAM1F.HMA the relative positions
of N, U, T are not changed but a new model (beta) is included in
the list (i.e. B, N, U, T). Some small changes in the orders of models
in Table 2 and Table 3 are due to the small variations in number of
intervals (cells) of the frequency distributions of the normalized and
non-normalized data sets.

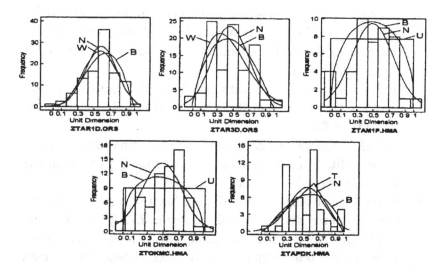

Figure 3 Frequency Distributions of Normalized Sample Measurement Sets and the Plot of the Best Three Distribution Functions (N=Normal, L=Log-normal, T=Triangular, U=Uniform, W=Weibull, E=Erlang, B=Beta)

Table 3 Chi-Square Test Results of Sample Hypothesized Distribution Functions for Normalized Distributions

Code of the Data Set	Number of Data	BETA								NORMAL						WEIBULL							FITNESS OF MODELS
		Model Parameters					Chi-square Test			Model Parameters					Chi-square Test		Model Parameters						Chi-square Test	
		α	β	l	k	υ	χ²	SL	x	s	l	k	υ	χ²	SL	α	β	l	k	υ	χ²	SL	
ZTAM1F.HMA	45	1.445	1.472	5	2	2	1.101	0.576	0.495	0.252	5	2	2	1.185	0.552	1.313	0.518	5	2	2	4.69	0.095	B, N,U, T
ZTAM2F.HMA	45	1.00	1.179	4	2	1	0.569	0.450	0.444	0.281	4	2	1	0.52	0.217	0.972	0.441	4	2	1	5.49	0.019	B, N,U, E+
ZFRTSD.HMA	125	1.00	1.298	7	2	4	90.34		0.434	0.273	7	2	4	74.81		1.034	0.437	7	2	4	107.21	0.00	**
ZTOLSD.HMA	120	1.692	3.896	7	2	4	14.14	0.007	0.276	0.177	8	2	5	20.44	0.001	1.375	0.297	8	2	5	28.60	1E-03	B+, N+, W+, E+
ZTOKMC.HMA	75	1.388	1.546	8	2	5	13.65	0.020	0.476	0.252	7	2	4	11.58	0.020	1.661	0.520	7	2	4	24.58		B=N,U+, T+
ZTAPDK.HMA	50	2.234	1.896	7	2	4	15.64	0.0035	0.540	0.220	7	2	4	16.59	0.0023	2.382	0.597	7	2	4	29.41		T+, B+, N+,U+
ZTARJD.ORS	100	3.316	2.297	6	2	3	7.36	0.068	0.580	0.190	6	2	3	6.18	0.068	3.177	0.637	6	2	3	8.11	0.043	B=N, W+, T+
ZTAR3D.ORS	100	1.834	2.068	7	2	4	15.66	0.0035	0.470	0.225	7	2	4	24.14		1.910	0.516	7	2	4	23.18	1E-04	B+, W+, T+, N+
ZTAR6Y.ORS	100	1.00	1.302	8	2	5	35.26		0.405	0.274	8	2	5	38.20		1.132	0.418	8	2	5	56.26		**
ZTAR2E.ORS	100	1.00	2.00	8	2	5	13.16	0.022	0.306	0.235	8	2	5	14.40	0.013	0.735	0.273	7	2	4	48.99		B+, N+, E+, W+
ZTOR4C.ORS	87	2.556	2.150	6	2	3	2.297	0.513	0.543	0.208	6	2	3	3.276	0.350	2.445	0.595	6	2	3	8.90	0.090	B, N, W+, T+
ZTOREN.ORS	88	2.573	1.882	6	2	3	27.58		0.577	0.211	6	2	3	27.40		1.761	0.606	7	2	4	58.23		B+, N+, T+, W+
ZTAEKB.TS	30	1.455	1	5	3	2	1.483	0.476	0.619	0.299	5	2	2	1.387	0.499	1.775	0.674	5	2	2	3.26	0.195	N, B, W, T
ZTAEMO.TS	32	1	1.15	5	2	2	0.936	0.626	0.461	0.317	5	2	2	1.66	0.435	1.319	0.495	5	2	2	1.467	0.480	B, N, W, E
ZTAEKM.TS	33	1.373	1.061	5	2	2	0.242	0.885	0.564	0.267	5	2	2	1.115	0.572	1.900	0.617	5	2	2	1.505	0.470	B, N, W, E
ZDESA1.ASE	70	2.34	2.15	6	2	3	2.23	0.525	0.522	0.213	6	2	3	5.91	0.115	2.431	0.578	6	2	3	51.47		B, N, T,U+
ZDESA2.ASE	140	1.918	1.431	6	2	3	5.46	0.141	0.572	0.237	6	2	3	16.33	0.001	2.52	0.638	6	2	3	18.40	3E-04	B, N+, T+, W+
ZDESA3.ASE	70	2.548	1.945	7	2	4	8.60	0.071	0.567	0.211	7	2	4	5.40	0.248	2.487	0.624	7	2	4	9.39	0.052	N, B+, W+, T+
ZGKT1C.MKE	120	1.15	1.00	8	2	5	6.57	0.254	0.546	0.299	9	2	6	14.75	0.022	1.445	0.585	9	2	6	30.32		B, U, N+, W+
ZGKT2C.MKE	95	3.39	2.29	7	2	4	8.69	0.069	0.596	0.189	7	2	4	6.45	0.167	2.964	0.648	7	2	4	10.44	0.033	N, B, W+, T+
ZTAEMM.HE	72	1.128	1.644	8	2	5	2.40	0.790	0.406	0.252	8	2	5	4.95	0.421	1.547	0.448	8	2	5	3.99	0.55	B, W, N, E
ZTOSBB.MAN	300	2.297	3.172	10	2	7	14.22	0.027	0.409	0.194	10	2	7	15.01	0.035	2.232	0.471	10	2	7	14.44	0.043	B+, W+, N+, E+
ZDIYPB.MAN	240	1.943	2.363	8	2	5	15.02	0.0102	0.451	0.216	9	2	6	31.69		2.134	0.506	9	2	6	9.84	0.131	W, E, T+, B+

** : Very small SL value

(+): SL less than 0.05

The results obtained from the chi-square tests for normalized measurement sets (Table 3) can be summarized as follows:

1. For all the proposed models the significance levels of ZFRTSD.HMA

and ZTARGY.ORS measurement sets are in the order of 10^{-6} – -10^{-15}. Therefore, these two normalized data sets are discarded from the analysis, which would possibly give unreliable and misleading results.

2. In 16 of 21 normalized data sets beta distribution function gives the best fit. In 5 of these 16 sets, the level of significance is less than 0.05.

3. In 3 of 21 normalized sets, normal distribution function gives the best fit. Normal distribution is the second model in the sets which beta is the best.

4. Weibull and triangular distribution functions are best in only one normalized set. In normalized sets of ZTAPDK.HMA and ZDIYPB.MAN in which the triangular and Weibull distributions are the best, respectively, the second best fits are beta distribution functions.

5. Erlang model generally gives a poor fit for the normalized data sets.

6. Although the Weibull model seems to produce a better fit than the Erlang, in 14 of 16 normalized data sets the Weibull gives a poorer fit than the beta and normal distributions.

7. Triangular and uniform models are generally placed in the 3rd or 4th place after beta and normal. Only in the ZTAPDK.HMA set does the triangular distribution give the best fit.

8. After the careful visual inspection of the normalized frequency distributions, it is observed that the right- and left-skewness characteristics of the dimension frequency distributions are best represented by the beta distribution model.

It is clear from the above given results and observations that the beta distribution function models the distribution of dimensions better than normal and the other commonly used statistical models. Since normalization technique only changes the scale of the non-normalized distributions without changing the frequency and shape characteristics, it could be stated that the beta distribution function is the best model for reflecting the behavior of the measurement sets. The model parameters α and β control the skewness, shape and scale of the model. Erlang and Weibull distribution functions are also known to be good in reflecting the right- or left-skewness. However, their model parameter estimators are found weak when compared with beta, which will eventually result in a poor chi-square fit.

The beta model is also proposed by He (1991) to model the dimension distributions. In his work, only the beta model is used and the superiority of the model with respect to the other distribution functions is not emphasized. In his work, only two data sets are used. One of the two sets of data is used to explain the use of the distribution for a small data set (16 measurements) and the other is taken from Bennet's (1964) study to explain the procedure for a large data set.

III. The Adoption of Beta Model to x-bar Control Charts

In this part of the study an attempt is made for the adoption of beta model to x-bar control charts.

A. Normal Distribution Based x-bar Charts

16 sets of measurements (Table 1) are used in the construction of x-bar charts. In this study, the x-bar charts used for the normal data are based on sample size of five. Schilling and Nelson (1976) showed that the Shewhart x-bar chart for modeling means works well with a sample size four or five. Author of this work believes that real motivation for improving the control chart methods for skewed data occurs when a small sample size is required, such as case where $n = 1$.

The Upper and Lower Control Limits of the charts are abbreviated as UCLs and LCLs. The letter s in the abbreviations indicates 'standard' (normal distribution based) control limits. Sample calculation of UCLs and LCLs for TAM1F.HMA data set is given below.

UCLs= $x + A_2R$ = 53,9814+ 0,577(0,008)= 53,986mm
LCLs= $x - A_2R$ = 53,9814 - 0,577(0,008)= 53,977mm

Here, $x = 53,9814$mm is the average of nine samples (53,979,

53, 980, 53, 981, 53, 982, 53, 982, 53, 983, 53, 980, 53, 983 and 53, 983 mm) with 5 measurements in every sample. R (the maximum deviation of measurements in a sample) values for the samples are 0,007, 0,008, 0,010, 0,013, 0,012, 0,005, 0,004, 0,005, 0,004 mm and their mean value $(\bar{R})$ is 0,008 mm. A_2 chart constants are dependent on sample size and can be found in statistical quality control books in tabulated forms (Bury, 1975; Bain, 1978). For sample size (n) of 5, A_2 value is 0.577. In this study, control limit calculations are made by using the formula CLs $= x \pm A_2R$ for the simplicity of the calculations. The control limit values for standard (normal model based) procedure is given in Table 4.

Table 4 UCL and LCL values for Beta and Normal Distributions and Their Differences for the Sample Measurement Sets

Data Set	α	β	ZUCLb	ZLCLb	UCLs [mm]	LCLs [mm]	UCLb [mm]	LCLb [mm]	DUCLsb [%]	DLCLsb [%]	RCLRsb
ZTAM1F.HMA	1.445	1.472	0.9912	0.0079	53.986	53.977	53.9878	53.9751	0.0033	-0.0035	1.41
ZFRTSD.HMA	1.000	1.298	0.9933	0.0012	6.86	6.82	6.859	6.82005	-0.014	0.0007	0.97
ZTOKMC.HMA	1.398	1.546	0.9888	0.0065	37.34	37.28	37.378	37.2509	0.1	-0.078	2.11
ZTAPDK.HMA	2.234	1.896	0.9831	0.0339	39.698	39.688	39.7007	39.6855	0.0068	-0.0063	1.52
ZTAR1D.ORS	3.316	2.297	0.9765	0.0843	25.9967	25.9949	25.9974	25.9938	0.0027	-0.0042	2
ZTAR3D.ORS	1.834	2.068	0.9729	0.0160	89.9953	89.9925	89.9959	89.9921	0.0007	-0.0004	1.36
ZTOR4C.ORS	2.556	2.150	0.9765	0.0459	29.081	29.045	29.098	29.024	0.06	-0.07	2.05
ZTOREN.ORS	2.573	1.882	0.9854	0.0517	11.814	11.648	11.894	11.52	0.68	-1.09	2.25
ZTAEKB.TS	1.455	1.000	0.9990	0.0115	116.998	116.983	116.999	116.9753	0.00085	-0.0066	1.58
ZTAEMO.TS	1.000	1.150	0.9965	0.0013	4.73	4.004	4.618	4.1506	-2.36	3.66	0.64
ZTAEKM.TS	1.373	1.061	0.9984	0.0083	8.934	8.915	8.9319	8.9161	-0.023	0.012	0.83
ZDESA1.ASE	2.340	2.150	0.9747	0.0355	0.133	0.127	0.1338	0.1253	0.6	-1.33	1.42
ZDESA2.ASE	1.918	1.431	0.9941	0.0255	0.136	0.124	0.1369	0.1214	0.66	-2.09	1.29
ZDESA3.ASE	2.548	1.945	0.9834	0.0491	0.431	0.425	0.4328	0.4225	0.42	-0.59	1.72
ZGKT1C.MKE	1.150	1.000	0.9987	0.0035	2.05	1.96	2.049	1.9504	-0.49	-0.48	1.1
ZGKT2C.MKE	3.398	2.296	0.9770	0.0890	2.55	2.45	2.566	2.415	0.62	-1.42	1.51

B. Beta Distribution Based x-bar Charts

For most of the manufacturing companies, "three defectives in thousand parts" is an acceptable limit (type I error rate of 3/1000). With the normal distribution this error rate is represented very nearly by $pm3\sigma$ distance from the distribution mean which gives symmetric tail probabilities of about 0.0015 on each side. Due to the skewed shape of the beta distribution, it is impossible to find equal symmetric distances from the distribution mean which would give the above mentioned equal tail probabilities. Therefore, it is almost impossible to find chart constants $(A_1, A_2, B_1, B_2$ etc. as in normal model) for the beta model based control charts. No references are available in the literature for adjusting control limits of a x-bar chart for skewed data by using the beta distribution and other distributions different from the normal. Upper and lower control limits of the beta model can be estimated by using the probability limit method. That is, it can be estimated by obtaining the percentiles of the beta distribution. In this study, the following procedure is applied to estimate upper and lower control limits for the beta model (UCLb and LCLb):

1. Normalize all the measurements (variables) in the data set.
2. Obtain the α and β values of the beta model by using the STATGRAPH package (If a set of measurements has the population mean μ and variance σ^2, the beta model parameters α and β are estimated from equations (IV.1) and (IV.2) given in the Appendix.).
3. Obtain the critical values, which would give 0,0015 and 0,9985 probabilities in the beta function. The critical value, which would give 0.0015 probability is the ZLCLb. Similarly, the critical value which would give 0.9985 is the ZUCLb. The obtained values are the normalized control limits of the beta model and the letter Z indicates that these values are calculated from normalized distributions (i.e. unit dimensions).
4. Use the following formulation to convert ZUCLb and ZLCLb values to UCLb and LCLb.

$$UCLb = ZUCLb(bb - ab) + ab, \qquad (III.1)$$

$$LCLb = ZLCLb(bb - ab) + ab. \qquad (III.2)$$

Here, ab and bb are the lower and upper limits of the beta distribution population and they are taken as the minimum and the maximum measurements (dimensions) in the set. It should be known that the original sample's maximum (especially for cases with a small original sample size) is not the maximum for the whole population.

If the maximum and minimum used to normalize a new sample are taken from the data used to generate the control limits, there is a possibility that the new sample will have a data point outside the maximum and minimum of the original sample.

The percent difference between upper control limits of standard procedure (normal model based) and beta distribution populations (DUCLsb) and the percent difference between lower control limits of standard procedure and beta distribution population (DLCLsb) are calculated by using the following equations:

$$DUCLsb = (UCLb - UCLs).100/UCLs, \qquad (III.3)$$

$$DLCLsb = (LCLb - LCLs).100/LCLs. \qquad (III.4)$$

To compare the range between UCL and LCL values obtained both from beta model and standard (normal distribution based) procedure (RCLRsb) the following equation is used:

$$RCLRsb = (UCLb - LCLb)/(UCLs - LCLs). \qquad (III.5)$$

The results obtained from equations (III.1)–(III.5) is summarized in Table 4 for the 16 sets of measurements.

C.   Sample Case

For the ZTAM1F.HMA data set, the α and β values are found from the STATGRAPH package as 1,445 and 1,472. For the 0,0015 and 0,9985 probabilities, the critical values are 0,0079 and 0,9912 (i.e. ZLCLb and ZUCLb). Equations (III.1) and (III.2) are used to find UCLb and LCLb. From equations (III.1) and (III.2):
UCLb = 0.9912(53.988-53.975)+53.975 = 53.9878mm.
LCLb = 0.0079(53.988-53.975)+53.975 = 53.9751mm.
The percent difference between UCL and LCL values for beta and standard values are (equations (III.3) and (III.4)):
DUCLsb = (53.9878-53.986).100/53.986 = 0.0033%
DLCLsb = (53.9751-53.977).100/53.977 = -0.0035%
Comparison of the range between beta model control limits and standard procedure control limits is performed by using equation (III.5):
RCLRsb = (53.9878-53,9751)/(53.986-53.977) = 1.41

From numerical results obtained for TAM1F.HMA measurement set, it is clear that the UCLb is higher than the UCLs and LCLb is lower than LCLs. The range between UCLb and LCLb values is 41% bigger than the range obtained from standard (normal model based) procedure.

D. Results and Discussion

The sample analysis given in Section C is repeated for sixteen sets of measurements and summary of the results is given in Table 4. Some sample x-bar charts showing both standard (normal model based) and beta distribution control limits are given in Figure 4. The following results can be deduced from Table 4:

1. For the twelve of the sixteen data sets, the UCLb is higher than UCLs, and LCLb is lower than LCLs.
2. The range between UCLb and LCLb values is generally wider than that of UCLs and LCLs. The ratio of the range of beta control limits to the range of standard procedure (RCLRsb) is in between 1 and 2.

The DUCLsb and DLCLsb values indicate that the beta model based control limits are not symmetric with respect to x value. The control limits of the standard procedure are closer to the x value than that of beta model (narrower control zone). From the above results it can be deduced that the normal model control limits provides closer control over sample means than that of beta model. It is possible that some sample averages (i.e. x-bar values) which fall beyond the control limits of the normal model will be considered acceptable (safe) by the beta model based control limits. By using the proposed, one can estimate the control limits of beta model to monitor the ongoing process.

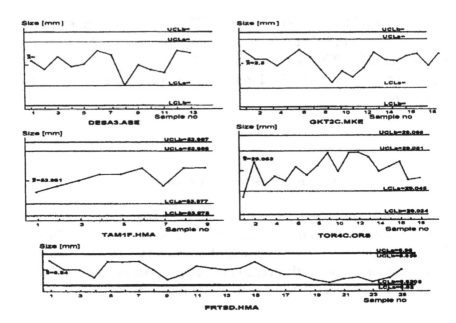

Figure 4 Some Sample $\bar{X}$ Control Charts for the Selected Measurement Sets

IV. Conclusions

In this study, experimental and theoretical efforts are spent to model the behavior of dimension distributions of machined workpieces. Twenty-three sets of dimension distributions of different parts are used in this study. In order to model the behavior of the collected data, seven different statistical distribution functions, namely, normal, log-normal, triangular, uniform, Weibull, Erlang and unit beta distributions, were used. The beta distribution is found the best statistical distribution function in representing the frequency distributions of the measurement sets by using chi-square goodness of fit tests.

In the second stage of the work, the upper and lower control limits of the beta model are estimated by obtaining the percentiles of the distribution. The tail probabilities of 0.0015 and 0.9985 are used to find the critical values for beta distribution function and these values are taken as the upper and lower control limits of the control charts by using type I error rate of 3/1000. It is found that mostly the UCL of the beta model is higher than that of normal model and LCL of the beta model is lower than that of the normal. So, the normal model based UCL and LCL provide closer control over sample means than that of beta model. It can be inferred that the beta based chart would result in fewer alarms, but it is difficult to know when it will detect true out of control points.

Many quality control engineers believe that in practice normality is not too much of a problem in the case of x-bar charts since the errors associated with its use are relatively small. The author believes that the real contribution for improving the control chart methods for skewed data occurs when a small sample is required, such as the case where $n = 1$. Development of an approach (or approaches) for adjusting the control limits of a control chart for skewed data, which can be modeled by using different models than normal, will be another contribution in the field.

Appendix: Unit Beta Distribution Function

The distribution function of the unit beta model is

$$f(x) = \begin{cases} \dfrac{\Gamma(\alpha + \beta)}{\Gamma(\alpha)\Gamma(\beta)} x^{\alpha-1}(1 - x)^{\beta-1}, & \text{if } 0 < x < 1, \\ 0, & \text{otherwise,} \end{cases}$$

where $\alpha > 0$ and $\beta > 0$ are the model parameters. The mean (μ) and variance (σ^2) of the model are calculated from

$$\mu = \frac{\alpha}{\alpha + \beta} \tag{IV.1}$$

and

$$\sigma^2 = \frac{\alpha\beta}{(\alpha + \beta)^2(\alpha + \beta + 1)}. \tag{IV.2}$$

The model parameter estimators (α and β) for data less than 21 and more than 21 are given in the works of Cooke (1979) and Bjorke (1978).

The model parameters α and β control the skewness and shape of the distribution. For different α and β values, the shapes of the unit beta functions are given in Figure 1.

Acknowledgments

The author would like to thank Engineer Bunyamin Kilinc (M.Sc.) for his efforts in this study.

References

1. Bai, D.S.; Chois, I.S. X bar-control and R-control charts for skewed populations, Journal of Quality Technology **1995**, *27*, 120-131.

2. Bain, L.J. *Statistical Analysis of Reliability and Life Testing*; Marcel Dekker: New York, 1978.

3. Balakrishnan, N.; Kocherlakota, S. Effects of non-normality on Xbar charts-assignable cause model, Sankhyā, B **1986**, *48*, 439-444.

4. Bennet, G. The application of probability theory to the allocation of engineering tolerances, Ph.D. Thesis, University of New South Wales, Australia, 1964.

5. Bjorke, Q. *Computer Aided Tolerancing*; Tapir Publishers: Trondheim, Norway, 1978.

6. Borror, C.M.; Montgomery, D.C.; Runger, G.C. Robustness of the EWMA control chart to non-normality, Journal of Quality Technology **1999**, *31*, 309-316.

7. Burr, I.W. Some theoretical and practical aspects of tolerances for mating parts, Quality Control **1958**, *15*, 18.

8. Bury, K.V. *Statistical Models in Applied Science*; John Wiley and Sons: New York, 1975.

9. Chan, L.K.; Hapuarachi, K.P.; McPherson, B.D. Robustness of Xbar and R-charts, IEEE Transactions on Reliability **1988**, *37*, 117-123.

10. Choobineh, F.; Ballard, J.L. Control limits of QC charts for skewed distributions using weighted variance, IEEE Transactions on Reliability **1987**, *36*, 473-477.

11. Cooke, P. Statistical inference for bounds of random variables, Biometrika **1979**, *66*, 367-374.
12. Crafts, J.W. Assembly tolerance problem, Engineer (London) **1952**, *206*, 918-928.
13. Doyle, L.E. Statistical aids for tool engineers, Tool Engineer **1951**, *26*, 48-51.
14. Fortini, E.T. Dimension control in design, Machine Design **1956**, *28*, 82-89.
15. Fortini, E.T. *Dimensioning For Interchangeable Manufacture*; Industrial Press Inc.: New York, 1967.
16. Gibson, J. *A New Approach to Engineering Tolerances*; The Machinery Publishing Co. Ltd.: Brighton and London, 1951.
17. Gladman, C.A. Techniques for applying probability to the tolerancing of -machined dimensions, C.S.I.R.O. Australian National Std. Lab. Tech. Paper No. 11, 1959.
18. Gladman, C.A. Applying probability in tolerance technology, The Institution of Engineers (Australia) **1980**, *5*, 82-88.
19. Haridy, A.M.A.; El-Shabrawy, A.Z. The economic design of cumulative sum charts used to maintain current control of non-normal process means, Computers in Industrial Engineering **1996**, *31*, 783-790.
20. He, J.R. Estimating the distributions of manufactured dimensions with the beta probability density function, Int. J. Mach. Tools Manuf. **1991**, *31*, 383-396.
21. Ishikawa, K. *Guide to Quality Control* (volume 9); Asia Productivity Organization: Tokyo, 1976.
22. Lashkari, R.S.; Rahim, M.A. An economic design of cumulative sum charts to control non-normal process means, Computers in Industrial Engineering **1982**, *6*, 1-18.
23. Mansoor, E.M. The dimension analysis of engineering designs, M.Sc. Thesis, University of Melbourne, 1960.
24. Mansoor, E.M. The application of probability to tolerances used in engineering designs, Proc. Instn. Mech. Engrs. **1964**, *178*, 29-51.
25. Nelson, P.R. Control charts for Weibull processes with standards given, IEEE Transactions on Reliability **1979**, *28*, 283-

288.

26. Nelson, L.S. The Shewhart control chart: tests for special causes, Journal of Quality Technology **1984**, *16*, 237-239.

27. Padget, W.J.; Spurrier, J.D. Shewhart type charts for percentiles of strength distributions, Journal of Quality Technology **1990**, *22*, 283-288.

28. Shainin, D. Cost-cutting chance laws can control design tolerances, Machine Design **1949**, *21*, 130-140.

29. Shilling, E.G.; Nelson, P.R. The effect of non-normality on the control limits of X-bar charts, Journal of Quality Technology **1976**, *8*, 183-188.

30. Shore, H. A new approach to analyzing non-normal quality data with application to process capability analysis, Int. J. Prod. Res. **1998**, *36*, 1917-1933.

31. STATGRAF *Statistical Graphics System by Statistical Graphics Corporation*; STSC Inc. - Plus Ware Product, version 5.0, 1993.

32. Zhang, H.C.; Huq, M.E. Tolerancing techniques: the state-of-the-art, Int. J. Prod. Res. **1992**, *30*, 2111-2135.

Applications of the Beta-Binomial Distribution to Animal Teratology Experiments

Dirk F. Moore and Woollcott Smith

Department of Statistics
Temple University
Philadelphia, Pennsylvania 19122

Teratology is the study of the causes of birth defects, and animal experiments provide a valuable method for studying the effects of suspected teratogens. Potential teratogens could be environmental contaminants or radiation, or they could be drug therapies which may have teratogenic side effects. Whatever the agent under study, the experimental design typically involves a series of pregnant female mice (or other animal species) that are divided into a control and one or more exposure groups. At some point in the pregnancy, the exposure groups are administered the agent under study. After a certain period of time, the pregnant mice are sacrificed, and the number of dead fetuses (resorptions), as well as the total number of implants, are counted. The statistical problem is to determine a dose-response relationship while accounting for the hierarchical structure of the binary data; fetuses within the same litter are more alike than are fetuses from different litters, resulting in a litter-to-litter variation that must be incorporated into the model.

The beta-binomial distribution provides a completely parametric model that can accommodate the among-litter variability (Williams, 1975). In this model, let r denote the number of resorptions ("successes") in a particular litter, and let n denote the total number of implants. Assume

$$\Pr(R = r) = \binom{n}{r} P^r (1 - P)^{n-r},$$

where P is a random variable with a beta density function,

$$p^{a-1}(1 - p)^{b-1}/B(a, b), \qquad a > 0, \qquad b > 0.$$

It is useful to re-parameterize the beta density in terms of a mean $\pi = a/(a + b)$ and variance parameter $\theta = 1/(a + b)$ (Griffiths, 1973). Then the mixture beta-binomial probability function is given by

$$g(r; \pi, \theta, n) = \binom{n}{r} \frac{\prod_{u=0}^{r-1}(\pi + \theta u) \prod_{u=0}^{n-r-1}(1 - \pi + \theta u)}{\prod_{u=0}^{n-1}(1 + \theta u)}, \qquad (.1)$$

where we adopt the convention that $\prod_{u=0}^{-1} = 1$. Note that when $\theta = 0$, the beta-binomial distribution reduces to a binomial with parameters n and π.

For a series of k litters, we may model the j'th litter by $\mathrm{logit}(\pi_j) = X_{(j)}\alpha$, where $X_{(j)}$ denotes the j'th row of a regression matrix X, and α is a vector of regression parameters to be estimated. The X matrix may consist of indicator variables for the control and experimental groups, or it may contain dose terms for a dose-response model. A likelihood may be constructed as a product of terms of form (.1), yielding the following expression for the log-likelihood:

$$l(\alpha, \theta) \propto \sum_{j=1}^{k} \left\{ \sum_{u=0}^{r_j-1} \log(\pi_j + u\theta) + \sum_{u=0}^{n_j-r_j-1} \log(1 - \pi_j + u\theta) \right.$$
$$\left. - \sum_{u=0}^{n_j-1} \log(1 + u\theta) \right\}$$

First and second derivatives may be readily obtained (Morgan, 1992, page 242), and these may be used to obtain maximum likelihood estimates for α and θ using, for example, Newton-Raphson iteration.

The data in Table 1 (Paul, 1982) are typical of many teratology animal experiments. Shown are the numbers of affected fetuses and the total numbers of fetuses for four groups of pregnant females: a control group and three experimental groups exposed to low, medium, and high doses. Maximum likelihood estimates and standard errors are presented in Table 2 for the regression parameters α and the overdispersion parameters θ for four nested models: *(i)* a common α_0 and θ for all groups combined, *(ii)* a dose-response model involving an intercept α_0 and slope term α_1 for the mean and an estimate for a common θ, *(iii)* separate estimates of α for each of the four groups and a common estimate of θ, and *(iv)* separate estimates of α and θ for each of the four groups. A likelihood-ratio test comparing Models *(i)* and *(ii)* yields twice the difference in the log-likelihoods equal to 6.3. This may be compared to a chi-square deviate with 1 degree of freedom, for a P-value of 0.012. But if one compares Models *(ii)* and *(iii)*, or *(iii)* and *(iv)*, neither comparison yields a statistically significant result. The best of the four models is thus the dose-response Model *(ii)*.

Some authors have discussed problems with bias and coverage probability of the maximum likelihood estimates of the beta-binomial parameters when the number of litters is small (Liang and McCullagh, 1993; Liang and Hanfelt, 1994). They show that the bias can be especially severe when a common variance parameter is fitted in the presence of heterogeneous group-to-group variability. Alternative approaches to modeling animal teratology data, based on quasi-likelihood and generalized estimating methods, are available (McCullagh and Nelder, 1989; Lefkopoulou, Moore, and Ryan, 1989; Moore and Tsiatis, 1991). These methods are robust to mispecifications of the variance function, and may be implemented using widely available software. A compromise method (Brooks, 1984) involves using an estimating equation to estimate the mean parameters, while the variance term is estimated using maximum likelihood with a beta-binomial distribution. Yamamoto and Yanagimoto (1994) propose factoring the beta-binomial likelihood into a marginal density for estimation of the mean and a conditional density for estimation of the overdispersion. Another approach, proposed by Moore, Park, and Smith (2001), involves fitting a mixture of beta-binomial distributions to each group. This method should be used if the assumption of a binomial mixed with a single beta is suspect, particularly if the numbers of proportions that are zeros or ones are inflated relative to what one would expect from a simple beta-binomial model.

Table 1. Data from the Shell Toxicology Laboratory taken from Paul (1982).

Control	1/12	1/7	4/6	0/6	0/7	0/8	0/10	0/7
	1/8	0/6	2/11	0/7	5/8	2/9	1/2	2/7
	0/9	0/7	1/11	0/10	0/4	0/8	0/10	3/12
	2/8	4/7	0/8					
Low dose	0/5	1/11	1/7	0/9	2/12	0/8	1/6	0/7
	1/6	0/4	0/6	0/7	1/5	5/9	0/1	0/6
	3/9							
Medium dose	2/4	3/4	2/9	1/8	2/9	3/7	0/8	4/9
	0/6	0/4	4/6	0/7	0/3	6/13	6/6	5/8
	4/11	1/7	0/6	3/10	6/6			
High dose	1/9	0/10	1/7	0/5	1/4	0/6	1/3	1/8
	2/5	0/4	4/4	1/5	1/3	4/8	2/6	3/8
	1/6							

Table 2. Maximum likelihood estimates for Model i ($\eta = \alpha_0$, common θ); Model ii ($\eta = \alpha_0 + \alpha_1 \cdot$ dose, common θ); Model iii ($\eta = \alpha_j$, common θ); and Model iv ($\eta = \alpha_j$, separate θ's).

parameter	log-like	α_0	α_1	α_2	α_3
Model i	-281.17	-1.134	–	–	–
		(0.162)	–	–	–
Model ii	-278.02	-1.816	0.335	–	–
		(0.258)	(0.133)	–	–
Model iii	-276.10	-1.823	-1.822	-0.732	-1.063
		(0.297)	(0.357)	(0.268)	(0.315)
Model iv	-274.42	-1.812	-1.926	-0.660	-1.160
		(0.315)	(0.335)	(0.304)	(0.301)

parameter	θ_0	θ_1	θ_2	θ_3
Model i	0.319	–	–	–
	(0.094)	–	–	–
Model ii	0.289	–	–	–
	(0.088)	–	–	–
Model iii	0.260	–	–	–
	(0.082)	–	–	–
Model iv	0.273	0.118	0.495	0.128
	(0.155)	(0.101)	(0.240)	(0.120)

References

1. Brooks, R.J. Approximate likelihood ratio tests in the analysis of beta-binomial data, Applied Statistics **1984**, *33*, 285-289.
2. Griffiths, D.A. Maximum likelihood estimation for the beta-binomial distribution and an application to the household distribution of the total number of cases of a disease, Biometrics **1973**, *29*, 637-648.
3. Lefkopoulou, M.; Moore, D.; Ryan, L. The analysis of multiple correlated binary outcomes: Application to rodent teratology experiments, Journal of the American Statistical Association **1989**, *84*, 810-815.
4. Liang, K.-Y.; Hanfelt, J. On the use of the quasi-likelihood method in teratological experiments, Biometrics **1994**, *50*, 872-880.
5. Liang, K.-Y.; McCullagh, P. Case studies in binary dispersion, Biometrics **1993**, *49*, 623-630.
6. McCullagh, P.; Nelder, J.A. *Generalized Linear Models* (second edition); Chapman and Hall: London, 1989.
7. Moore, D.F.; Park, C.-K.; Smith, W. Exploring extra-binomial variation in teratology data using continuous mixtures, Biometrics **2001**, *57*, 490-494.

8. Moore, D.F.; Tsiatis, A. Robust estimation of the variance for moment methods for extra-binomial and extra-Poisson variation, Biometrics **1991**, *47*, 383-401.

9. Morgan, B.J.T. *Analysis of Quantal Response Data*; Chapman and Hall: London, 1992.

10. Paul, S.R. Analysis of proportions of affected fetuses in teratological experiments, Biometrics **1982**, *38*, 361-370.

11. Williams, D.A. The analysis of binary data from toxicological experiments involving reproduction and teratogenicity, Biometrics **1975**, *31*, 949-952.

12. Yamamoto, Y.; Yanagimoto, T. Statistical methods for the beta-binomial model in teratology, Environmental Health Perspectives Supplements **1994**, *102*, Supplement 1, 25-31.

Applications of the Beta Distribution in Soil Science

Gerrit H. de Rooij[1] and Frank Stagnitti[2]

[1]Wageningen University
Department of Environmental Sciences
Nieuwe Kanaal 11
6709 PA Wageningen
The Netherlands
[2]Deakin University
School of Ecology and Environment
P. O. Box 423
Warrnambool, Victoria 3280
Australia

I. Solute Movement in Soils

Modern high input-high output agriculture relies heavily on fertilizers and pesticides, and industrial production and the use of fossil fuels cause widespread deposition of various substances onto soil. As a consequence, infiltrating water, supplied through natural rainfall or through irrigation, carries these substances from the soil surface downwards through the soil and, eventually, into the groundwater (e.g., Flury, 1996). Once in the groundwater, pollutants may experience little to no biodegradation due to oxygen-limiting conditions and consequently can remain for hundreds and even thousands of years. Pollutants in groundwater may also reenter surface water bodies such as streams, rivers and lakes as a result of seepage through the aquifer-interface with the water body. The pollutants may also percolate further downward into the vast bodies of deep groundwater that constitute major sources of reliable, high-quality potable water for domestic drinking water supplies and irrigation. Whether in the soil, the surface water, or in the groundwater, these contaminants can threaten human and animal health, the viability of natural and agricultural ecosystems, and the long-term sustainability of communities that rely on unpolluted soil and water resources.

For these reasons, the travel times and pathways of contaminants moving through the soil have been intensively studied within soil science (Nielsen et al., 1986; Jury and Flühler, 1992, Vanderborght et al., 2000). Historically, leaching at a given depth as a function of time has received much attention. For an inert solute applied uniformly as a pulse to the surface of a well-defined soil volume, a plot of the fraction of the total amount of applied solute leached below a predefined depth as a function of time gives the cumulative breakthrough curve (BTC_{cumt}). This is the traditional tool of soil scientists and solute transport theorists. The soil volume can be a soil column in the laboratory (ranging from ~ 100 cm^3 to ~ 1 m^3) from which outflow is collected and analyzed, or a field plot (ranging from ~ 1 m^2 to > 100 m^2) where solute movement is monitored at different locations and/or drainage water is sampled and analyzed. In the case of field experiments or transient experimental conditions, the time coordinate is routinely replaced by cumulative drainage (Figure 1).

To understand water flow and solute transport in soils, the soil may be viewed as a bundle of stream tubes (Jury et al., 1986). A stream tube is a volume of soil defined by the flow lines of the water: thus, water that enters a stream tube moves downward through that tube without leaving it. A stream tube widens or narrows in regions of diverging and converging flow, respectively (Figure 2). Each stream tube starts at the soil surface and reaches down to the groundwater level; the stream tubes are tortuous, and have different shapes, lengths, and travel times. When travel distances are short (e.g., less then a few meters), the exchange of solutes between stream tubes

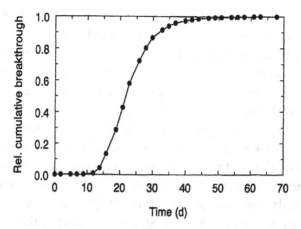

Figure 1. An example of a cumulative breakthrough curve of an inert solute that was applied uniformly to the soil surface and subsequently sampled at a given depth.

as a result of lateral diffusion, dispersion, etc. is often limited. In this contribution, the solute exchange between stream tubes will be neglected. For a discussion of the role of lateral solute movements refer to Flühler et al. (1996).

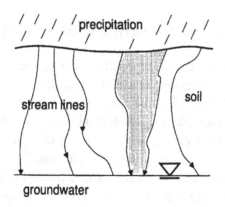

Figure 2. Two-dimensional view of the conceptualization of the soil as a population of stream tubes with varying properties. The flow diverges in the areas to the left and the right, while the center area has converging flow. The shaded area indicates a stream tube with converging flow.

Field soils exhibit massive spatial variation of properties such as the hydraulic conductivity. This has a marked effect on the shape of the stream tubes in soils, and on the distribution of travel times in the population of stream tubes. In addition, phenomena such as macropores (large, continuous voids produced by soil fauna, decayed roots, or by shrinkage of clay soils) and wetting front instability (where infiltrating water rapidly moves downward on a few locations while the remainder of the soil does not participate in the flow process) also affect the pattern and velocity of solute movement. This prompted the development of instruments in which water percolating through the soils was collected simultaneously in different compartments of equal size (e.g., Poletika and Jury, 1994; Quisenberry et al., 1994; Wildenschild et al., 1994; Buchter et al., 1995; de Rooij, 1996; Boll et al., 1997; Stagnitti et al., 1998; Strock et al., 2001). The percolate could be analyzed for desired solutes. Field versions of these multiple sample percolation systems (MSPSs) produce data that demonstrate the large spatial variation of drainage fluxes and solute leaching (e.g., Stagnitti et al., 1999b). Recent laboratory experiments for the first time provided detailed information not only about the temporal variation in solute and contaminant transport in soils but also as well as the spatial distribution of the solute leaching.

II.   Mathematical Description of Solute Leaching

The discussion is limited to solutes that are not subject to adsorption or decay processes. The BTC_{cumt} of these inert solutes at a variety of scales can often be fitted successfully by a solution to a second-order linear partial differential equation, which produces smooth breakthrough curves with a degree of asymmetry that depends on the initial and boundary conditions. In cases with pronounced macropore flow or other causes for irregular travel time variations, a better fit can often be obtained by assuming two or more parallel flow domains with a simple mechanism to exchange solutes (van Genuchten and Wierenga, 1976; Gerke and van Genuchten, 1993; Flühler et al., 1996). Each domain, when uncoupled, produces its own smooth breakthrough curve (BTC), and the coupled domains can reproduce a wide variety of breakthrough curves.

To characterize the spatial redistribution of solutes during their downward movement in the soil, the sampling compartments of a multiple sample percolation experiment can be ranked in order of decreasing amount of total collected solute over the entire period during which solutes were leached. The fraction of total solute leached can then be plotted against the fraction of the total sampling area (e.g., Quisenberry et al., 1994). This produces a curve with a continuously decreasing slope that, by definition, must always be non-negative (compartments cannot contain negative solute amounts). In analogy to the breakthrough curve, this can be termed the cumulative spatial solute distribution curve (SSDC$_{\text{cumx}}$). A continuous function able to describe these observations must be bounded in the interval between 0 and 1. It must also be skewed and the level of the skew determined by parameters of the function. The Beta distribution has these properties, and Stagnitti et al. (1999b) recognized its potential to parameterize the SSDC$_{\text{cumx}}$:

$$P(x; \alpha, \eta) = [B(\alpha, \eta)]^{-1} x^{\alpha-1} (1-x)^{\eta-1} \qquad (\text{II.1})$$

where x denotes the fraction of the total sampling area ($0 \leq x \leq 1$), P denotes a probability density, B is the Beta function, and α and η are fitting parameters. The Beta distribution reduces to the uniform distribution when $\alpha = \eta = 1$. This would be the case if the soil was perfectly uniform, with all stream tubes running parallel and having equal travel times. The cumulative form can be used to analyze the spatial redistribution of solutes in soils: the fraction of total collected solute leached through the section $[0, x]$ of the sampling area is given by $\int_0^x P(\psi; \alpha, \eta) d\psi$, with ψ an integration variable.

Given the utility and simplicity of equation (II.1), Stagnitti et al. (1999b) could produce accurate fits to observed spatially non-uniform solute leaching as well as drainage amounts for several soils, with only two fitting parameters, α and η, the shape fitting parameters of the Beta distribution. Furthermore, the mean μ and the standard deviation σ of equation (II.1) are functions of the fitting parameters only:

$$\mu = \frac{\alpha}{\alpha + \eta}$$

and

$$\sigma = \sqrt{\frac{\alpha\eta}{(\alpha+\eta)^2(\alpha+\eta+1)}}$$

Based on this, Stagnitti et al. (1999b) proposed an expression to capture the degree of heterogeneity of solute leaching elegantly in a single number, which they termed the Heterogeneity Index (HI):

$$HI(\alpha,\eta) = \sqrt{3}\frac{\sigma}{\mu} = \sqrt{\frac{3\eta}{\alpha(\alpha+\eta+1)}}. \tag{II.2}$$

This heterogeneity index is a scaled expression of the coefficient of variation (σ/μ). The scaling factor $3^{1/2}$ ensures that $HI = 1$ for uniform flow $(\alpha = \eta = 1)$, and increases for an increasing deviation from uniformity.

III. A Physical Interpretation of the Spatial Solute Distribution Curve

If stream tubes are assumed to be immobile (a requirement that may be satisfied to a reasonable degree for the duration of a typical solute transport experiment, even though stream tubes can be very dynamic through the seasons), any arbitrary area on the soil surface is the top of a soil volume whose shape is determined by the combined volume of the stream tubes emerging from it. If stream tubes do not exchange solutes, i.e. each solute particle remains in the stream tube in which it entered upon infiltration at the soil surface, solutes enter this soil volume at the soil surface, and leave at its bottom (the groundwater, or any arbitrary depth above it).

If a solute pulse was applied uniformly to the soil surface, the assumptions above allow the cumulative Beta distribution that was fitted to the observed $SSDC_{cumx}$ to serve as a descriptor of the geometry of the population of stream tubes in the soil volume under investigation (Figure 3). Each point on the fitted curve is associated with a stream tube with an infinitesimal outflow area at the sampling depth. At arbitrary x, the slope $P(x)$ of the fitted curve gives the ratio of the area of infiltration (A_i, L^2) over the area of outflow (A_o, L^2) at the sampling depth, for an individual stream tube. Thus,

$P(x)$ quantifies the degree of convergence or divergence of the flow in one stream tube. In principle, this reasoning is also valid for an observed curve, but its slope at any point is sensitive to experimental error and noise.

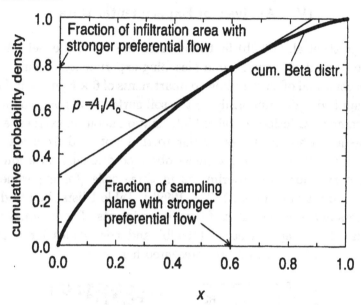

Figure 3. The cumulative Beta distribution, fitted to data of the fraction of solute leaching (vertical coordinate) as a function of the fraction of the area (at a given depth below the soil surface) where leaching was collected (horizontal coordinate). The slope $P(x)$ at any point on the curve (represented by the straight line) equals the ratio of the infiltration area, A_i, over the outflow area, A_o, of the stream tube represented by that point. The coordinates define the fractions of the sampling plane (x) and the soil surface (cumulative probability density) occupied by stream tubes with more strongly converging flow than that corresponding to $P(x)$. (After de Rooij and Stagnitti, 2000, Figure 1).

As indicated in Figure 3, the horizontal coordinate x of $P(x)$ gives the fraction of the sampling plane that is occupied by stream tubes that converge more strongly than the stream tube at x, and the vertical coordinate (cumulative probability density; $\int_0^x P(\psi)d\psi$) gives the fraction of the soil surface occupied by these stream tubes. Hence, $\int_0^x P(\psi)d\psi$ cannot be smaller than x. Finally, the fraction of the total leaching (sum of the cumulative leaching of all stream tubes when all solute has leached) carried by a subset of stream tubes between x_1 and x_2 is given by $\int_{x_1}^{x_2} P(\psi)d\psi$.

IV. Analysis of Experimental Data

Stagnitti et al. (1999b) first illustrated the use of the Beta distribution to multiple sample solute leaching experiments. Their instruments consisted of 25 sampling compartments of 6 × 6 cm each. Since they applied water uniformly to the soil surface (without ponding), the drainage variation contained information about convergence and divergence of stream tubes similar to that obtained from leaching data of an inert tracer. Fig 4 shows observed drainage variation and fitted curves for five experiments in three soils (for experimental details see Stagnitti et al., 1999a, 1999b). Table 1 lists the fitted parameters and the HI-values for all experiments. These values were calculated in Stagnitti et al. (1999b) and new values for another experiment (Tower Hill-2) are presented here.

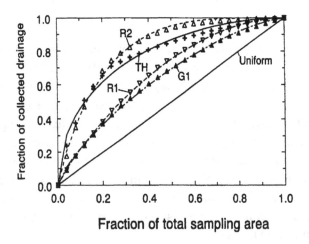

Figure 4. Observed spatial distribution curves of cumulative drainage of uniformly applied water for five different locations in three soils in Victoria (Australia): Rutherglen (1) and Rutherglen (2), denoted R1 and R2, Grassmere (1) and (2), denoted G1 and G2, and Tower Hill-2 (TH). Details of the experiments and the soils are in Stagnitti et al. (1999b).

Table 1. Parameter values of equation (II.1) fitted to drainage data from five multisampler experiments in three soils in Victoria, Australia (Stagnitti et al., 1999a, 1999b). The Heterogeneity Index (HI) was calculated according to equation (II.2).

Soil (experiment)	α	η	HI
Rutherglen (1)	0.8885	1.835	1.29
Rutherglen (2)	0.8477	3.993	1.56
Grassmere (1)	0.7639	1.366	1.31
Grassmere (2)	0.7775	1.582	1.35
Tower Hill (2)	0.4648	1.761	1.88

Table 2 gives leaching characteristics derived from the observed curves and the parametric fits for the soils with the smallest (Rutherglen 1) and largest HI (Tower Hill-2). The relative contribution of the highest, intermediate, and lowest draining sampling compartment gives an indication of the non-uniformity of the flow.

Table 2. Characteristics of the spatial distribution of water flow in the soils with the smallest and largest HI-value in Table 1.

	Rutherglen (1)	Tower Hill
$0 \leq x < 0.04$	0.092	0.24
$0.48 \leq x < 0.52$	0.036	0.018
$0.96 \leq x \leq 1.0$	0.0069	0.0058
	0.72	0.69
	0.47	0.27

The values in the first three rows are the fractions of total drainage flowing through the indicated interval of x while those in the last two rows are the fraction of the soil surface accessing converging flow paths and the fraction of the sampling plane draining converging flow paths, respectively.

The areas accessing and draining converging stream tubes were calculated by determining the x-value for which $P(x) = 1$. There, $A_i = A_o$. All x smaller than that are locations where stream tubes exit for which $A_i > A_o$, i.e. converging stream tubes, larger x-values correspond to diverging stream tubes. The value of the fitted cumulative Beta distribution at the x-value for which $P(x) = 1$ gives the fraction of the soil surface that gives access to the converging stream tubes (see Figure 3). The ratio of the soil surface area accessing converging stream tubes to the sampling area where these stream tubes exit indicates the degree of flow concentration. This type of information allows a quantitative view of non-uniform flow and its effect on solute transport.

The observed deviation from uniformity of the spatial distribution of solutes or drainage depends on the size of the individual sampling compartments: larger sampling areas tend to underestimate the degree of flow heterogeneity. In view of the importance of preferential flow (high flow rates occurring in a small fraction of the soil) for rapid transport of contaminants, sampling compartments should be kept as small as practically feasible to provide accurate data. De Rooij and Stagnitti (2000) reported on a experiment in which a cylindrical core (1 m^2 area and 0.85 m high) of a sandy soil with a hydrophobic top layer was subjected to a series of 20 mm artificial rain showers. Drainage and solutes were collected in 360 sampling compartments, of which the 300 compartments (0.05 $\times$ 0.05 m) in the center were used in the analysis. The estimated fraction of the sampling area occupied by strongly converging stream tubes ($A_i/A_o \geq 2$) dropped from 0.05 to 0.003 when the sampling compartments increased from 0.05 $\times$ 0.05 m to 0.15 $\times$ 0.15 m (Figure 5).

The large number of samples available to de Rooij and Stagnitti (2000) allowed them to assess the required number of sampling compartments. They randomly selected 2^n samples ($n = 1, 2, \ldots, 8$) from the population of 300, sorted them as explained above, and fitted the cumulative Beta distribution to the resulting sets of $2, 4, \ldots, 256$ data points. Fitted values of α, η, and HI became almost invariant to changes in n for $n \geq 4$ (≥ 16 samples, i.e. $\geq 5\%$ of the outflow area). A visual comparison of the fitted curves for 16 and 300 samples showed they were very nearly identical. This suggests that the 25 sampling compartments of many MSPSs currently in use are adequate to quantify spatial heterogeneity of leaching at the plot scale ($< 1m^2$). To aid the design of future experiments and MSPSs, it appears recommendable to estimate the required number of sampling compartments for other soils and experimental conditions by carrying out a similar analysis whenever data from a sufficient number

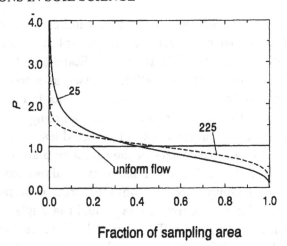

Figure 5. Fitted curves of the Beta distribution to experimental data from a 1 m² undisturbed core of a sandy soil. The fits were obtained for a single chloride leaching experiment, but for different sampling compartment sizes. The actual compartments were 25 cm² (solid curve). By combining the compartments in sets of 3 × 3, artificial 225 cm² compartments were obtained (dashed curve). The vertical coordinate indicates the degree of convergence or divergence of the stream tubes (A_i/A_o). Adapted from de Rooij and Stagnitti (2000).

of sampling compartments is available. This approach illustrates the power of the parameterization provided by the Beta distribution and its value for designing soil percolate sampling experiments.

V. Relation Between the Beta Distribution and the Spatio–Temporal Distribution of Solute Leaching

When solute leaching at a given depth is monitored in a multiple sample percolation experiment, breakthrough curves are obtained for all sampling compartments. These describe the temporal distribution of solute leaching at each sample location. Similarly, the spatial distribution of solute leaching over the total sampling area is obtained for every sampling interval. De Rooij and Stagnitti (2002a, 2002b) introduced the concept of the leaching surface as a tool to analyze both the temporal and the spatial aspect of solute leaching. The approach hinges on the fact that ranking the sampling compartments in order of decreasing amounts of total collected leaching as outlined above makes the two spatial axes of the sampling plane collapse into a single axis with dimension L^2. This, in turn, allows leaching from all compartments comprising the sampling area to be plotted as a function of time (or cumulative drainage) and the single transformed spatial coordinate (Figure 6). The curved leaching surface thus obtained describes the full spatio-temporal behavior of solute leaching.

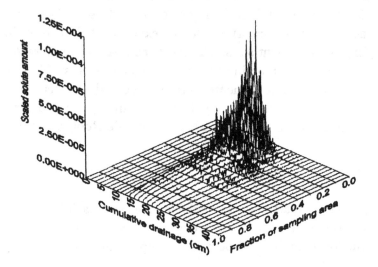

Figure 6. The leaching surface observed during the chloride leaching experiment of Figure 5. The leaching surface was scaled to integrate to unity. It comprises leaching observations from 300 sampling compartments (0.05×0.05 m) on 29 sampling times.

The leaching surface can be scaled to make the volume underneath it equal to one. This scaled leaching surface S can be modeled as a bivariate probability density function. The associated marginal distributions are the scaled SSDC of cumulative leaching at the time when all solute has leached, which is given by

$$S_x(x) = \int_0^\infty S(x,t)dt, \qquad x \in [0,1], \qquad \text{(V.1)}$$

and the scaled breakthrough curve of the entire sampling area (all sampling compartments combined), which is

$$S_t(t) = \int_0^1 S(x,t)dx, \qquad t \in [0,\infty],$$

where x is the transformed spatial coordinate, scaled by dividing it by the total sampling area, $t[T]$ represents the time since the solute application, S is the scaled solute flux density, and S_x and S_t are the marginal pdfs of S along the x-axis and the t-axis, respectively. The Beta distribution can generally adequately fit the SSDC of cumulative leaching defined by equation (V.1), and thus represents one of the marginal pdfs of the leaching surface.

The concept of a leaching surface represents a fundamental new step in understanding and using solute transport theory. The BTC, the classical description of temporal variation of solute transport in soils, can now be integrated with the concept of a spatial solute distribution curve SSDC in a single analysis. Therefore not only can the solute breakthrough at any time for any fixed depth in the soil be described but the proportion of the sample area contributing the bulk of the breakthrough can also be quantified. The Beta distribution is pivotal for the concepts of the SSDC and S since it is the only suitable continuous pdf for describing the SSDC. Thus the Beta distribution has an important role to play in soil science.

References

1. Boll, J.; Selker, J.S; Shalit, G.; Steenhuis, T.S. Frequency distribution of water and solute transport properties derived from pan sampler data, Water Resources Research 1997, *33*, 2655-2664.

2. Buchter, B.; Hinz, C.; Flury, M.; Flühler, H. Heterogeneous flow and solute transport in an unsaturated stony soil monolith, Soil Science Society of America Journal **1995**, *59*, 14-21.

3. de Rooij, G.H. Preferential flow in water-repellent sandy soils − Model development and lysimeter experiments. Doctoral thesis, Wageningen Agricultural University, The Netherlands, 1996, 229 pp.

4. de Rooij, G.H.; Stagnitti, F. Spatial variability of solute leaching: experimental validation of a quantitative parameterization, Soil Science Society of America Journal **2000**, *64*, 499-504.

5. de Rooij, G.H., Stagnitti, F. Spatial and temporal distribution of solute leaching in heterogeneous soils: analysis and application to multisampler lysimeter data, Journal of Contaminant Hydrology **2002a**, *54*, 329-346.

6. de Rooij, G.H.; Stagnitti, F. The solute leaching surface as a tool to assess the performance of multidimensional unsaturated solute transport models. In: *Computational Methods in Water Resources* (editors S.M. Hassanizadeh, R.J. Schotting, W.G. Gray, and G.F. Pinder), pp. 639-646, Vol. 1. Proceedings of the XIVth International Conference, June 23-28, Delft, The Netherlands. Elsevier, Developments in water science, no. 47: Amsterdam, 2002b.

7. Flühler, H.; Durner, W.; Flury, M. Lateral solute mixing processes - A key for understanding field-scale transport of water and solutes, Geoderma **1996**, *70*, 165-183.

8. Flury, M. Experimental evidence of transport of pesticides through field soils - a review, Journal of Environmental Quality **1996**, *25*, 25-45.

9. Gerke, H.H.; van Genuchten, M.Th. Evaluation of a first-order water transfer term for variably saturated dual-porosity flow models, Water Resources Research **1993**, *29*, 1225-1238.

10. Jury, W.A.; Flühler, H. Transport of chemicals through soil: mechanisms, models, and field applications, Advances in Agronomy **1992**, *47*, 141-201.

11. Jury, W.A.; Sposito, G.; White, R.E. A transfer function model of solute transport through soil. 1. Fundamental concepts, Water Resources Research **1986**, *22*, 243-247.

12. Nielsen, D.R.; van Genuchten, M.Th.; Biggar, J.W. Water flow and solute transport processes in the unsaturated zone, Water Resources Research **1986**, *22*, 89S-108S.

13. Poletika, N.N.; Jury, W.A. Effects of soil surface management on water flow distribution and solute dispersion, Soil Science Society of America Journal **1994**, *58*, 999-1006.

14. Quisenberry, V.L.; Phillips, R.E.; Zeleznik, J.L. Spatial distribution of water and chloride macropore flow in a well-structured soil, Soil Science Society of America Journal **1994**, *58*, 1294-1300.

15. Stagnitti, F.; Allinson, G.; Sherwood, J.; Graymore, M.; Allinson, M.; Turoczy, N.; Li, L.; Phillips, I. Preferential leaching of nitrate, chloride and phosphate in an Australian clay soil, Toxicological and Environmental Chemistry **1999a**, *70*, 415-425.

16. Stagnitti, F.; Li, L.; Allinson, G.; Phillips, I.; Lockington, D.; Zeiliguer, A.; Allinson, M.; Lloyd-Smith, J.; Xie, M. A mathematical model for estimating the extent of solute- and water-flux heterogeneity in multiple sample percolation experiments, Journal of Hydrology **1999b**, *215*, 59-69.

17. Stagnitti, F., Sherwood, J.; Allinson, G.; Evans, L.; Allinson, M.; Li, L.; Phillips, I. An investigation of localised soil heterogeneities on solute transport using a multisegment percolation system, New Zealand Journal of Agricultural Research **1998**, *41*, 603-612.

18. Strock, J.S.; Cassel, D.K.; Gumpertz, K.L. Spatial variability of water and bromide transport through variably saturated soil blocks, Soil Science Society of America Journal **2001**, *65*, 1607-1617.

19. Vanderborght, J.; Timmerman, A.; Feyen, J. Solute transport for steady-state and transient flow in soils with and without macropores. Soil Science Society of America Journal **2000**, *64*, 1305-1317.

20. van Genuchten, M.Th.; Wierenga, P.J. Mass transfer studies in sorbing porous media. I. Analytical solutions. Soil Science Society of America Journal **1976**, *40*, 473-480.

21. Wildenschild, D.; Jensen, K.H.; Villholth, K.; Illangasekare, T.H. A laboratory analysis of the effect of macropores on solute transport. Ground Water **1994**, *32*, 381-389.

Limited-Range Distributions with Informative Dropout

Alan H. Feiveson

National Aeronautics and Space Administration
Lyndon B Johnson Space Center, Mail Code SK3
Houston, Texas 77058

A mixed discrete-continuous modification of a beta distribution can be used to model range-limited performance data that contains dropout that is known to be associated with poor performance capability. In this scenario, normalized performance capability is modeled by a beta distribution, but scores from this distribution are only observed if a subject completes the test protocol. When a subject fails to complete the protocol, no quantitative test score is observed, however it is assumed that the conditional probability of non-response given the non-observable performance capability y^* is a function $g(y^*)$ of known form. If low values of y^* reflect poor performance capability and high values reflect good performance capability, it is reasonable to expect that $g(y^*)$ would be a decreasing function of y^* such that $g(0) = 1$ and $g(1) = 0$. More specifically, when $g(y^*)$ is in the form of a power law; e.g. $g(y^*) = (1 - y^*)^r$, and y^* has the Beta density

$$f(y^*) = \frac{\Gamma(p+q)}{\Gamma(p)\Gamma(q)} (y^*)^{p-1} (1 - y^*)^{q-1},$$

then it can be shown that α, the unconditional probability of non-response is equal to $\dfrac{\Gamma(r+q)\Gamma(p+q)}{\Gamma(q)\Gamma(r+p+q)}$. If a mixed discrete-continuous random variable $\tilde{y}$ is equal to y^* for completed tests and is assigned the value 0 for uncompleted tests, it follows that the conditional density of $\tilde{y}$ given $\tilde{y} > 0$ is

$$f(y \mid y > 0) = (1 - g(y))f(y)/(1 - \alpha) \tag{.1}$$

and that $F(y)$, the unconditional CDF of $\tilde{y}$ is equal to α for $y = 0$ and is given by

$$F(y) = \alpha + I_y(p,q) - \frac{\Gamma(r+q)\Gamma(p+q)}{\Gamma(q)\Gamma(r+p+q)} I_y(p, r+q) \tag{.2}$$

for $y > 0$, where $I_y(p,q)$ is the incomplete Beta function with parameters p and q evaluated at y. Given a set of observations on $\tilde{y}$, one may use (.1) and the expression for α to construct a likelihood and subsequently estimate the parameters p, q, and r (which may be functions of other parameters, as in a regression model). One may then use (.2) to obtain estimates of percentage points of the distribution of $\tilde{y}$.

An example of this type of modeling is given in Feiveson et al. (2002), in which the authors study the effect of age on computerized dynamic posturography (CDP) equilibrium scores, widely used as part of a neurological assessment of balance control (Asai et al., 1993; Nashner, 1993). These equilibrium scores are inversely related to the maximum amount of sway that a subject attains in an effort to maintain an erect stance on apparatus designed to nullify visual or proprioceptive sensory input. Scores are normalized to a 0 - 1 range with 1 representing perfect balance (no sway). However a CDP trial is terminated if a patient loses balance and takes a step or has to be restrained form falling. In these situations, the equilibrium score $\tilde{y}$ is assigned the lowest possible score, zero. As a result, $\tilde{y}$ has a mixed discrete-continuous distribution. In Feiveson et al. (2002), a regression model is fit to observations of $\tilde{y}$, examining the effect of age on the Beta parameters p and q as well as on the dropout parameter r.

References

1. Asai, M.; Watanabe, Y.; Ohashi, N; Mizukoshi, K. Evaluation of vestibular function by dynamic posturography and other equilibrium examinations, Acta Otolaryngol (Stockh) **1993**, Suppl *504*, 120-124.
2. Feiveson, A.H.; Metter, E.J.; Paloski, W.H. A statistical model for interpreting computerized dynamic posturography data, IEEE Transactions in Biomedical Engineering **2002**, *49*, 300-309.
3. Nashner, L.M. Computerized dynamic posturography, in *Handbook of Balance Function Testing* (editors G.P. Jacobsen, C.W. Newman, J.M. Kartush) **1993**, pp. 309, 323, Mosby-Year Book, Inc.: Chicago, IL.

New Binomial and Multinomial Distributions From Graph Theory

Yontha Ath[1] and Milton Sobel[2]

[1]Department of Mathematics
California State University, Dominguez Hills
Carson, California 90747
[2]Department of Statistics and Applied Probability
University of California, Santa Barbara
Santa Barbara, California 93106

I. Introduction

Let $G(N, E)$ or G be a simple connected, undirected graph with N vertices and E edges. A *random walk* on G is the process, which occurs in a sequence of discrete steps. At $t = 0$ one (or an object) starts at a certain vertex, say at vertex v_0 (SP). At $t = 1$ the object moves from this vertex (SP) with probability $\frac{1}{d_i}$ to one of its neighboring vertices where d_i is the degree of vertex i ($i = 1, 2, \ldots, N$). This movement is repeated for $t = 2, 3, 4, \ldots$ until two or more (or all) vertices have been visited. When all vertices have been visited, we called it the mean cover time for random walk on graphs or simply coverage. An introduction to cover times for graph can also be found in Blom and Sandell (1992), Barnes and Feige (1996) and Ball et al. (1997).

In this paper, our main goal is to set up an analogy between independent observations in the usual binomial or multinomial distribution and the steps from one node to a neighboring node in the given graph. The analogy for a fixed number of observations is a fixed number of steps. The cells corresponding to nodes and the frequency in a cell corresponds to the number of visits to the corresponding node. Most of our discussion deals with equal probabilities for each edge that lead out of the current node position. Stopping rules can be of at least 3 different types:

(1) Fixed # of steps (corresponding to fixed # of observations)

(2) Sequential Problems: For example, we can stop when we have visited j different nodes.

(3) We can stop when somebody rings an unpredictable bell.

Since we need a starting point, it is necessary to decide whether the starting point is a gratis visit (G) or is not counted as a visit (NG). It is our desire to open up new areas for statisticians and probabilists to explore.

II.   Problems with a Fixed Number of Steps (S = 5 and 10)

In this section we would like to find the GF(and results of the GF) for (1) a specified NSP (non-starting point) node, (2) for the SP (starting point) itself, (3) binodal GF for a pair of NSP nodes (4) binodal GF for a mixed pair of nodes; one NSP and the other is the SP; the GF is given as a matrix(or quadratic form) in all binomal cases.

The $S = 5$-Step GF Results for one NSP (respectively, one SP) node for the complete graph family: K_N (i.e. graphs with N vertices in which there is exactly one edge joining every pair of vertices, $N = 2, 3, 4, \ldots$)

The general formulas for any N are given below

$$GF_1(5, N) = \Big\{ (N-2)^5 + (5N-6)(N-2)^3 s$$
$$+ 3(2N-3)(N-2)(N-1)s^2$$
$$+ (N-1)^2 s^3 \Big\} / (N-1)^5, \quad \text{(NSP)}$$

$$\mu_1(5, N) = \frac{(5N^4 - 24N^3 + 45N^2 - 40N + 15)}{(N-1)^5},$$

$$\sigma_1^2(5, N) = \Big\{ (5N^9 - 62N^8 + 347N^7 - 1155N^6 + 2530N^5$$
$$- 3795N^4 + 3907N^3 - 2667N^2$$
$$+ 1099N - 210) \Big\} / (N-1)^{10}$$

and

$$GF_0(5, N) = \Big\{ (N-2)^4 + (4N-5)(N-2)^2 s$$
$$+ (3N-5)(N-1)s^2 \Big\} / (N-1)^4, \quad \text{(SP)}$$

$$\mu_1(5, N) = \frac{(4N^3 - 15N^2 + 20N - 10)}{(N-1)^4},$$

$$\sigma_1^2(5, N) = \Big\{ 4N^7 - 41N^6 + 184N^5 - 471N^4 + 744N^3$$
$$- 725N^2 + 404N - 100) \Big\} / (N-1)^8.$$

The first formula $GF_1(5, 3)$ reduces to $\dfrac{1 + 9s + 18s^2 + 4s^3}{32}$ for $N = 3$ and $S = 5$ steps. It is easy to check that (starting from x_0 in a complete triangle graph the possible orderings of 0, 1, 2 (including the starting point x_0) are show below.

# of no visits to x_1	# of 1 visit to x_1	# of 2 visits to x_1	# of 3 visits to x_1
020202	010202 021202	010102 021210 021021	
	021020 020120	012102 021212 021201	010101
	020102 020212	010120 010210 010201	012101
	020210	012102 010212 012021	010121
	020201	021010 012010 020101	012121
	012020	021012 012012 020121	

For $GF_0(5,3)$ the result is $\dfrac{1 + 7s + 8s^2}{16} = \dfrac{2 + 14s + 16s^2}{32}$

# of no visits to x_0	# of 1 visits to x_0	# of 2 visits to x_0
021212 012121	010121 021021 010212 021201 012012 021202 012021 012101 020121 012102 020212 021210 021012 012120	010101 010120 010102 010210 010201 020120 010202 020210 020101 012010 020102 021010 020201 012020 020202 021020

For example, with $N = 10$ nodes:

$$GF_1(5,10) = \frac{32768 + 22528s + 3672s^2 + 81s^3}{59049} \quad \text{(NSP)},$$

$$GF_0(5,10) = \frac{4096 + 2240s + 225s^2}{6561} \quad \text{(NSP)}.$$

Similarly for the $S = 10$-steps we get

$$
\begin{aligned}
GF_1(10,N) = \Big\{ &(N-2)^{10} + (10N-11)(N-2)^8 s \\
&+4(9N-11)(N-2)^6(N-1)s^2 \\
&+7(8N-11)(N-2)^4(N-1)^2 s^3 \\
&+5(7N-11)(N-2)^2(N-1)^3 s^4 \\
&+(6N-11)(N-1)^4 s^5 \Big\} \Big/ (N-1)^{10}, \quad \text{(NSP)}
\end{aligned}
$$

$$
\begin{aligned}
GF_0(10,N) = \Big\{ &(N-2)^9 + (9N-10)(N-2)^7 s \\
&+7(4N-5)(N-2)^5(N-1)s^2 \\
&+5(7N-10)(N-2)^3 s^3 \\
&+5(3N-5)(N-2)(N-1)^3 s^4 \\
&+(N-1)^4 s^5]/(N-1)^9, \quad \text{(SP)}
\end{aligned}
$$

For example with $N = 5$,

$$GF_1(10,5) = \left\{59049 + 255879s + 396576s^2 + 263088s^3\right.$$
$$\left. +69120s^4 + 4864s^5\right\}\big/1048576, \text{ (NSP)}$$

$$GF_0(10,5) = \left\{19863 + 76545s + 102060s^2 + 54000s^3\right.$$
$$\left. +9600s^4 + 256s^5\right\}\big/262144. \text{ (SP)}$$

The table below gives the expectation (Exp) and Variance (σ^2) of the number of visits to one specified NSP (respectively, the SP node) in $S = 5$ (respectively,10) steps for K_N, $3 \leq N \leq 10$ (all obtained from $GF_1(5, N)$, $GF_0(5, N)$, $GF_1(10, N)$ and $GF_0(10, N)$).

	S = 5 Steps		S = 10 Steps	
	NSP	SP	NSP	SP
N	Exp (σ^2)	Exp (σ^2)	Exp (σ^2)	Exp (σ^2)
3	1.78125 (0.48340)	1.43750 (0.37109)	3.44434 (0.83870)	3.11133 (0.76690)
4	1.31276 (0.55239)	1.06173 (0.45298)	2.56249 (1.01952)	2.31250 (0.92581)
5	1.04004 (0.54430)	0.83984 (0.44701)	2.03999 (1.02400)	1.84000 (0.92800)
6	0.86112 (0.51319)	0.69440 (0.42021)	1.69444 (0.97608)	1.52778 (0.88349)
7	0.73470 (0.47732)	0.59182 (0.38972)	1.44898 (0.91462)	1.30612 (0.82716)
8	0.64063 (0.44264)	0.51562 (0.36054)	1.26563 (0.85278)	1.14063 (0.77075)
9	0.56790 (0.41092)	0.45679 (0.33407)	1.12346 (0.79500)	1.01235 (0.71818)
10	0.51000 (0.38250)	0.41000 (0.31049)	1.01000 (0.91000)	0.91000 (0.67050)

For example, if we consider 2-Node GF's for $S = 5$, $N = 3$: (1) with 2 NSP nodes (i.e. x_1 and x_2) and (2) with mixed type (1 NSP and the SP). The mixed matrix is for the pair x_0 and x_1, where x_0 is the starting point (SP). It is not the same as for x_1 and x_2 (both NSP). If S denotes $(1, s, s^2, s^3)$ and T denotes $(1, t, t^2, t^3)$, we need only display the matrix of the quadratic form:

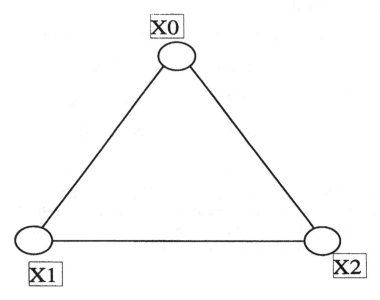

Figure 1. K_3

Thus for 5 steps on K_3

$$M_{(1,2)}(5,3) = \begin{array}{c} \\ 1 \\ s \\ s^2 \\ s^3 \end{array} \overset{\begin{array}{cccc} 1 & t & t^2 & t^3 \; t^4 \end{array}}{\begin{pmatrix} 0 & 0 & 0 & 1 \\ 0 & 0 & 7 & 2 \\ 0 & 7 & 10 & 1 \\ 1 & 2 & 1 & 0 \end{pmatrix}} \; /32 \quad \text{(NSP)};$$

Here $\frac{7}{32}$ is the probability that one NSP node x_1 gets $i = 1$ visit and the other NSP x_2 gets $j = 2$ visits, all within $S = 5$ steps.

$$M_{(1,0)}(5,3) = \begin{pmatrix} 0 & 0 & 0 & 0 \\ 0 & 0 & 7 & 0 \\ 1 & 10 & 7 & 0 \\ 1 & 2 & 1 & 0 \end{pmatrix} /32 \quad \text{(Mixed Type)}$$

The notation above (which is no longer symmetric; rows are for NSP and columns are for the SP) indicates that the probability is $\frac{10}{32}$ that x_1 get $i = 2$ visits and the $SP = x_0$ gets 1 visit, all within 5 steps.

When $S = 10$ steps,

$$M_{(1,0)}(10,3) = \begin{pmatrix} 0 & 0 & 0 & 0 & 0 & 1 \\ 0 & 0 & 0 & 0 & 14 & 5 \\ 0 & 0 & 0 & 38 & 80 & 0 \\ 0 & 0 & 40 & 182 & 132 & 10 \\ 0 & 17 & 116 & 182 & 80 & 5 \\ 2 & 17 & 40 & 38 & 14 & 1 \end{pmatrix} /1024 \quad \text{(Mixed Type)}$$

Thus $\frac{116}{1024}$ is the probability that x_1 gets 4 visits and the SP gets 2 visits, all within $S = 10$ steps.

$$M_{(1,2)}(10,3) = \begin{pmatrix} 0 & 0 & 0 & 0 & 0 & 1 \\ 0 & 0 & 0 & 0 & 5 & 14 \\ 0 & 0 & 0 & 10 & 80 & 38 \\ 0 & 0 & 10 & 132 & 182 & 40 \\ 0 & 5 & 80 & 182 & 116 & 17 \\ 1 & 14 & 38 & 40 & 17 & 2 \end{pmatrix} /1024 \quad \text{(NSP)}$$

The computation of the above matrix probabilities can be complicated when #N and #Steps are increasing. For example K_4 in 10 steps.

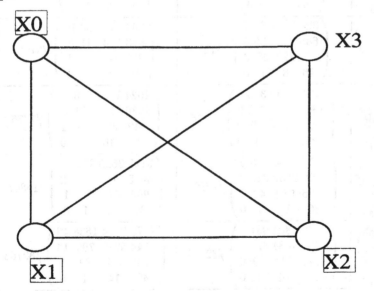

Figure 2. K_4

$$M_{(1,2)}(10,4) = \begin{pmatrix} 1 & 19 & 128 & 364 & 400 & 112 \\ 19 & 290 & 1499 & 3064 & 2200 & 352 \\ 128 & 1499 & 5642 & 7891 & 3664 & 376 \\ 364 & 3064 & 7891 & 7346 & 2327 & 176 \\ 400 & 2200 & 3664 & 2327 & 554 & 35 \\ 112 & 352 & 376 & 176 & 35 & 2 \end{pmatrix} /59049 \text{ (NSP)}$$

$$M_{(1,0)}(10,4) = \begin{pmatrix} 2 & 36 & 226 & 584 & 560 & 128 \\ 34 & 484 & 2290 & 4184 & 2624 & 368 \\ 196 & 2092 & 7054 & 8768 & 3698 & 368 \\ 440 & 3260 & 7382 & 6256 & 1942 & 160 \\ 320 & 1472 & 2168 & 1336 & 346 & 28 \\ 32 & 80 & 80 & 40 & 10 & 1 \end{pmatrix} /59049 \quad \text{(Mixed Type)}$$

$S = 5$-steps Results for $M(1,0)$ and $M(1,2)$ with K_N, $(N = 5, 6, 7, 8, 9, 10)$

Graphs	M(1,0)	M(1,2)
K_5	$\begin{pmatrix} 48 & 132 & 63 & 0 \\ 156 & 282 & 75 & 0 \\ 111 & 120 & 21 & 0 \\ 9 & 6 & 1 & 0 \end{pmatrix}$ /1024	$\begin{pmatrix} 32 & 112 & 90 & 9 \\ 112 & 268 & 127 & 6 \\ 90 & 127 & 34 & 1 \\ 9 & 6 & 1 & 0 \end{pmatrix}$ /1024
K_6	$\begin{pmatrix} 324 & 540 & 160 & 0 \\ 648 & 752 & 136 & 0 \\ 292 & 220 & 28 & 0 \\ 16 & 8 & 1 & 0 \end{pmatrix}$ /3125	$\begin{pmatrix} 243 & 513 & 252 & 16 \\ 513 & 774 & 241 & 8 \\ 252 & 241 & 46 & 1 \\ 16 & 8 & 1 & 0 \end{pmatrix}$ /3125
K_7	$\begin{pmatrix} 1280 & 1530 & 325 & 0 \\ 1840 & 1570 & 215 & 0 \\ 605 & 350 & 35 & 0 \\ 25 & 10 & 1 & 0 \end{pmatrix}$ /7776	$\begin{pmatrix} 1024 & 1536 & 540 & 25 \\ 1536 & 1688 & 391 & 10 \\ 540 & 391 & 58 & 1 \\ 25 & 10 & 1 & 0 \end{pmatrix}$ /7776
K_8	$\begin{pmatrix} 3750 & 3450 & 576 & 0 \\ 4200 & 2832 & 312 & 0 \\ 1086 & 510 & 42 & 0 \\ 36 & 12 & 1 & 0 \end{pmatrix}$ /16807	$\begin{pmatrix} 3125 & 3625 & 990 & 36 \\ 3625 & 3130 & 577 & 12 \\ 990 & 577 & 70 & 1 \\ 36 & 12 & 1 & 0 \end{pmatrix}$ /16807
K_9	$\begin{pmatrix} 9072 & 6804 & 931 & 0 \\ 8316 & 4634 & 427 & 0 \\ 1771 & 700 & 49 & 0 \\ 49 & 14 & 1 & 0 \end{pmatrix}$ /32768	$\begin{pmatrix} 7776 & 7344 & 1638 & 49 \\ 7344 & 5220 & 799 & 14 \\ 1638 & 799 & 82 & 1 \\ 49 & 14 & 1 & 0 \end{pmatrix}$ /32768

K_{10}	$\begin{pmatrix} 19208 & 12152 & 1408 & 0 \\ 14896 & 7072 & 560 & 0 \\ 2696 & 920 & 56 & 0 \\ 64 & 16 & 1 & 0 \end{pmatrix} /59049$	$\begin{pmatrix} 16807 & 13377 & 2520 & 64 \\ 13377 & 8078 & 1057 & 16 \\ 2520 & 1057 & 97 & 1 \\ 64 & 16 & 1 & 0 \end{pmatrix} /59049$

III. Expected Number of Steps (for NG Coverage)

The GF and Expectation for the # of steps needed for non-gratis (NG) Coverage for the complete graph K_N

$$GF = \frac{(N-2)!\,t^N}{\displaystyle\prod_{j=1}^{N-2}(N-jt-1)},$$

$$Exp. = 1 + (N-1)\sum_{j=1}^{N-1}\frac{1}{j}$$

The coefficient of t^α is the probability that coverage will be completed in α steps i.e. that every node is visited at least once within α steps. Below is the general formula for any pair (ν, N) with $2 \le \nu \le N$. It is for the number of steps needed to visit ν new nodes on the complete graph. The formula is a further generalization of the coverage since it gives the coverage for $\nu = N$.

$$GF = \frac{t^\nu \displaystyle\prod_{j=1}^{\nu-2}(N-j-1)}{\displaystyle\prod_{j=1}^{\nu-2}(N-jt-1)}$$

For coverage on the complete graph K_N we simply set $\nu = N$. The table with results for $(v =) N = 2$ to 10 from the GF Formula is given below.

Table for Expected # of Steps for NG Coverage

N	GF (for Coverage)	$Exp.$	$Variance$

2	t^2	2	0
3	$t^3/(2-t)$	4	2
4	$2!t^4/\prod_{j=1}^{2}(3-jt)$	6.5	6.75000
5	$3!t^5/\prod_{j=1}^{3}(4-jt)$	9.33333	14.44444
6	$4!t^6/\prod_{j=1}^{4}(5-jt)$	12.41667	25.17361
7	$5!t^7/\prod_{j=1}^{5}(6-jt)$	15.70000	38.99000
8	$6!t^8/\prod_{j=1}^{6}(7-jt)$	19.15000	55.92806
9	$7!t^9/\prod_{j=1}^{7}(8-jt)$	22.74285	76.01215
10	$8!t^{10}/\prod_{j=1}^{8}(9-jt)$	26.46071	99.26047

IV. 95% Confidence Interval

In this section we are interested in the # of steps needed to reach the probability at least 95% of visiting all the nodes in the complete family K_N. Below are results for N up to 10.

N # of steps needed
3   7 gives .96875
4   12 gives .96533
5   17 gives .96000
6   22 gives .95410
7   28 gives .95659
8   33 gives .95000
9   39 gives .95045
10 45 gives .95004

V.   Sequential Stopping Rules

One of the stopping rule that we are interested in is the GF for seeing α new nodes before returning to the starting point (BRSP). First we consider the complete bipartite graphs or $K_{X,Y}$, where $K_{X,Y}$ is a simple graph in which the set of vertices can be partitioned into two sets X and Y such that there is an edge between every vertex in X and every vertex in Y. This $K_{X,Y}$ family has $N = X + Y$ nodes and $E = XY$ edges. The GF's for seeing α new nodes BRSP for $K_{X,Y}$ are given below

Complete Bipartite Family Results

Graphs (N,E)	GF, Exp. and Variance
(4,4)	$GF = \dfrac{2t^3 + t^2 + 3t}{6}$ $Exp. = \dfrac{11}{6} \approx 1.83333$ $Var. = \dfrac{29}{36} \approx 0.80555$

(6, 9)	$GF = \dfrac{81t^5 + 69t^4 + 95t^3 + 35t^2 + 140t}{420}$ $Exp. = \dfrac{14}{5} \approx 2.80000$ $Var. = \dfrac{1214}{525} \approx 2.31238$
(8, 16)	$GF = \dfrac{250t^7 + 238t^6 + 256t^5 + 205t^4 + 325t^3 + 91t^2 + 455t}{1820}$ $Exp. = \dfrac{53}{14} \approx 3.78571$ $Var. = \dfrac{4393}{980} \approx 4.48265$

where the coefficient of t^α = probability of seeing exactly α new nodes (BRSP).

The second family of graphs that we are interested in is the wheel family (W_N). The wheel graph is the graph that has $N - 1$ nodes on the circle and one more node in the center of the circle as SP. This wheel family has N nodes and $2(N - 1)$ edges. The table below contains results for the wheel family.

Wheel Family Results

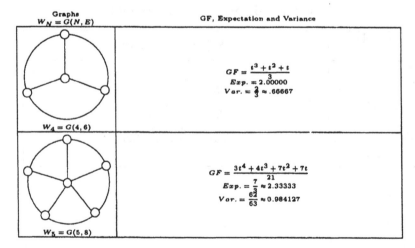

Graphs $W_N = G(N, E)$	GF, Expectation and Variance
$W_4 = G(4, 6)$	$GF = \dfrac{t^3 + t^2 + t}{3}$ $Exp. = 2.00000$ $Var. = \frac{2}{3} \approx .66667$
$W_5 = G(5, 8)$	$GF = \dfrac{3t^4 + 4t^3 + 7t^2 + 7t}{21}$ $Exp. = \frac{7}{3} \approx 2.33333$ $Var. = \frac{62}{63} \approx 0.984127$

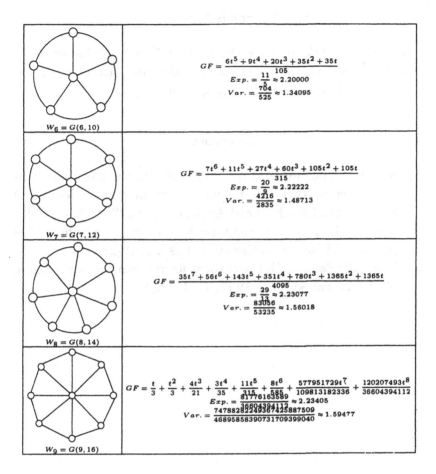

$$W_6 = G(6, 10)$$

$$GF = \frac{6t^5 + 9t^4 + 20t^3 + 35t^2 + 35t}{105}$$
$$Exp. = \frac{11}{5} \approx 2.20000$$
$$Var. = \frac{704}{525} \approx 1.34095$$

$$W_7 = G(7, 12)$$

$$GF = \frac{7t^6 + 11t^5 + 27t^4 + 60t^3 + 105t^2 + 105t}{315}$$
$$Exp. = \frac{20}{9} \approx 2.22222$$
$$Var. = \frac{4216}{2835} \approx 1.48713$$

$$W_8 = G(8, 14)$$

$$GF = \frac{35t^7 + 56t^6 + 143t^5 + 351t^4 + 780t^3 + 1365t^2 + 1365t}{4095}$$
$$Exp. = \frac{29}{13} \approx 2.23077$$
$$Var. = \frac{83056}{53235} \approx 1.56018$$

$$W_9 = G(9, 16)$$

$$GF = \frac{t}{3} + \frac{t^2}{3} + \frac{4t^3}{21} + \frac{3t^4}{35} + \frac{11t^5}{315} + \frac{8t^6}{585} + \frac{5779517729t^7}{109813182336} + \frac{1202074936t^8}{36604394112}$$
$$Exp. = \frac{81776163589}{36604394112} \approx 2.23405$$
$$Var. = \frac{74788282249367425887509}{46895858390731709399040} \approx 1.59477$$

It is interesting to note that as N increases only two's terms change (but no general expression for this change was found). Thus the sum of the first two fractions for $N = 9$ is $\frac{1}{117}$, which is the same as $\frac{35}{4095}$.

VI. Conclusion

Graph theory contains many more families that are not included here. Many of these families have a multitude of applications. We feel that by bringing graph theory into the statistical realm, we are opening up new areas of research for statisticians and probabilities.

Acknowledgment

The authors wish to thank Dr. Saralees Nadarajah for his assistance in translating our paper into the the publisher's format and for his suggestions to put all the graphs in the paper.

References

1. Ball, F.; Dunham, B.; Hirschowitz, A. On the mean and variance of cover times for random walks on graphs, Journal of Mathematical Analysis and Applications **1997**, *207*, 506-514.
2. Barnes, G.; Feige, U. Short random walks on graphs, SIAM Journal of Discrete Mathematics **1996**, *9*, 19-28.
3. Blom, G.; Sandell, D. Cover times for random walks on graphs, Mathematical Scientist **1992**, *17*, 111-119.

Index

Printed in the United States
by Baker & Taylor Publisher Services